T0176753

Reversible Ligand Binding

Reversible Ligand Binding

Theory and Experiment

Andrea Bellelli
*Department of Biochemical Sciences,
Sapienza University of Rome, Italy*

Jannette Carey
Department of Chemistry, Princeton University, USA

Registered Office(s)
John Wiley & Sons, Inc., 111 River Street, Hoboken, NJ 07030, USA
John Wiley & Sons Ltd, The Atrium, Southern Gate, Chichester, West Sussex, PO19 8SQ, UK

Editorial Office
The Atrium, Southern Gate, Chichester, West Sussex, PO19 8SQ, UK

For details of our global editorial offices, customer services, and more information about Wiley products visit us at www.wiley.com.

Wiley also publishes its books in a variety of electronic formats and by print-on-demand. Some content that appears in standard print versions of this book may not be available in other formats.

Library of Congress Cataloging-in-Publication Data

Names: Bellelli, Andrea, 1958– author. | Carey, Jannette, author.
Title: Reversible ligand binding : theory and experiment / by Andrea Bellelli, Jannette Carey.
Description: First edition. | Hoboken, NJ : John Wiley & Sons, 2018. |
    Includes bibliographical references and index. |
Identifiers: LCCN 2017026931 (print) | LCCN 2017040990 (ebook) | ISBN 9781119238478 (pdf) |
    ISBN 9781119238492 (epub) | ISBN 9781119238485 (cloth)
Subjects: LCSH: Ligand binding (Biochemistry)
Classification: LCC QP517.L54 (ebook) | LCC QP517.L54 B45 2017 (print) | DDC 572/.33–dc23
LC record available at https://lccn.loc.gov/2017026931

Cover design by Wiley
Cover Image: Courtesy of RCSB Protein Data Bank (entry 1vyf, deposited by Angelucci et al. 2004)

Set in 10/12pt Warnock by SPi Global, Pondicherry, India
Printed and bound in Malaysia by Vivar Printing Sdn Bhd

10 9 8 7 6 5 4 3 2 1

A.B. dedicates this book to Maurizio Brunori, teacher, mentor and friend for thirty-eight years until now and with more to come.

Contents

Preface

Ligand binding is a crucial event in virtually every biological phenomenon. Detailed understanding of many biologically relevant events including enzymatic catalysis, transport, and molecular recognition requires quantitative description of ligand binding. Such description may prove exquisitely complex because biological macromolecules may bind multiple ligands at once or alternatively and their reactions may present several types of thermodynamic linkage. The scope of this book is to provide a comprehensive view of the various biochemical considerations that govern reversible as well as irreversible ligand binding. Special attention is devoted to enzymology, a field usually treated separately from ligand binding, but actually governed by identical thermodynamic relationships.

This book is intended for PhD students and researchers, and aims at providing the understanding necessary to interpret ligand-binding experiments, formulate plausible reaction schemes, and analyze the data according to the chosen model(s). Attention is given to the design of the experiment because a properly designed experiment may provide clear evidence of biochemical features that can otherwise escape notice. Classical experiments are reviewed in order to further highlight the importance of the design of the experiment.

The book includes treatment of thermodynamic relationships that are most often left to the specialized literature, for example, ligand-linked dissociation. To make the book accessible to a general audience, we simplified the analysis of these relationships to the maximum possible extent, for example, whenever possible we adopted homodimeric proteins as model systems instead of more complex macromolecular assemblies. This choice allowed us to explore a large range of effects with minimally complex equations. Our scope throughout the book has been to present all the essential and distinguishing aspects of the phenomena we describe rigorously, but at the same time in the simplest possible form. Indeed, we are confident that every reader having competence in elementary algebra may take advantage of our work.

Acknowledgments

The authors of scientific books always contract debts with many people who, directly or indirectly, contributed ideas, suggestions, and comments. This book contains considerations and ideas that we elaborated over many years, long before our decision to write, and the list of people we would like to thank is very long. Moreover, the background of the authors being different, our two lists sum without overlapping.

We acknowledge the innumerable hours of education on the subject matters discussed in this book by our teachers and mentors. A.B. expresses his thanks to Maurizio Brunori, Eraldo Antonini, Bruno Giardina, Gino Amiconi, and Quentin H. Gibson.

We enjoyed many long and fruitful discussions with our colleagues, that often shaped our views as expressed in this book. Thanks are due to W.A. Eaton, S.J. Gill, T. Yonetani, M. Coletta, P. Ascenzi, G. Antonini, P. Sarti, F. Malatesta, E. Henry, P. Brzezinski, J.S. Olson, W.F. Xue, S. Linse, G.K. Ackers, and many others.

We express our thanks to our marvelous students, whose questions made us grow. Many of them are now well reputed researchers and close friends: R. Ippoliti, E. Lendaro, G. Boumis, A. Brancaccio, S. Santanché, A. Arcovito, F. Angelucci, L. Jin, B. Harish and R. Grandori. Special thanks are due to F. Saccoccia and A. Di Matteo who read the manuscript more than once and pointed out errors and unclear passages: their contribution has been invaluable.

Part I

Ligand Binding to Single Binding Site Targets

1

Theory of Ligand Binding to Monomeric Proteins

1.1 Importance of Ligand-Binding Phenomena in Biology

Reversible interactions between or among molecules underlie nearly every aspect of biology. To understand these interactions in a chemical way means to describe them quantitatively. To do so we must be able to determine their affinity, stoichiometry, and cooperativity by carrying out ligand-binding experiments. We use the term "ligand" in a way distinct from its use to discuss coordination complexes within inorganic chemistry. In ligand-binding theory we use this term to mean any interacting partner. Although many people consider a ligand to be a small molecule that interacts with a macromolecule, in fact, either partner can be considered to be the ligand of the other. In a typical experiment the concentration of one partner is held fixed while the concentration of the other partner is incremented. In ligand-binding theory and practice we define the ligand operationally as the partner whose concentration is incremented during the experiment. Such experiments resemble pH titrations both practically and theoretically, and thus are referred to as titrations. The partner whose concentration is held fixed is referred to as the "target," and again the definition is strictly operational, that is, either a small molecule or a macromolecule can be the target, depending on how the titration is set up.

As we will show later, there are good reasons to carry out experiments with first one partner, then the other, treated as the ligand; however, depending on the chemical properties of the two partners in the reaction, there may be practical limitations to, or conceptual difficulties in, the possibility of interchanging their respective roles as the ligand and the target. *In particular, if either of the reaction partners has multiple binding sites for the other, one may not obtain superimposable binding isotherms when exchanging the target and the ligand.*

Affinity refers to the strength of interaction between partners. Affinity is quantitatively expressed by an equilibrium constant that we measure in our experiment or, equivalently, a free energy difference between the bound and free states of the system that we calculate from the equilibrium constant.

Stoichiometry refers to the number of molecules of each partner that participate in the binding process, and it must also be determined by our measurements. In practice, what we really mean by stoichiometry is more often *molar ratio*. For example, in a process involving four molecules of one kind with two of another kind, the stoichiometry

Reversible Ligand Binding: Theory and Experiment, First Edition. Andrea Bellelli and Jannette Carey.
© 2018 John Wiley & Sons Ltd. Published 2018 by John Wiley & Sons Ltd.

is 4:2 but the molar ratio is 2:1. The determination of true stoichiometries usually requires additional information from sources other than a binding experiment (e.g., molecular weights, and state of aggregation of the target and ligand in solution).

Thermodynamic linkage is a general term that applies to ligand-binding experiments in which the same target binds two or more molecules of the same or different ligands, and each ligand modulates the affinity of the target for the other. There are at least three different types of linkage, called identical, homotropic, and heterotropic (Wyman and Gill, 1990). *Identical linkage* occurs when two different ligands compete for the same binding site on the target, and their binding is mutually exclusive. This type of linkage is discussed in Section 1.8 for single binding site targets, and in Section 4.8 for multiple binding site targets. Competitive enzyme inhibition is a very important case of identical linkage and is described in Section 8.6. *Homotropic linkage* occurs when the target can bind more than a single molecule of the same ligand, with different affinity. Homotropic linkage can occur only in targets with multiple binding sites, thus its analysis is deferred to Chapter 4. Finally, *heterotropic linkage* occurs when the target can bind two different ligands in a non-exclusive manner and the binding of one ligand alters the affinity of the other. It is described in Section 1.9, and for targets with multiple binding sites in Section 4.9. Important examples of heterotropic linkage are uncompetitive enzyme-inhibition (Section 8.7), and regulation of the oxygen affinity of hemoglobin by effectors, including protons (Bohr effect), diphosphoglycerate, or inositol hexaphosphate, dealt with in Chapter 7.

Homotropic and heterotropic linkage are typically regarded as an emergent property unique to proteins, but some non-protein molecules of ~500–1000 Da have been shown to exhibit cooperative binding of their ligands (Rebek, 1985). An interesting and biologically relevant example is provided by the axial ligands of iron-porphyrins (Traylor and Sharma, 1992).

Homotropic and heterotropic linkages may be either positive (if each ligand increases the affinity of the other) or negative (if each ligand decreases the affinity of the other). *Cooperativity* has often been used as a synonym of linkage, but unfortunately not always with the necessary precision. Often, cooperativity, or positive cooperativity, is used to indicate positive homotropic linkage, but the terms of negative cooperativity or anti-cooperativity may be used to indicate a negative homotropic or heterotropic linkage. The definition of cooperativity is sufficiently general to encompass cases in which even monomeric proteins can respond cooperatively to two different ligands, for example, the physiological ligand and an ionic component of the solution (Weber, 1992). Note that in such cases the ion must be also considered as a physiological effector. Because the general applicability of a definition is inversely related to its precision, in this book, we shall prefer the terms positive or negative homotropic or heterotropic linkage whenever precision is required. Positive cooperativity occurs in several proteins and has special relevance in physiology. For example, the binding of oxygen to hemoglobin is cooperative in that oxygen affinity becomes stronger as binding progresses, as described in detail in Chapter 7.

In this chapter we describe the theoretical bases of ligand binding under equilibrium conditions for protein:ligand complexes with 1:1 stoichiometry; in the next chapter we discuss the kinetics of the same system, and in Chapter 3 we consider some practical aspects of experimental design, and some common sources of errors.

1.2   Preliminary Requirements for Ligand-Binding Study

Whether or not a given ligand binds to a given macromolecular target may be known from prior experiment or inferred from physiological or chemical data. When no such information is available, or when the equilibrium constant for the reaction is required, then the interaction must be evaluated from titration, using a method of analysis that is suitable for the solution conditions of interest and the concentrations of partners that can be practically achieved. Although specific methods will not be detailed here, all useful methods have in common that they offer some observable that changes during, and thus reports on, the binding process. Any observable is referred to as the *signal*, and its change relates to a shift between the bound and free states of the system.

Whether or not a ligand binds reversibly and without any chemical transformation must also be established. Ligand-binding theory and practice generally refers only to such cases; but we shall cover in this book also some common non-conformant cases, for example, thiol reagents and enzyme-substrate reactions, that of course do begin with a ligand-binding process followed by a chemical transformation. The most common approach to determining reversibility is to simply assume it; this is not appropriate for rigorous scientific work. Reversibility is generally established by showing that the signal change is reversed when concentrations are reduced, for example, by dilution or dialysis. A more stringent criterion is to show that the separated ligand and target are recovered unchanged after their interaction, although it can be difficult to rule out a small extent or minor degree of change. One of the best ways to do so is to repeat the binding measurement itself with the recovered materials to evaluate whether the affinity is the same. However, this method can fail with labile partners.

Finally, the molar ratio and stoichiometry must be determined from the same kinds of binding experiments that are used to determine affinity and cooperativity, using strategies that will be outlined later. Without knowing the correct molar ratio the interpretation of the affinity may be plagued by high uncertainty. Another common, unfortunate, and sometimes untested, assumption is that an interaction has a 1:1 molar ratio. As we will show in Chapter 3, the experimental design required to establish molar ratio is not difficult conceptually, and usually is not practically difficult either. Thus there is no reason whatsoever to leave this very important feature to untested assumption. The analysis of binding data is simpler for the 1:1 stoichiometry case, thus we will treat this case first before expanding the treatment to cases of any molar ratio.

1.3   Chemical Equilibrium and the Law of Mass Action

Every chemical reaction, *if allowed enough time*, reaches an equilibrium condition in which the rate of product formation from reactants equals the rate of product degradation to reactants. When this condition is reached the concentrations of reactants and products undergo no further net change. The specific chemical composition of the reaction mixture at equilibrium depends on experimental conditions such as temperature, pH, solution composition, and so on, but at any given set of conditions it obeys the law of mass action. This law states that, under the equilibrium condition, the ratio between the product of product concentrations and the product of reactant concentrations, each raised to a power corresponding to its stoichiometric coefficient, equals a constant, the *equilibrium constant*. Technically, *molar activities should be used*, but under most experimental conditions the concentrations of ligand and protein are low enough to allow the researcher to neglect this distinction.

The equilibrium constant of a chemical reaction is independent of the initial concentrations of reagents and products, but varies with the solution conditions, as discussed further below. The equilibrium constant of a chemical reaction has the units of molarity elevated to a positive or negative, usually integer, power factor that corresponds to the difference between the stoichiometric coefficients of the products and reagents. If the sum of the stoichiometric coefficients of the products equals that of reagents, the equilibrium constant is a pure number.

Ligand binding to proteins usually conforms to the above description, and can be described as:

$$P + X \Leftrightarrow PX \qquad \text{(eqn. 1.1)}$$

where P represents the unliganded protein, X the molecule that binds to it and PX the protein-molecule complex. For practical reasons, it is often convenient to keep constant the protein concentration and to vary in the course of the experiment the concentration of the molecule X, in which case P is the target and X the ligand. Thus, unless differently specified, we shall assume that X is the ligand.

When the above reaction reaches its equilibrium condition, the law of mass action dictates:

$$K_a = [PX]/[P][X] \qquad \text{(eqn. 1.2)}$$

where the square brackets indicate the molar concentrations of the chemical species involved, and the subscript "a" indicates the association direction of the reaction.

A typical example of the above reaction is that of respiratory proteins that reversibly bind oxygen, for example, myoglobin (Mb):

$$Mb + O_2 \Leftrightarrow MbO_2$$

Reversible chemical reactions may be written, and observed experimentally, in either forward or reverse direction. Thus one may write reaction 1.1 in the form of the dissociation of the protein-ligand complex:

$$PX \Leftrightarrow P + X$$

with

$$K_d = [P][X]/[PX] \qquad \text{(eqn. 1.3)}$$

where the subscript "d" refers to the dissociation direction. From equations 1.2 and 1.3 we observe that the equilibrium constant of the dissociation reaction is the reciprocal of that of the combination reaction. The dissociation equilibrium constant has units of molar concentration, whereas K_a has units of reciprocal concentration. Thus K_d can be directly compared to the concentration of the free ligand. Exceptions to this rule do exist (see Section 1.11) and can be a cause of confusion.

A consistent and unique formulation for ligand-binding experiments, using either the association or the dissociation reactions and equilibrium constants, would probably be desirable, but is not likely to be universally adopted. Indeed, we usually design an experiment considering the association reaction, because adding the ligand to the unliganded protein is more obvious and easier to do than dissociating an already-formed complex (although dissociation is possible, e.g., by dilution, chemical transformation of the unbound ligand, or phase extraction). Once the association experiment has been carried out, however, we often switch our reasoning to the dissociation reaction because the dissociation constant, having the units of a molar concentration, corresponds to a point on the ligand concentration axis of the binding plot, and actually the ligand concentration itself can be expressed as a multiple or sub-multiple of K_d. Thus, in the literature one finds experiments, analyses, and models developed in both ways and must be familiar with both.

Unless one directly measures [P], [X], and [PX] after the equilibrium condition has been reached, one only knows the total concentrations of the ligand and the protein, that is, $[X]_{tot} = [X] + [PX]$ and $[P]_{tot} = [P] + [PX]$ (for a 1:1 reaction). If this is the case, eqn. 1.3 should be rewritten as:

$$K_d = \left([P]_{tot} - [PX]\right)\left([X]_{tot} - [PX]\right)/[PX] \qquad \text{(eqn. 1.4)}$$

Although eqn. 1.4 can be easily solved for [PX], yielding a second-degree equation, the procedure is not completely straightforward. Indeed, *the more [PX] approaches [X]$_{tot}$, the greater the uncertainty in [X] and, consequently, in K_d.*

A great simplification can be achieved under conditions in which either (i) both the free and bound ligand (i.e., [X] and [PX]) can be measured directly (e.g., by using equilibrium dialysis); or (ii) $[X]_{tot} \gg [PX]$. As we shall demonstrate in Section 1.4, the latter condition implies $[P]_{tot} \ll K_d$. If the experiment can be run under this condition, only a small fraction of the total ligand will be bound and $[X] \approx [X]_{tot}$, making the use of eqn. 1.4 unnecessary, and allowing direct use of eqns. 1.2 or 1.3. It may happen that, depending on K_d and the experimental method chosen, this condition cannot be met, as it would require protein concentrations too low to be detected. We shall discuss experimental approaches that may overcome this limitation in Sections 1.8 and 1.9.

1.4 The Hyperbolic and Sigmoidal Representations of the Ligand-Binding Isotherms

The graphical representation of binding measurements is important because it is usually difficult to visualize equations like those in the above paragraph or the more complex ones we shall encounter in the following chapters. Some graphical representations may offer clear indications of some property of the system, but may distort or alter other properties. Thus some caution in their use is in order, especially when we want not only to look at them, but to use them for quantitative analysis, that is, to determine the values of the parameters describing the binding reaction. In the present section we shall describe the simplest graphical representations of the ligand-binding isotherm, that is, the hyperbolic plot of [PX] versus [X] and its variants. These representations do not usually distort the experimental error, and should be preferred for quantitative analysis. More complex, and distorting, plots will be considered in a following section.

The soundest and statistically least biased way to represent ligand-binding data is to plot the signal recorded in the experiment, whatever it may be, as a function of the free ligand concentration. The signal will be discussed further in Chapter 3; it is a detectable physical property of the system that depends on the concentrations of P, PX, or both. Thus the graph of the signal is essentially equivalent to a plot of [P] or [PX] versus [X]. However, for the sake of clarity, the researcher may decide to calculate and report the concentration of the bound protein ([PX]) or its fraction ([PX]/[P]$_\text{tot}$) rather than the experimentally recorded signal. The graph of the fraction of bound protein, that is, the ratio between bound and total binding sites, versus the concentration of the free ligand, or total ligand if [X]$_\text{tot}$ >> [PX], is probably the most commonly adopted representation in the biological context.

In order to define the relationship between [PX] and [X], we need to define the *binding polynomial* of the reaction. The binding polynomial expresses the sum of all species of the target as a function of one of them that is adopted as a reference species. For example, if the experiment conforms to eqn. 1.3 and we adopt as a reference the concentration of the unliganded protein P, we can write:

$$[PX] = [P][X]/K_d$$
$$[P]_\text{tot} = [P] + [PX] = [P](1 + [X]/K_d)$$

(eqn. 1.5a)

Eqn. 1.5 represents the binding polynomial of reaction 1.1. Notice that the direction in which the reaction is written, either combination or dissociation, is irrelevant because the binding polynomial does not distinguish reactants and products. The definition of the binding polynomial for such a simple chemical mechanism is obvious, but we write it explicitly in view of its pedagogical value for more complex reaction schemes, to be described in the following chapters.

From eqn. 1.5 we obtain:

$$[P] = [P]_\text{tot} K_d/(K_d + [X])$$

(eqn. 1.5b)

or

$$[PX] = [P]_\text{tot} - [P] = [P]_\text{tot}[X]/(K_d + [X])$$

(eqn. 1.5c)

Both equations correspond to rectangular hyperbolas, which asymptotically tend either to zero (1.5b) or to [P]$_\text{tot}$ (eqn. 1.5c).

Eqn. 1.5c is most often employed and can be rearranged to represent the fractional saturation, defined as the fraction of bound over total ligand binding sites:

$$\bar{X} = [PX]/[P]_\text{tot} = [X]/(K_d + [X])$$

(eqn. 1.6)

The fractional saturation is called in the ligand-binding literature $\bar{Y}$, $\bar{X}$, θ, or ν. We prefer $\bar{X}$ since in reactions involving two or more ligands this allows us to call $\bar{X}$ the fractional saturation of the target for ligand X, $\bar{Y}$ that for ligand Y and so on (see below).

Eqn. 1.6 describes a rectangular hyperbola with unitary asymptote (Figure 1.1A). For many biological systems this is the most meaningful representation of experimental data and it has been preferred traditionally by physiologists studying systems as different as oxygen carriers, hormone receptors, and enzymes. This representation has also the advantage of introducing minimal distortions in the experimental data and their errors. It is important to stress that the ligand concentration that appears in the law of

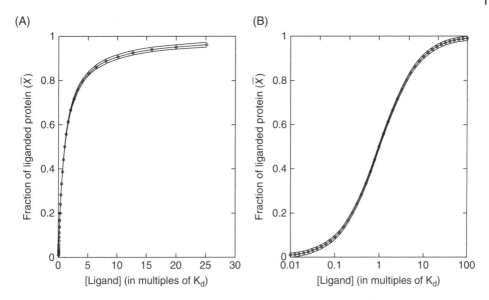

Figure 1.1 Two common representations of ligand-binding isotherms. Panel A: the $\bar{X}$ versus [X] plot; Panel B: same as A but [X] is logarithmically scaled. In both panels the points are calculated from eqn. 1.6. The two lines above and below the points on each panel are obtained by adding and subtracting a constant error of 0.01 $\bar{X}$ units, and demonstrate that in these representations there is no distortion of the error.

mass action, and hence in all the equations we developed thus far, except eqn. 1.4, refers to the free ligand, that is, $[X]_{tot}-[PX]$.

A variant of the hyperbolic plot $\bar{X}$ versus [X] is obtained if one reports $\bar{X}$ as a function of the logarithm of the free ligand concentration. Chemists may prefer this representation over the hyperbolic one because the chemical potential of the ligand is proportional to the logarithm of its concentration (or activity), and thus the plot emphasizes some properties of the system that do not appear clearly in the hyperbolic plot. However, it is unusual for the physiological ligand concentrations to vary over several orders of magnitude; thus, this plot compresses the physiological range of ligand concentrations in a small region of the curve and physiologists seldom use this representation. The $\bar{X}$ versus log([X]) plot has the same overall accuracy as $\bar{X}$ versus [X] plot, and does not distort the experimental error. Actually, the error on ligand concentration, which is usually assumed to be negligible, may in some cases be better represented on a logarithmic than a linear scale (see Chapter 3).

From eqn. 1.6 one may calculate the free ligand concentration (expressed as a multiple or sub-multiple of K_d, as shown on Figure 1.1) required to achieve any value of $\bar{X}$. As an example, we may calculate that at [X]=1/3 K_d we have $\bar{X}=0.25$ and at [X]=3 K_d, $\bar{X}=0.75$; or that to raise $\bar{X}$ from 0.1 to 0.9 an increase of the logarithm of [X] of 1.91 units is required. Generalizing, if we consider ligand concentrations that are multiples or sub-multiples of the K_d by the same factor i, we have that $\bar{X}_{[X]=K_d/i}=1-\bar{X}_{[X]=iK_d}$, which demonstrates that *the $\bar{X}$ versus log [X] plot is symmetric* (at least in the cases of 1:1 and 1:2 stoichiometries).

An important property of the $\bar{X}$ versus log([X]) plot is that its shape is perfectly invariant with respect to K_d: that is, if one explores a set of experimental conditions that cause K_d to vary, one obtains a series of symmetric sigmoidal curves of identical shape, shifted right or left according to their different K_ds. The shape of the plot can be quantified as the approximate slope of the straight line passing through two points symmetric with respect to the center of the curve. For example, the points $\bar{X}_{[X]=Kd/3} = 0.25$ and $\bar{X}_{[X]=3Kd} = 0.75$ are symmetric in the sense defined above, and are joined by a straight line with slope ~0.52 (for a single-site macromolecule). The slope of the $\bar{X}$ versus log ([X]) plot was called the *binding capacity* of the macromolecule by Di Cera and Gill (1988), in analogy with the physical concept of heat capacity. This terminology is potentially confusing because the term capacity is also used to indicate the total amount of some substance that the sample may contain (e.g., the maximum amount of oxygen that a given volume of blood can transport is commonly referred to as its oxygen capacity), which is not what these authors meant. Nevertheless, the concept conveyed by the slope is important because changes in slope from the canonical value given above provide important clues on the binding properties of the protein, and/or the composition of the experimental system. In particular, as we shall demonstrate further on, positive cooperativity increases the slope, whereas chemical heterogeneity of the protein or negative cooperativity decreases it (see Chapters 3 and 4).

In the present chapter we deal only with single-site targets, whose changes in ligand affinity shift the position of the $\bar{X}$ versus log([X]) curve along the X axis but do not change its shape, and in particular do not change its slope; this will be true for all types of linkage considered in this chapter (Sections 1.8–1.10) and for the effect of temperature (Section 1.7). However we call the attention of the reader to this point, because in later chapters we shall describe systems having steeper or shallower $\bar{X}$ versus log[X] plots, and we shall develop an interpretation of the increased or decreased slope of these plots, which may constitute the first indication that the protein:ligand stoichiometry differs from 1:1.

The derivative of the $\bar{X}$ versus log([X]) plot, $\Delta\bar{X}$ versus log([X]), has been used by S.J. Gill to describe the ligand-binding isotherms of hemoglobin, as recorded using the thin layer dilution method (Figure 1.2) (Gill *et al.*, 1987). This is a very specialized

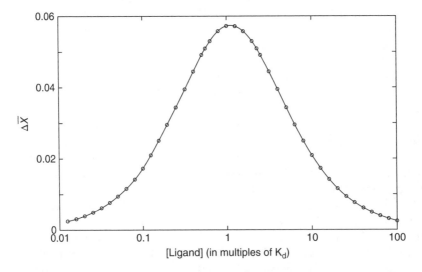

Figure 1.2 Plot of $\bar{X}$ increments versus [X] (on a log scale). Same data set as in Figure 1.1; because the errors added to calculated $\bar{X}$ values are constant and independent of $\bar{X}$, in this plot the points and the lines exactly overlap. Notice that the values on the ordinate axis depend on the dimension of the intervals of log [X] explored and that to obtain a readable plot, these should be kept constant.

plot justified by the experimental method used, and true to the experimental data in that it aims to minimize the error distortion implicit in other types of plots. However, this graphical representation is strongly linked to the experimental method, and greatly benefits from constant logarithmic steps of increase or decrease of the ligand concentration, which is why it never gained widespread use. A more general use of the derivative plot of the sigmoid-binding curve, applicable independently of the specific experimental setup, is to reveal the inflection point (i.e., the midpoint, $X_{1/2}$, whose significance is discussed in the next Section) even when the sigmoid is incomplete.

1.5 The Important Concept of $X_{1/2}$

An important point of the $\bar{X}$ versus [X] or the $\bar{X}$ versus log [X] plots is their midpoint. We may calculate that $\bar{X}$ equals 0.5 (saturation of half the available binding sites) when [X]=K_d, consistent with the symmetry of the $\bar{X}$ versus log [X] plot for monomeric proteins.

The free ligand concentration required to achieve $\bar{X} = 0.5$ is called $X_{1/2}$ (or X_{50}) and is always a function of K_d. However, the simple equation $X_{1/2}=K_d$ applies only to the simple cases of monomeric, single-site proteins or multiple, identical, non-interacting sites; we shall consider below some cases in which this relationship is more complex and we shall devote Chapters 4, 5 and 7 to proteins with multiple interacting ligand binding sites.

In view of the widespread use and misuse of the term $X_{1/2}$, and its equivalence with IC_{50} in enzymology (Section 8.6), it is important to provide a clear and unequivocal definition. *$X_{1/2}$ is the free ligand concentration required to saturate half the available binding sites under the chosen experimental conditions, provided that chemical equilibrium has been reached.* This definition is consistent with that of IC_{50} (Cheng and Prusoff, 1973) and stresses the fact that $X_{1/2}$ is a thermodynamic parameter that may depend on the experimental conditions, but must be independent of the direction in which the equilibrium condition is approached (e.g., whether the experiment was carried out by successive additions of the ligand to the target, as in a titration, or by removing the ligand by dilution or dialysis). Moreover, $X_{1/2}$ cannot depend on the time required by the mixture to reach its equilibrium state.

A related, but perhaps less intuitive, concept is that of Xm, the ligand concentration required to express half the binding free energy change. If the $\bar{X}$ versus log [X] plot is symmetric, as always occurs in monomeric, single-site proteins, $Xm = X_{1/2}$ (Wyman 1963; Ackers *et al.*, 1983).

Due to its necessary relationship with K_d, $X_{1/2}$ is commonly used as an overall empirical parameter to define the apparent affinity of the protein for its ligand. However, $X_{1/2}$ may depend not only on K_d, but also on the concentration of components of the mixture other than X. If this happens, the type of dependence, whether linear, hyperbolic, or other, is a clue to the type of thermodynamic linkage between X and these components (Sections 1.8 and 1.9).

1.6 Other Representations of the Ligand-Binding Isotherm

Before the widespread use of electronic computers, the analysis of binding isotherms was carried out using graphical methods, and representations aimed at linearizing the [PX] versus [X] hyperbola were widely employed. Unfortunately, the transformations required to linearize the hyperbola entail significant distortion of the magnitude of the experimental error, and are statistically unsound, thus they should not be used to obtain a quantitative estimate of the thermodynamic parameters (i.e., K_d or $X_{1/2}$). They are nevertheless discussed here because a thorough understanding of these representations (and their weaknesses) is required to read and understand many classical papers on ligand binding; moreover, these representations may still have some limited usefulness because they may provide visual evidence of some property of the system that must be confirmed using statistically sounder methods.

A classical and widely employed plot was proposed by the Nobel laureate Archibald Vivian Hill to represent his hypothesis on hemoglobin cooperativity. The Hill plot is based on a simple re-elaboration of eqn. 1.6, which yields:

$$[PX]/[P] = \bar{X}/(1-\bar{X}) = [X]/K_d$$

and in logarithmic form:

$$\log\left[\bar{X}/(1-\bar{X})\right] = \log([X]) - \log(K_d)$$

The Hill plot for a single site macromolecule is a straight line with unitary slope and intercept $-\log(K_d)$. Although this representation of experimental data lacks a sound thermodynamic basis and was based on a hypothesis for the function and structure of hemoglobin that was later proven wrong (Bellelli, 2010), it had considerable success due to its simplicity and apparent information content. Indeed, it allows the researcher to measure $\log(K_d)$ using a ruler, by drawing a straight line through the experimental points in the range $0.1 < Y < 0.9$ (outside this range the experimental error affects the data very significantly, as shown in Figure 1.3), an important

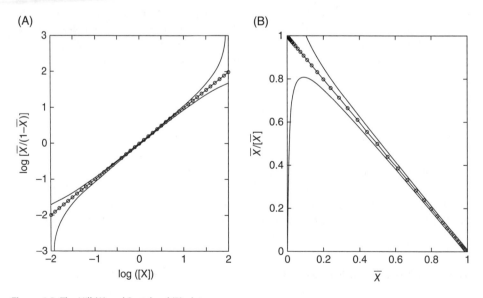

Figure 1.3 The Hill (A) and Scatchard (B) plots.
Ligand concentration is expressed as multiples of K_d (i.e., $[X]/K_d$).
These representations severely distort the experimental errors (continuous lines are calculated to represent an error of 0.01 on $\bar{X}$, i.e., $\bar{X} + 0.01$ and $\bar{X} - 0.01$), and should never be used for quantitative analyses. However, they may be helpful to visualize some qualitative details of the ligand-binding curve (e.g., the presence of heterogeneity or cooperativity), and may maintain some value for figurative purposes.

advantage at a time when personal computers able to implement sounder statistical methods were not available. Moreover, deviations from the unitary slope may indicate either heterogeneity or cooperativity, consistent with Hill's original objective (see Chapter 7).

Like several other graphical representations, the Hill plot severely distorts the experimental errors of the original data set. Thus it can be used, at most, as an instrument to depict a property of the system, but should not be used for quantitative analysis (Forsén and Linse, 1995). Some authors advocated a quantitative use of the Hill plot for the special cases of instruments with demonstrated greater precision (and smaller errors) at very low and very high values of ligand saturations (e.g., Imai *et al.*, 1970), but this suggestion cannot be generalized.

Figure 1.3A reports the Hill plot of the data set of Figure 1.1, together with the lines representing the error range of $+/-0.01$ independent of $\bar{X}$. The actual amplitude of the error of the experimental points cannot be defined *a priori*: it is a function of the experimental method used, of the value of the variable one measures, and often of $\bar{X}$ itself (e.g., the error on absorbance measurements is usually a small fraction of the actual value recorded by the instrument, which is in turn a function of $\bar{X}$, the protein concentration, and the extinction coefficients of the liganded and unliganded derivatives). Thus, the $+/-0.01$ confidence interval in Figure 1.3A is only representative of how severely a constant error is distorted in the Hill plot. The figure shows that the central region of the Hill plot, in the range $-1 < \log(\bar{X}/(1-\bar{X})) < 1$ (approximately equivalent to $0.1 < \bar{X} < 0.9$) is relatively free from error distortion, whereas, outside this range, the distortion becomes very pronounced.

The Hill plot has been widely used in the past to empirically evaluate the cooperativity of multimeric proteins (as described in Chapters 4 and 7); however, the advent of personal computers has made it obsolete. Moreover, its slope fails to provide a thermodynamically interpretable quantitative measure of the magnitude of cooperativity (Forsén and Linse, 1995) and remains an empirical parameter essentially devoid of any direct physical meaning. However, since the Hill plot has been widely employed in the past, interpretation of a host of literature data entails an understanding of its parameters (log $(X_{1/2})$ and the slope, often called the Hill coefficient n).

The Scatchard plot is another type of graphical representation of great historical value that is unsuitable for quantitative data analysis, despite continuing to be in use. It aims to highlight different properties of a binding system than the Hill plot, but shares the same concerns for error distortion. This graphical representation was proposed by G. Scatchard (Scatchard, 1949) and is based on the following linear rearrangement of eqn. 1.6:

$$[PX]/[X] = [P]/K_d = [P]_{tot}/K_d - [PX]/K_d$$

or:

$$\bar{X}/[X] = 1/K_d - \bar{X}/K_d$$

Thus, a plot of $[PX]/[X]$ versus $[PX]$ (or $\bar{X}/[X]$ versus $\bar{X}$) yields a straight line with slope $-1/K_d$ and two intercepts: with the abscissa at $[P]_{tot}/K_d$ (or $1/K_d$) and with the ordinate at $[P]_{tot}$ (for the case of 1:1 binding) (Figure 1.3B).

The Scatchard plot may seem to have little to offer: (i) it severely distorts the experimental errors; (ii) it uses [PX], which is error-prone, in both the dependent and independent variables; and (iii) it is strongly counterintuitive, as can be seen from a comparison of Figure 1.3B with the same data plotted in Figure 1.1.

The Scatchard plot was used to graphically detect heterogeneity of the protein preparation (i.e., the presence of isoforms or non-equivalent binding sites; see Section 3.10); however, it is obsolete, because the same information can be obtained by sounder statistical methods at the expense of minimal computing power.

1.7 Effect of Temperature: Thermodynamic Relationships

The reaction scheme of eqn. 1.1 neglects any energy contribution to the formation or dissociation of the protein-ligand complex; a more reasonable scheme would be:

$$P + X + Q \Leftrightarrow PX \quad \text{or} \quad P + X \Leftrightarrow PX + Q \qquad \text{(eqn. 1.7)}$$

In which energy (Q, which may indicate the free energy, the enthalpy or the entropy change) appears as a reactant or as a product of the reaction. Many ligand-binding reactions are exothermic, that is, they release heat as a consequence of the formation of the protein ligand bond(s); however, there are also examples of endothermic reactions, in which the amount of heat released upon bond formation is exceeded by the heat absorbed because of changes within the protein or the ligand (eqn. 1.7).

The fundamental relationship between the reaction free energy and the equilibrium constant is:

$$\Delta G^0 = -RT \ln K \qquad \text{(eqn. 1.8)}$$

The $\Delta G^0 (= G^0_{\text{products}} - G^0_{\text{reagents}})$ is defined for standard thermodynamic conditions (1 M for every species in solution, 1 atm for gases), and measures the free energy that is absorbed or released during the process in which the system, starting from the standard conditions, reaches its chemical equilibrium state (at constant temperature). Biological systems rarely or never can be equilibrated under standard conditions, thus their free energy is corrected for their actual conditions and is called their $\Delta G'$; the correction is such that a system under equilibrium conditions has $\Delta G' = 0$. $\Delta G'$, if different from zero, defines the direction in which the system will evolve: if $\Delta G' < 0$ then evolution is in the direction of a net increase of products at the expense of the reactants; if $\Delta G' > 0$ then evolution is in the opposite direction. Progression of the reaction changes the concentrations of reactants and products, hence the $\Delta G'$ that becomes zero when equilibrium is reached.

Eqn. 1.8 allows the researcher to calculate the binding energy of the protein-ligand pair, or, to be more precise, the free-energy difference between the PX complex and the dissociated P and X pair. Thus, one might completely replace the equilibrium constant with the binding free energy. Although we do not suggest the reader to take this step, it is useful to keep in mind the order of magnitude of the binding energies commonly encountered in biological systems; thus a K_d of 1 mM corresponds to a free-energy difference of approximately 4 kcal/mole of complex at 25 °C, a K_d of 1 uM to 8 kcal/mole, and a K_d of 1 nM to 12 kcal/mole. These values have a positive sign in the direction of dissociation, that is, free energy is released when the complex forms and is absorbed when the complex dissociates.

One should resist the simplistic idea of equating ΔG^0 to the protein-ligand bond energy, since this parameter also includes the energy of any ligand-linked structural rearrangement of the protein and the ligand, and any solvation-desolvation processes of the ligand and the binding site.

The equivalence of equilibrium constants and energies clarifies that no chemical reaction can be truly irreversible, as this would imply an infinite ΔG^0 and would violate the first principle of thermodynamics. What we usually call irreversible is a chemical reaction whose equilibrium constant is so large that the concentration of reagents at equilibrium escapes detection by any practical means. For example, the solubility product of mercuric sulfide is so low that the precipitate is in equilibrium with less than one of the constitutive ions in essentially any volume of water one may practically use, the equilibrium concentrations of mercuric and sulfide ions being of the order of 10^{-27} M. In cases like this one, thermodynamic relationships are used instead of direct measurements to infer the equilibrium constants, and we should imagine "concentrations" as the probability of at least one molecule of the reagent to be present in solution at any given time in the reaction mixture.

Protein-ligand interactions rarely approach these affinities, but may still reach levels that may make it difficult or impossible to detect the free ligand. For example, the K_d of the avidin-biotin complex is in the order of 10^{-15} M and that of heme-hemopexin is $<10^{-12}$ M, as is that of the Mb-NO complex. In these and similar cases competition (see below and Chapter 3) or kinetic (Chapter 2) experiments may be required to determine equilibrium constants.

The study of the dependence of the equilibrium constant on the temperature allows the researchers to better quantify the components of its ΔG^0. Indeed, if we combine eqn. 1.8 with the Gibbs' free energy equation:

$$\Delta G = \Delta H - T\,\Delta S$$

we obtain:

$$\ln K = -\left(\Delta H^0/RT\right) + \left(\Delta S^0/R\right) \qquad \text{(eqn. 1.9)}$$

In many simple chemical equilibria ΔH^0 is independent of temperature (at least on small temperature intervals) and the plot ln K versus 1/T yields a straight line, with slope $-\Delta H^0/R$ and intercept $\Delta S^0/R$ (the van t'Hoff plot; Figure 1.4). In such cases one may derive all three fundamental thermodynamic parameters for the reaction, ΔG, ΔH, and ΔS from determinations of the equilibrium constant at different temperatures. When ΔH is not temperature independent, the van t'Hoff plot is curved, and the dependence of ΔH on temperature conveys information about the heat capacity change for the system.

We stress once more that when we say that a function is a straight line whose slope and intercept correspond to certain parameters, we do not imply that the statistically soundest method to determine those parameters is by linear regression: our aim is to offer to the reader the description of a function or a concept in a way that is easy to visualize and remember. Actually, the soundest method to determine the thermodynamic

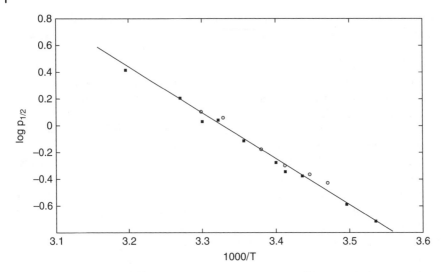

Figure 1.4 van't Hoff plot of the reaction of sperm whale myoglobin with oxygen. The ordinate reports the logarithm of the oxygen partial pressure required to half saturate the protein ($P_{1/2}$) that for a gaseous ligand is equivalent to $\log X_{1/2}$. *Source:* Adapted from Antonini and Brunori, 1971; original data by G. Amiconi.

parameters of a protein-ligand couple is to fit all the experimental points ($\overline{X}$ values as functions of [X] and T) globally with a function that includes the lowest possible number of relevant parameters. In the present case, this would mean K_d at a given temperature, chosen as the reference state (usually 298 K, independently of the fact that this temperature has been actually explored), ΔH^0, and ΔS^0 (see eqn. 1.9). The ΔG^0 will be derived from Gibbs' free energy equation.

An interesting practical application of the van t'Hoff plot is to extrapolate the value of ln K_d to temperatures at which it is too large or too small to be determined directly. For example, if K_d is too small to allow precise determination, for example, because the condition $[P]_{tot} \ll K_d$ cannot be achieved (see Section 1.3), but the reaction has a large ΔH, the researcher can measure ln K_d in a range of temperatures where it is greater, calculate ΔH and extrapolate ln K_d value at the temperature of interest. Since this approach assumes that the ΔH is temperature independent, a condition that is not necessarily true, some caution in the use of the final value is required. For this use, the integrated form of eqn. 1.9 (van't Hoff equation) may be useful:

$$\Delta H = R\,T_1\,T_2\left(\ln K_2 - \ln K_1\right)/\left(T_1 - T_2\right) \qquad\qquad \text{(eqn. 1.10)}$$

where K_1 and K_2 are the equilibrium constant measured at temperatures T_1 and T_2, respectively. Needless to say, ΔH is lower than zero (negative) in the case of exothermic reactions (heat appearing as a product in eqn. 1.7) and greater than zero (positive) in the case of endothermic reactions (heat appearing as a reactant).

An important point to be considered when determining the apparent ΔH from van t'Hoff plots is that temperature changes may affect the ligand-binding equilibrium in more aspects than the one described by eqn. 1.9, and their effect on the equilibrium constant is the sum of a large number of contributions, only one of which is the reaction heat. The heat of solubilization of the ligand, of its dilution, and so on, all contribute; the reaction heat of the buffer may cause pH variations (and in these measurements it is a good idea to use a buffer with as small as possible ΔH of protonation, e.g., phosphate), and so on. If the contributions other than the reaction heat cannot be neglected, they should be measured by careful control experiments carried out in the absence of the protein. Moreover, the true reaction ΔH, after correction for the above contributions, is itself the sum of several components that include the heat of bond formation, the heat of desolvation of the ligand and of the binding site, the heat of eventual structural changes in the protein, and the reaction heat of titratable groups on the protein may cause changes in their protonation state. These components are to be considered because they represent important properties of the biological system under study.

A clear-cut example of the above-mentioned effects is the following. ΔH for the reaction of oxygen binding to heme model compounds is -14 kcal/mol (Traylor and Berzinis, 1980). Overall, ΔH for the same reaction in hemoproteins from different sources varies over a significant range, from -14.9 for sperm whale myoglobin (Antonini and Brunori, 1971), to -11 kcal/mol for human hemoglobin, down to -2 kcal/mol for musk ox and reindeer hemoglobins (Coletta *et al.*, 1992). The chemical reaction in all the above cases is the same, namely the reversible binding of oxygen to the histidine/imidazole coordinated heme iron. However, when the heme is embedded in the protein moiety, oxygenation is coupled to other reactions (e.g., structural changes and proton release, see Chapter 7), whose ΔH adds to that of heme oxygenation, thus explaining the variations.

1.8   Replacement Reactions: Competitive Ligands

A biologically important problem is that of the specificity of ligand binding, that is, the ability of a protein to discriminate between two ligands of similar molecular structure that may happen to be simultaneously present in the biological or experimental system. Common examples of this condition are steroid hormones and their receptors, substrates and competitive inhibitors in enzymology, and competition between CO and oxygen for the heme iron in hemoglobin and myoglobin, which leads to carbon monoxide poisoning. The set of relationships that describe a protein and two ligands competing for the same binding site has been called identical linkage by Wyman and Gill (1990). The same authors recognized an analogy with the partition of a solute between two phases and called this experiment the partition of the protein between two ligands (or two complexes), a term that we shall use as synonymous with ligand replacement.

When two similar molecules are present in solution and the protein can bind either but not both of them at the same binding site, the two ligands effectively compete with each other. The intrinsic affinity of each ligand is not affected by the presence of the other, but the apparent affinity is, depending on their relative affinities and concentrations. A detailed understanding of this feature is relevant because one may turn it to advantage. Let us describe the system as follows:

$$Y + PX \Leftrightarrow Y + X + P \Leftrightarrow X + PY$$

with:

$$K_{d,X} = [P][X]/[PX]$$

and

$$K_{d,Y} = [P][Y]/[PY]$$

The binding polynomial, which we write using P as the reference species, results:

$$[P]_{tot} = [P] + [PX] + [PY] = [P](1 + [X]/K_{d,X} + [Y]/K_{d,Y})$$ (eqn. 1.11)

From eqn. 1.11 we derive the fraction of saturation for both ligands:

$$\bar{X} = [X]/K_{d,X} / (1 + [X]/K_{d,X} + [Y]/K_{d,Y})$$ (eqn. 1.12)

$$\bar{Y} = [Y]/K_{d,Y} / (1 + [X]/K_{d,X} + [Y]/K_{d,Y})$$ (eqn. 1.13)

and

$$X_{1/2} = K_{d,X}(1 + [Y]/K_{d,Y})$$
$$Y_{1/2} = K_{d,Y}(1 + [X]/K_{d,X})$$ (eqn. 1.14)

Eqn. 1.14 demonstrates that the apparent $X_{1/2}$ is increased by the factor $[Y]K_{d,X}/K_{d,Y}$ in the presence of the competing ligand Y; that is, the competing ligand shifts the binding curves to the right, as shown in Figure 1.5. Conversely, the experimental finding that the $X_{1/2}$ of a ligand is directly proportional to the concentration of another component of the solution strongly suggests identical linkage.

Unless both ligands are present at very low concentration relative to their K_ds, the fraction of the unliganded species P may be neglected, and the reaction scheme may be simplified.

$$Y + PX \Leftrightarrow X + PY$$

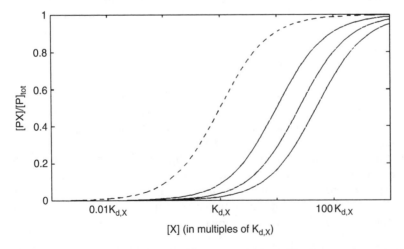

Figure 1.5 A typical ligand replacement experiment. The dashed line represents the binding isotherm of ligand X in the absence of competing ligand Y; the three lines on the right represent the replacement of ligand Y by ligand X at constant fixed concentrations of ligand Y (from left to right: [Y]=10 $K_{d,Y}$, 25 $K_{d,Y}$, 50 $K_{d,Y}$). Notice the right shift of each curve, that is, the decrease of the apparent affinity for ligand X, as a response to successive increases of the concentration of ligand Y, and their shape invariance, anticipated in Section 1.4.

with:

$$\bar{X} = [X]/K_{d,X}/([X]/K_{d,X} + [Y]/K_{d,Y}) = [X]/([X] + [Y]K_{d,X}/K_{d,Y}) \qquad \text{(eqn. 1.15)}$$

Under experimental conditions where the total protein concentration is significantly lower than those of X and Y, eqn. 1.15 describes (and eqn. 1.12 approximates) a binding function analogous to eqn. 1.6, with the apparent constant:

$$K_p = [PY][X]/[PX][Y] = K_{d,X}/K_{d,Y}$$

K_p, the ligand partition constant, is a measure of the relative specificity of the protein for the two ligands. The value of K_p is easily determined, provided that one can discriminate the two protein-ligand complexes PX and PY from each other (and from the unliganded protein P if present at all under the chosen experimental conditions).

If the experiment is carried out at constant concentration of either one of the ligands (under experimental conditions in which the amount of ligand bound to or released by the protein during the experiment causes negligible variations of its concentration), the above equation reduces to a simple hyperbola, fully analogous to Figure 1.1A, and described by the same equation except that in place of the unliganded protein concentration we find the complex of the protein with the constant ligand. For example, in the case of the concentration of ligand Y being constant:

$$K_{p,[Y]=const} = K_{d,X}[Y]/K_{d,Y} = [PY][X]/[PX] \qquad \text{(eqn. 1.16)}$$

We finally remark that the interpretation of ligand replacement experiments in the case of proteins having more than a single ligand-binding site differs from the one presented above, thus the interested reader is encouraged to compare this section with Section 4.8.

Two practical problems may take great advantage of a ligand replacement experiment: that of ligands that provide poor or no signal, and that of high-affinity ligands, as we shall detail in Chapter 3.

The reason high-affinity ligands benefit from replacement experiments is evident from eqn. 1.15 and Figure 1.5. When $K_{d,X}$ is small with respect to the protein or ligand concentration required to produce a measurable signal (a condition that we recognized above as undesirable), addition of a competing ligand will increase the apparent equilibrium constant by the factor $(1+[Y]/K_{d,Y})$, which the experimenter controls at will (eqn. 1.14 and Figure 1.5). $K_{d,X}$ is then calculated from K_p or, better, from a series of experiments carried out at different concentrations of ligand Y fitted globally to eqn. 1.12, or to eqn. 1.15 if the concentration of unliganded protein can be neglected, to find the statistically soundest estimates of $K_{d,X}$ and $K_{d,Y}$. This experimental approach demands great attention to the incubation time required to reach the equilibrium condition after each addition of the variable ligand: indeed high-affinity ligands often have very low dissociation rate constants and replacement equilibration may require many minutes or hours (see Chapter 2).

Replacement reactions can also be used to determine ligand affinities when formation or dissociation of their complexes does not provide a detectable signal. For example, the complex of fatty acid binding protein (FABP) with its physiological ligands is indistinguishable from the mixture of the unbound protein and fatty acids. However, FABP binds the fluorescent probe anilinonaphthalene sulfonate (ANS) at the same site as the fatty acid, and the reaction is associated with a large increase of its quantum yield. Thus, the affinity of FABP for ANS can be measured directly. In an experiment in which ANS-saturated FABP is titrated with the physiological ligand palmitic acid (or vice versa), the ligand replacement can be followed as a decrease in fluorescence, caused by the release of ANS, and the partition constant of the two ligands can be measured (see Angelucci *et al.*, 2004). From K_d of ANS and the partition constant for the ANS/fatty acid, K_d of the fatty acid can be calculated.

1.9 Heterotropic Linkage: Non-Competitive Binding of Two Ligands

Proteins are complex macromolecules and the case of one and the same protein having two binding sites for two different ligands is by no means uncommon. The two different ligands may or may not affect each other's affinity. When they do, one has the condition called heterotropic linkage by J. Wyman (Wyman, 1948; Wyman, 1964; Wyman and Gill, 1990). Examples of ligands involved in this type of linkage include uncompetitive enzyme inhibitors, allosteric effectors in hemoglobin, enzymes, and many other proteins, and regulatory ligands such as lactose for the repressor protein in the Lac operon.

The reaction scheme is as depicted in Figure 1.6.

The reaction scheme of Figure 1.6 is a thermodynamic cycle, and energy conservation dictates the important property:

$$K_X {}^X K_Y = K_Y {}^Y K_X = [P][X][Y]/[PXY] \qquad \text{(eqn. 1.17)}$$

Thus any equilibrium constant can be determined from the other three.

In the absence of linkage, the equilibrium constants for each ligand are independent of the presence of the other ligand, that is, $K_X = {}^Y K_X$ and $K_Y = {}^X K_Y$. In the presence of linkage, the following relationships apply:

$$^Y K_X = K_X {}^X K_Y / K_Y$$
$$^X K_Y = K_Y {}^Y K_X / K_X$$

These relationships may be summarized as follows: if X increases the affinity of P for Y (i.e., $^X K_Y < K_Y$) then Y also increases the affinity of P for X (i.e., $^Y K_X < K_X$), and we have a case of positive heterotropic interaction. Conversely, if X decreases the affinity of P for

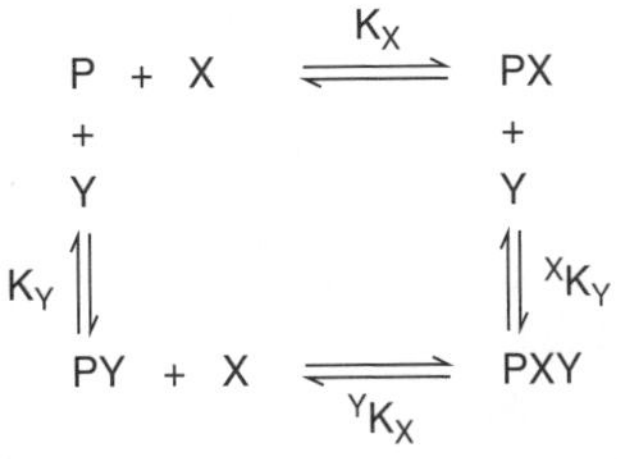

Figure 1.6 Reaction scheme for a protein binding two ligands at different binding sites. Notice that the ternary complex PXY is possible in this case, whereas it was impossible in the case of identical linkage.

Y (i.e., $^{X}K_Y > K_Y$) we have a case of negative heterotropic interaction (and $^{Y}K_X > K_X$). Negative heterotropic interactions are particularly common in biology and may play important regulatory roles, typically in negative feedback mechanisms.

The effect of ligand Y on the apparent affinity of the protein for ligand X may be derived from the above relationships (Wyman, 1948, 1964). The binding polynomial for the reaction scheme of Figure 1.6 is:

$$\begin{aligned} \left[P\right]_{tot} &= \left[P\right] + \left[PX\right] + \left[PY\right] + \left[PXY\right] \\ &= \left[P\right]\left(1 + \left[X\right]/K_X + \left[Y\right]/K_Y + \left[X\right]\left[Y\right]/K_Y{}^{Y}K_X\right) \end{aligned}$$

(eqn. 1.18)

If the experiment is carried out at constant concentration of one ligand, for example, Y, the fractional saturation of the other ligand results:

$$\begin{aligned} \bar{X} &= \left(\left[PX\right] + \left[PXY\right]\right)/\left(\left[P\right] + \left[PX\right] + \left[PY\right] + \left[PXY\right]\right) \\ &= \left[X\right]\left(K_Y + \left[Y\right]K_X/{}^{Y}K_X\right)/\left\{K_X\left(K_Y + \left[Y\right]\right) + \left[X\right]\left(K_Y + \left[Y\right]K_X/{}^{Y}K_X\right)\right\} \end{aligned}$$

(eqn. 1.19)

Note that if $K_X = {}^{Y}K_X$ (absence of linkage between the two ligands) or $[Y]=0$, this case reduces to eqn. 1.6.

$X_{1/2}$ (i.e., the concentration of free ligand X required to half saturate the protein in the presence of ligand Y) results:

$$X_{1/2} = K_X\left(K_Y + \left[Y\right]\right)/\left(K_Y + \left[Y\right]K_X/{}^{Y}K_X\right)$$

(eqn. 1.20)

which, as expected, is limited between $X_{1/2} = K_X$ in the case $\left[Y\right] << \left(K_Y, {}^{X}K_Y\right)$ and $X_{1/2} = {}^{Y}K_X$ in the case $\left[Y\right] >> \left(K_Y, {}^{X}K_Y\right)$. Contrary to the direct proportionality of $X_{1/2}$ to [Y] observed in the ligand replacement experiment described in Section 1.8, in this case the dependence of $X_{1/2}$ on [Y] is hyperbolic and essentially analogous to a binding curve; its midpoint has coordinates $[K_Y{}^{Y}K_X/K_X, 0.5(K_X + {}^{Y}K_X)]$ (Table 1.1 and Figure 1.7).

An important consequence of the relationships described above is that *one can measure K_Y from the effect of [Y] on $X_{1/2}$*. This feature may turn out very useful if either ligand yields a poor signal: for example, one can measure the affinity of Hb for its heterotropic effectors, whose binding is spectroscopically silent, from the changes they induce in its affinity for oxygen, which is associated with large changes in spectroscopic signal (Figure 1.8).

Table 1.1 Relationships between $X_{1/2}$ and K_d in monomeric single-site proteins.

$X_{1/2}$ **independent of effectors**	absence of linkage; $X_{1/2} = K_d$
$X_{1/2}$ **depends linearly on effector concentration**	competitive binding; identical linkage. $X_{1/2}$ as in eqn. 1.14.
dependence of $X_{1/2}$ on effector concentration is hyperbolic	non-competitive binding; heterotropic linkage. $X_{1/2}$ as in eqn. 1.20.

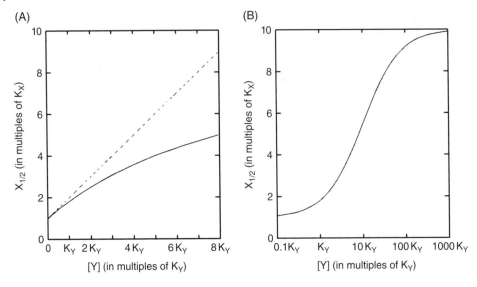

Figure 1.7 A comparison of identical and heterotropic linkages.

Panel A: Identical linkage between ligands X and Y causes $X_{1/2}$ to be a linear function of [Y] (dashed line); heterotropic linkage causes $X_{1/2}$ to depend hyperbolically on [Y]. Panel B: dependence of $X_{1/2}$ on [Y] for the case of heterotropic linkage plotted on a logarithmic scale. $^{Y}K_X = 10\,K_X$; $^{X}K_Y = 10\,K_Y$. The figure demonstrates that dependence of $X_{1/2}$ on the thermodynamic parameters of the system (K_X, $^{Y}K_X$ and K_Y) and the concentration of the other ligand unequivocally identifies the type of linkage between the two ligands X and Y. However, $X_{1/2}$ should not be used for determination of those parameters, since a sounder statistical procedure is to fit the original experimental data (fractional saturation $\bar{X}$ as a function of [X] and [Y]) directly with eqn. 1.14 or 1.19, and recover the three thermodynamic parameters from the fit.

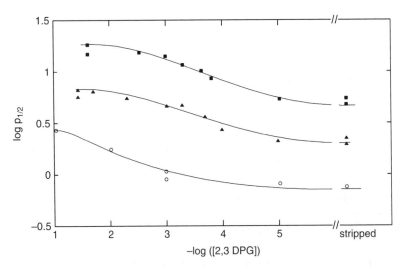

Figure 1.8 Measurement of the affinity of human hemoglobin for glycerate 2,3 bis-phosphate from its effect on the affinity for oxygen. Source: Adapted from Antonini *et al.* (1982).

Experimental conditions and symbols: 0.1 M Bistris or Tris HCl buffers, containing 0.1 M NaCl, pH 6.5 (squares), 7.5 (triangles), or 8.6 (circles); T = 20°C; Hb concentration 60 μM /tetramer.

1.10   Allostery and Allosteric Phenomena in Monomeric Proteins

The term allostery (a neologism from the Greek words for "other structure") was created by J. Monod to describe a possible interpretation of the functional behavior of
enzymes (or other macromolecules) presenting cooperativity and/or heterotropic
linkage (Monod *et al.*, 1963). Two years later Monod, Wyman, and Changeux (Monod
et al., 1965] elaborated the underlying formal concepts in what became known as the
two-state allosteric model (often nicknamed the MWC model). This model suggests
that allosteric proteins can sample two (or more) spatial configurations, each characterized by its own set of equilibrium constants for binding to each of its ligands.
Monod's original idea applied to proteins made up by multiple identical subunits,
each having a ligand-binding site, and required the oligomer to be symmetric
(Monod *et al.*, 1965). However, it was successively demonstrated that also monomeric proteins may exist as equilibrium mixtures of conformational isomers (Austin
et al., 1975). This idea, which when formulated was ahead of its time, has tremen-
dous implications, first of all the thermodynamic necessity that all protein isomers
bind all ligands, albeit with different affinities (Figure 1.9). The reaction scheme of
a monomeric allosteric protein capable of binding two ligands is quite complex see
Figure 1.9.

The important features of this reaction scheme are as follows:

i) The protein has access to (at least) two conformational isomers P and P*, independently of the presence of its ligands; the equilibrium constant of the isomerization
reaction is L = [P*] / [P].
ii) The two isomers have different affinities for the protein's ligands, which bind to different sites (the case of two or more molecules of the same ligand binding to identical sites will be considered in Chapter 4).
iii) The binding of a ligand biases the equilibrium between the isomeric forms of the
protein and favors the form for which it has higher affinity, increasing its relative
population at the expense of the other(s).

The two-state model was meant to explain not only the heterotropic interactions
between two ligands of a monomeric protein, but also the homotropic interactions of
the same ligands binding to multiple identical sites in an oligomeric protein such as

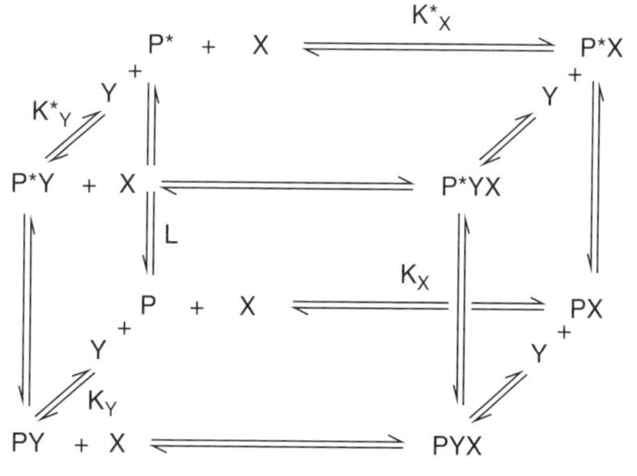

Figure 1.9 Reaction scheme of an allosteric
monomeric protein binding two ligands.
Under the limiting condition where each
ligand has significant affinity for only one of
the different conformational isomers of the
protein, some intermediates may be
neglected.

hemoglobin. A full description of allostery in ligand binding is presented in Chapter 4. However, it is important to stress that allosteric phenomena have been described in monomeric proteins (Cui and Karplus, 2008), and actually some crucial experiments on this subject were carried out on myoglobin (Austin *et al.*, 1975).

A fundamental, yet elusive, tenet of the concept of allostery is the presence of the two (or more) conformational isomers of the protein in the absence of the ligand. Although today the existence of protein conformational dynamics is widely accepted, there was little evidence supporting the MWC viewpoint in its day. Indirect evidence for protein dynamics generally, and for multiple conformational states in particular, became available for selected systems within the decade following publication of the MWC model. Among the earliest and best known of these two categories of evidence are the following. Lakowicz and Weber (1973) demonstrated that compounds as large and chemically diverse as molecular oxygen, acrylamide, or iodide ion could quench the native fluorescence of myoglobin during the lifetime of the excited state of its tryptophan residues (i.e., they showed that the quenching was dynamic rather than static). This result was unexpected because the crystal structure of myoglobin clearly showed the absence of any channels that could allow such compounds to reach the buried tryptophan residues from the protein exterior. Shortly thereafter, Frauenfelder and coworkers, working at the same institution as Lakowicz and Weber (the University of Illinois at Urbana), demonstrated that the photodissociation product of the myoglobin-CO complex at low temperature contains a mixture of conformational isomers (Austin *et al.*, 1975). This result implied that the conformational isomers are present in the bound complex, before the photodissociating light pulse, and made it difficult to argue that unliganded Mb does not also sample multiple states.

We shall not discuss the two-state model further in this section since its formulation is clearly redundant if applied to a monomeric, single ligand-binding site protein, and shall defer its analysis to Chapter 4.

1.11 The Special Case of Cys Ligands (and Similar Reactions)

There are cases in which the protein-ligand reaction may not conform to the equations developed above, and great caution is in order before the researcher commits himself or herself to concepts developed for the standard case. Probably the most common of these is that of ligands of Cys residues. Thus in this section we shall devote our attention to these, and delay the analysis of other special cases to the next section. However, the reaction schemes considered in this section are of general relevance, Cys residues and their reagents being only the most typical examples of a class.

Eqns. 1.1 through 1.20 refer to reversible reactions in which *the ligand and the protein have the same chemical formula in the bound and free states*. This is the case for the majority of ligands forming weak bonds (e.g., steroid hormones and serum albumin) or coordination complexes with the protein (e.g., the ferrous heme iron of hemoglobin and its gaseous ligands). There are cases, however, in which *the ligand has a different chemical formula in its bound and free state*, and these cannot be treated like the preceding cases. It is important to firmly state this point because in these cases usage of the equations developed in this chapter may lead to errors.

This phenomenon commonly occurs with *covalent ligands*. An example is given by the reaction of Cys residues with alkylating agents, metal donors, or other reagents capable of transferring a reactive group. The reaction scheme may be described as follows:

$$P-SH+RX \Leftrightarrow P-SX+HR(\text{or } H_3O^+ + R^-)$$ (eqn. 1.21)

where P is the protein, RX the free ligand used, and HR the leaving group. Some -SH reagents do not present a leaving group, and thus their reaction, if reversible, conforms to eqn. 1.1 instead of 1.21; these are not considered in this Section.

The reaction mechanism and leaving group(s) may vary. Examples of such ligands include gold organometallic complexes like aurothiomalate and auranofin, iodoacetamide, iodoacetate, and their derivatives, and so on (see Cai *et al.,* 2012 for a review). Cys residues are not the only sites where covalent ligand binding may occur, and in all cases where the leaving group is different from water a similar treatment of the experimental data is required.

The reaction described by eqn. 1.21 may be irreversible or reversible. The former case is probably more common, a typical example being provided by alkylating agents. However, the latter case cannot be excluded. An important example of a physiologically relevant reversible ligand of Cys is provided by nitrosothiols. The reaction of these compounds follows the scheme:

$$P-SH+R-SNO \Leftrightarrow P-SNO+R-SH$$

with R-SNO being nitroso-cysteine or nitroso-glutathione. This reaction is reversible and excess free Cys or GSH effectively removes the protein-bound NO (Meyer *et al.,* 1994).

If the reaction is reversible, a dimensionless equilibrium constant can be defined and determined.

$$K = [P-SH][RX]/[P-SX][HR]$$ (eqn. 1.22)

The following considerations apply:

i) *Reversibility should be carefully assessed.* A covalent protein-ligand complex is often irreversible for all practical purposes, and its equilibrium constant is indistinguishable from zero. However, covalent binding does not necessarily imply irreversibility since the reaction's ΔG^0 reflects the algebraic sum of the free energies of all the chemical bonds that break and form in the reaction, and breakage of the RX bond may require an amount of energy similar to that released by the formation of the P-SX bond. As a consequence, in spite of the fact that the protein-ligand complex entails covalent bond(s), we cannot take for granted that the complex is irreversible, even though this is often the case. We suggest that a system involving the formation of covalent bonds is considered irreversible until proven otherwise by extensive dialysis under the appropriate experimental conditions, but we also recommend that reversibility is carefully assessed in all cases. Because of the very high affinity of most Cys reagents and their slow reactivity true thermodynamic reversibility has been demonstrated in few cases, the best example being probably that of nitrosothiols.

ii) *Dialysis removes equally RX and HR and thus is ineffective in dissociating the complex, unless one takes the precaution of adding HR to the dialysis buffer.* As a consequence, if one does not take the necessary precautions these complexes may appear irreversible even when they are not.

iii) *The system is insensitive to dilution* and does not obey Ostwald's law unless the concentration of HR (or H_3O^+ and R^-) is made constant. In fact, if HR dissociates into H_3O^+ and R^- in an unbuffered solution, and no external R^- or HR is added, dilution would be expected to promote complex formation!

iv) If the reaction proves reversible and allows the researcher to define an equilibrium constant, *the equilibrium constant is a pure number* or has units of M^{-1} rather than M. $X_{1/2}$ (in this case $RX_{1/2}$), the concentration of free RX necessary to half-saturate the protein, depends on the protein concentration, unless one forces the system to behave in a way consistent with eqn. 1.1 by carrying out the measurement in the presence of added HR (or R^- ions) at high and constant concentration (see below). (In the buffered solutions usually employed in biochemistry one need not worry about changes in hydrogen ion concentration.)

v) *Binding may be slow* because of the high activation energy for covalent bond breaking and formation, and reaching the true equilibrium condition may require long incubation times. If the incubation time is not long enough, one obtains an $\bar{X}$ versus [RX] plot that may superficially resemble a hyperbola, but is actually an exponential (see Section 3.4).

vi) A phenomenon that for lack of a better name we call *pseudo-reversibility* may be observed. Pseudo-reversibility occurs when the complex dissociates according to a reaction scheme different from that of association. As a consequence, only the native state of the protein is recovered, and not the native state of the ligand. An example of pseudo-reversibility is provided by the case of thioredoxin reductase covalently alkylated by nitrosoureas (Schallreuter *et al.*, 1990; Saccoccia *et al.*, 2014). In this case thioredoxin, the substrate of the enzyme, can extract the ligand from the Cys residues in the active site (and can also catalyze a redox reaction, which releases a reduced state of X).

The consequence pseudo-reversibility is recovery of the unmodified protein, via a reaction that is not the inverse of the binding reaction, and the compound RX is not recovered at the end of the process. We refer the reader to the review by Saccoccia and co-workers (Saccoccia *et al.*, 2014) for a more detailed discussion of sulfhydryl ligands. The reason pseudo-reversibility is important is that, besides its possible biological relevance, it may simulate true thermodynamic reversibility, and may induce the researcher to assume that a reaction is reversible when in fact it is not.

Obviously, the correct identification of the reaction scheme and application of the appropriate formula for the equilibrium constant is necessary and sufficient to determine the equilibrium constant from the experimental data. Provided that the reaction has a signal, that reversibility has been carefully checked by dialysis against excess HR, and that the time required for the complex to reach its equilibrium condition has been determined (see points 2 and 5, above), a binding isotherm may be recorded by successive additions of RX to the protein. The experiment may be carried out in two different ways.

In the presence of added HR at a concentration significantly higher than that of the protein, one may assume that the concentration of this component is nearly constant throughout the measurement, leading to a rearrangement of eqn. 1.22

$$K' = K[HR] = [P-SH][RX]/[P-SX]$$

which is analogous to eqn. 1.3. Under these conditions all the equations developed in this Chapter may be applied, using K' instead of K.

In the absence of added HR, equal amounts of P-SX and HR are produced (i.e., [HR]=[P-SX]), which leads to the following rearrangement of eqn. 1.22

$$K = [P-SH][RX]/[P-SX]^2$$

Introducing $[P]_{tot} = [P-SH] + [P-SX]$ leads to:

$$K[P-SX]^2 + [P-SX][RX] - [P]_{tot}[RX] = 0$$

from which [P-SX] (and $\bar{X} = [P-SX]/[P]_{tot}$) can be calculated.

The $\bar{X}$ vs log[RX] plot has a symmetric sigmoidal shape, broader than that one would obtain from eqn. 1.6. The midpoint of the plot, that is, the concentration of RX required to half-saturate the protein, is defined by the condition: [P-SX]=[HR]=[P-SH]=0.5[P]$_{tot}$, and results

$$\mathbf{RX}_{1/2} = 0.5K[P]_{tot}$$

We call the reader's attention on the fact that the concept of $\mathbf{X_{1/2}}$ does not usually depend on the target's concentration, and thus the above definition of the $\mathbf{RX_{1/2}}$ is atypical. It is common that enzymes having reactive Cys residues in their active site are inhibited by -SH reagents having reaction mechanisms similar to those described in this Section, for which $\mathbf{IC_{50}}$ values (fully analogous to $\mathbf{RX_{1/2}}$) are of interest. In these cases one must pay attention to the necessary controls and considerations (points 1 to 6, above), and to the atypical form of $\mathbf{RX_{1/2}}$ ($\mathbf{IC_{50}}$). This point is further discussed under Section 11.5.

1.12   Other Special Cases

Other special cases not fully conformant to eqns. 1.1–1.20 include reactions having semi-stable intermediates, and those cases that, unknown to the researcher, are replacement reactions rather than binding equilibria, and obey eqns. 1.11–1.16, even though the researcher expects them to obey eqns. 1.1–1.6.

Monomeric proteins having more than a single binding site for the same ligand will not be considered here and are deferred to Chapter 4. This case is often observed when the ligand is a low molecular weight ion: for example, the monomeric protein calmodulin binds four calcium ions. We only remark here that these cases highlight the importance of direct determination of binding stoichiometry and reaction mechanism, even in the case of monomeric proteins.

A ligand-binding reaction having a stable (equilibrium populated) intermediate is as follows:

$$P + X \Leftrightarrow PX \Leftrightarrow PX^*$$

where PX and PX* are structural isomers, whose ratio is independent of ligand concentration. In this case the bound state of the protein is a mixture of two conformers in fixed proportions, that may or may not be recognized depending on whether the signal associated with PX and PX* is the same or not.

If the researcher assumes that the full signal change corresponds to the complete titration of the protein with the ligand, this case may be unnoticed. However, if the signal quantitatively measures the concentration of only one of the two bound states (e.g., PX*), while the other is not distinguished from the unbound state, incomplete binding may be erroneously suspected. Kinetics may often provide the clues to correctly identify this case (see Chapter 2).

Reactions that, unknown to the researcher, are replacements rather than ligand binding equilibria are not uncommon. A typical example is the reaction of several ferric hemoproteins their ionic ligands, for example, the reaction of ferric myoglobin with azide, cyanide or other ligands:

$$MbFe^{+3} - N_3^- + H_2O \Leftrightarrow MbFe^{+3} - OH^- + HN_3$$

The heme iron of several ferric hemoproteins, including vertebrate hemoglobins and myoglobins, is usually six-coordinated, either a water molecule or a hydroxyl ion occupying the sixth coordination position, depending on pH. When these proteins are titrated with their typical ligands (azide, cyanide, etc.), the water or hydroxyl ion must be displaced. Since the concentration of these species is constant (water because of its abundance and OH^- because of the buffer), the equilibrium experiments fully conform to a titration; but the kinetics does not (see Chapter 2). Moreover, the ionic ligands of ferric hemoproteins are Lewis bases and the reaction above may be read as a double ligand replacement: the azide ion is exchanged between two Lewis acids (iron and the hydrogen ion) at the same time as it is replaced at the iron by water.

A similar type of problem may occur in the case of the binding of metal ions to proteins, since upon dissociation these react with solution components, and may possibly precipitate. For example, copper(II), dissociated from a carrying protein, forms copper hydroxide $Cu(OH)_2$, which has a solubility product of $4.8 \times 10^{-20} M^3$.

As a general rule, the problem in these cases is to unravel the reaction mechanism and stoichiometry, in order to apply the analysis appropriate to the experimental data one collects.

References

Ackers G.K., Shea M.A., and Smith F.R. (1983) Free energy coupling within macromolecules. The chemical work of ligand binding at the individual sites in co-operative systems. J Mol Biol, 170: 223–242.

Angelucci *et al.* (2014) Schistosoma mansoni fatty acid binding protein. Biochemistry, 43: 13000–13011.

Antonini E., Condò S.G., Giardina B., Ioppolo C., and Bertollini A. (1982) The effect of pH and D-glycerate 2,3-bisphosphate on the O2 equilibrium of normal and SH(β 93)-modified human hemoglobin. Eur J Biochem, 121: 325–328.

Antonini E. and Brunori M. (1971) *Hemoglobin and Myoglobin in Their Reactions with Ligands*. North Holland, Amsterdam.

Austin R.H., Beeson K.W., Eisenstein L., Frauenfelder H., and Gunsalus I.C. (1975) Dynamics of ligand binding to myoglobin. Biochemistry, 14(24): 5355–5373.

Bellelli A. (2010) Hemoglobin and cooperativity: Experiments and theories. Curr Protein Pept Sci, 11: 2–36.

Cai W., Zhang L, Song Y, Wang B, Zhang B, Cui X, Hu G, Liu Y, Wu J, and Fang J. (2012) Small molecule inhibitors of mammalian thioredoxin reductase. Free Radic Biol Med, 52: 257–265.

Cheng Y., and Prusoff W.H. (1973) Relationship between the inhibition constant (K1) and the concentration of inhibitor which causes 50 per cent inhibition (I50) of an enzymatic reaction. Biochem Pharmacol. 22: 3099–3108.

Coletta M., Clementi M.E., Ascenzi P., Petruzzelli R., Condò S.G., and Giardina B. (1992) A comparative study of the temperature dependence of the oxygen-binding properties of mammalian hemoglobins. Eur J Biochem, 204: 1155–1157.

Cui Q., and Karplus M. (2008) Allostery and cooperativity revisited. Protein Sci, 17: 1295–1307.

Di Cera E., and Gill S.J. (1988) A new method for the determination of equilibrium constants through binding capacity measurements. Biophys Chem, 32: 149–152.

Gill S.J., Di Cera E., Doyle M.L., Bishop G.A., and Robert C.H. (1987) Oxygen binding constants for human hemoglobin tetramers. Biochemistry, 26: 3995–4002.

Lakowicz J.R. and Weber G. (1973) Quenching of protein fluorescence by oxygen. Detection of structural fluctuations in proteins on the nanosecond time scale. Biochemistry, 12: 4171–4179.

Forsén S., and Linse S. (1995) Cooperativity: over the Hill. Trends Biochem Sci, 20: 495–497.

Meyer D.J., Kramer H., Ozer N., Coles B., and Ketterer B. (1994) Kinetics and equilibria of S-nitrosothiol-thiol exchange between glutathione, cysteine, penicillamines and serum albumin. FEBS Lett, 345: 177–180.

Monod J., Changeux J.P., and Jacob F. (1963) Allosteric proteins and cellular control systems. J Mol Biol, 6: 306–329.

Monod J., Wyman J., and Changeux J.P. (1965) On the nature of allosteric transitions: a plausible model. J Mol Biol, 12: 88–118.

Rebek J. Jr., Costello T., Marshall L., Wattley R., Gadwood R.C., and Onant K. (1985) Allosteric effects in organic chemistry: Binding cooperativity in a model for subunit interactions. J Am Chem Soc, 107: 7481–7487.

Saccoccia *et al.* (2014) Thioredoxin reductase and its inhibitors. Curr Prot Pept Sci, 15(6): 621–646.

Scatchard G. (1949) The attraction of proteins for small molecules and ions. Ann NY Acad Sci, 51: 660–672.

Schallreuter K.U., Gleason F.K., and Wood J.M. (1990) The mechanizm of action of the nitrosourea anti-tumor drugs on thioredoxin reductase, glutathione reductase and ribonucleotide reductase. Biochim Biophys Acta, 1054: 14–20.

Traylor T.G., and Berziniz A/P. (1980) Binding of O2 and CO to hemes and hemoproteins. Proc Natl Acad Sci USA, 77: 3171–3175.

Traylor T.G., and Sharma V.S. (1992) Why NO? Biochemistry, 31: 2847–2849.

Weber G. (1992) *Protein Interactions*. Elsevier: Chapman and Hall.

Wyman J. (1948) Heme proteins. Adv Protein Chem, 4: 407–531.

Wyman J. (1963) Allosteric effects in hemoglobin. Cold Harbor Spring Symposia on Quantitative Biology, XXVIII: 483–489.

Wyman J. (1964) Linked functions and reciprocal effects in hemoglobin: a second look. Adv Protein Chem; 19: 223–286.

Wyman, J. and Gill, S.J. (1990) *Binding and Linkage*. Mill Valley, CA: University Science Books.

2

Ligand-Binding Kinetics for Single-Site Proteins

As explained in the preceding chapter, it is essential to the determination of ligand-binding isotherms that the system is allowed enough time to reach its equilibrium condition: this implies *per se* that the researcher has some estimate of the rate at which the system moves towards the equilibrium. Thus, determination of the rate constant and reaction order is important even though the researcher may only be interested in determining K_d or $X_{1/2}$. Neglecting the kinetic aspects of the reaction may lead to errors in design of the experiment or in interpretation of the experimental data.

Kinetics, however, is not limited to this ancillary role: it provides important information about the reaction mechanism. Thus, in this chapter we shall describe the kinetic aspects of the phenomena described in Chapter 1. Wherever possible, we follow in this chapter the excellent treatment of the subject matter by Antonini and Brunori (1971). More specialized subjects are referenced to the original research papers.

2.1 Basic Concepts of Chemical Kinetics: Irreversible Reactions

All chemical reactions are reversible, at least to some extent; however, some key concepts of chemical kinetics are better introduced under the assumption that reversibility can be at least temporarily neglected. The reason for this lies in the fact that kinetic reversibility stems from the coexistence of a forward and a backward reaction, proceeding independently of each other. The overall rate of the reaction results from the sum of the forward and backward rates, and chemical equilibrium is the condition in which the rate of the forward reaction equals that of the backward reaction. A great simplification of the analysis occurs when either the forward or backward reaction is negligible with respect to the other. In these cases the equilibrium condition is indistinguishable from the exhaustion of at least one of the reagents, and the reaction rate is dominated by either the association or the dissociation process. Thus, for pedagogical purposes, we shall introduce our analysis by referring to experimental conditions where either the dissociation of the protein-ligand complex or the association of the ligand with its target may be neglected.

The rate of the association reaction may be made negligible either because one suddenly changes the experimental conditions in a way that dramatically lowers the affinity of the ligand for its target, for example, by a pH change, or because one chemically removes

Reversible Ligand Binding: Theory and Experiment, First Edition. Andrea Bellelli and Jannette Carey.
© 2018 John Wiley & Sons Ltd. Published 2018 by John Wiley & Sons Ltd.

the free ligand from the solution. A typical example of the latter is provided by the dithionite reduction of dioxygen, a reaction used to induce the irreversible dissociation of oxy-myoglobin. The rate of the dissociation reaction will be practically negligible in the case of high-affinity complexes and high-ligand concentrations, leading to reaction time courses that approximate irreversible association. A typical example is provided by the association of NO to myoglobin, where the association rate at micromolar ligand concentration exceeds the dissociation rate by over 100,000-fold.

The irreversible dissociation of the target-ligand complex is described by the chemical equation:

$$PX \rightarrow P + X \qquad \text{(eqn. 2.1)}$$

The reaction rate is defined as the ratio between the change (decrease) of the concentration of the reagent and the time interval in which it occurred:

$$-\Delta[PX]/\Delta t = -\left([PX]_2 - [PX]_1\right)/(t_2 - t_1)$$

In the above equation $t_2 > t_1$, and $[PX]_2 < [PX]_1$; thus we adopt a negative sign for $\Delta[PX]$ in order to have a positive rate. Moreover, for reasons analyzed below, we want $\Delta[PX]$ and Δt to be as small as possible.

Since every PX complex has the same probability as any other to dissociate during the time interval Δt, the differential may be equated to the product of a probability constant (the *kinetic constant*) times the instantaneous concentration of the reagent at time t_1; the kinetic differential equation results:

$$-\delta[PX]/\delta t = k_d[PX] \qquad \text{(eqn. 2.2)}$$

By convention lower case k is used for kinetic constants (and capital case for equilibrium constants) and we use the suffix d to indicate the dissociation reaction. Because the reaction is irreversible, its time course is not affected by the presence of its products, and the kinetic law contains only the concentration of the reactant PX. The kinetic constant k_d in this case has the units of s^{-1}.

A differential equation like eqn. 2.2 allows us to calculate $[PX]_2$ if we know $[PX]_1$, δt, and k_d. This is not a little feat in itself, and the use of numerical integration routines and algorithms on modern computers can very quickly calculate stepwise the entire time evolution of the reaction from t=0 to the virtual disappearance of the reactant PX. However, eqn. 2.2 can be easily integrated analytically, to yield:

$$\bar{X}_t = [PX]_t / [PX]_0 = e^{-k_d t} \qquad \text{(eqn. 2.3)}$$

where $\bar{X}_t$ represents the fractional saturation of protein P with ligand X after time t from the start of the reaction.

In most (but not all) experimental designs, only one species of the protein is present at the start of the reaction (time=0). If this is the case, the initial concentration of the protein-ligand complex is identical to the total protein concentration, that is, $[PX]_0 = [P]_{tot}$.

Eqn. 2.3 demonstrates that the concentration of PX (or the fractional ligand saturation) at time t is an exponential function of the time elapsed from the start of the reaction.

A chemical reaction in which only one molecule of a reactant is converted to (any number of) products, as in eqn. 2.1, is defined as a *monomolecular* reaction, and the *molecularity* of a reaction equals the number of reactant molecules it requires. A reaction whose kinetic law is a function of only one reactant concentration, as in eqn. 2.2, is defined a *first-order* reaction, and the order of a reaction equals the degree of the monomial (or polynomial) that expresses its time-dependence (Cornish-Bowden, 1995). The order and molecularity of a reaction are two different concepts; they are expressed by the same number if the reaction mechanism is simple and does not imply long lived intermediates (as in eqns. 2.1 and 2.2, whose molecularity and order are both unity); by different numbers otherwise. It follows that when we write a kinetic model of a chemical reaction with the ambition of describing all relevant reaction steps, the molecularity and order should be expressed by the same number for every step; if they do not, the model requires additional steps.

An irreversible ligand association reaction may be expressed by the following chemical equation.

$$P + X \rightarrow PX \qquad \text{(eqn. 2.4)}$$

whose kinetic law is:

$$-\delta[P]/\delta t = -\delta[X]/\delta t = k_a[P][X] \qquad \text{(eqn. 2.5)}$$

This reaction is bimolecular and, if its mechanism does not include intermediates, it is second order, as its velocity depends on two concentrations (eqn. 2.5). It is important to remark that while we can with some confidence infer the molecularity from knowledge of the formulas of the reactants and products (though exceptions may occur), we cannot infer the reaction order. We need to determine the reaction order by experiment, from the dependence of the reaction velocity on the concentration(s) of the reactant(s).

Here again the reaction time course can be calculated by numerical integration, or eqn. 2.5 can be analytically integrated to yield:

$$[X]_t / [P]_t = [X]_0 / [P]_0\, e^{([X]0-[P]0)\,k_a\, t} \qquad \text{(eqn. 2.6)}$$

Where $[P]_0$ and $[X]_0$ refer to the protein and ligand concentrations at the start of the reaction, and $[P]_t$ and $[X]_t$ to their concentrations at time t.

Eqn. 2.6 may seem quite awkward to use; however, it can be rearranged to:

$$([X]_0 - [PX]_t)/([P]_0 - [PX]_t) = ([X]_0 / [P]_0)\, e^{([X]_0-[P]_0)\,k_a\, t} \qquad \text{(eqn. 2.7)}$$

which yields:

$$\bar{X}_t = [PX]_t / [P]_{tot} = [X]_0 \left(e^{([X]_0-[P]_0)\,k_a\, t} - 1 \right) / \left([X]_0\, e^{([X]_0-[P]_0)\,k_a\, t} - [P]_0 \right) \qquad \text{(eqn. 2.8)}$$

A considerable simplification is obtained if the ligand concentration is so much higher than that of the protein that the amount of the bound ligand at the end of the reaction is negligible, that is, $[X]_{t\rightarrow\infty} \approx [X]_{tot}$. If this condition applies, we can consider the concentration of the ligand X to be constant throughout the time course, and define the apparent *pseudo-first order* combination rate constant k_a':

$$k_a' = k_a[X]_{tot}$$

and the kinetic law reduces to:

$$-\delta[P]/\delta t = k'_a\,[P]$$

$$\bar{X}_t = 1 - e^{-[X]_{tot}\,ka't}$$

(eqn. 2.9)

which are analogous to eqns. 2.2 and 2.3. The difference between eqn. 2.3 and 2.9 depends on the different direction and starting point, PX for the former, P+X for the latter. The reader may also verify that eqn. 2.9 can be derived from eqn. 2.8 under the assumption $[X]_0 e^{([X]_0 - [P]_0)\,ka\,t} >> [P]_0$.

An important kinetic concept is the half-time or half-life, indicated as $t_{1/2}$, and defined as *the time required for the concentration of the reactant to decrease to half its initial value*. The $t_{1/2}$ of a first-order reaction is independent of the initial concentration of the reactant and equals $\ln(2)/k$. The $t_{1/2}$ of a second-order reaction of the type $P + X \rightarrow PX$ depends on the initial concentration of the reactant that is in excess; under conditions of pseudo-first order it is: $t_{1/2,P} = \ln(2)/k[X]_{tot}$ (assuming that $[X]_{tot}>>[P]_{tot}$). Notice that only the half time of the reactant that is in lower amount can be defined, whereas the half time of the reactant that is in excess is not meaningful, and may not exist at all, if $[X]_{tot}>2[P]_{tot}$, because under these conditions P will have been entirely used up before the concentration of X has been halved.

Figure 2.1 reports two ligand-binding time courses for the second-order and pseudo-first order conditions, adjusted so as to have approximately the same $t_{1/2}$.

The kinetic rate constant expresses the probability that a reactant acquires, via collisions with other molecules, a kinetic energy high enough to break some chemical bonds,

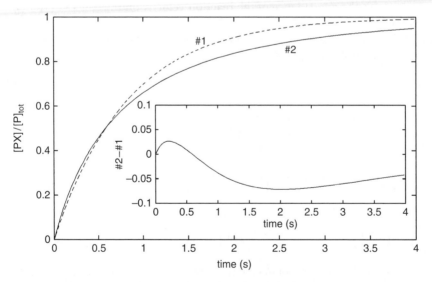

Figure 2.1 Ligand-binding time courses for the second order (continuous line) and pseudo-first order (dashed line) conditions adjusted to have approximately the same $t_{1/2}$. ($\approx$0.6 s) Time courses were calculated using eqns. 2.9 (k=800 M^{-1}s^{-1}, [P]$_0$ = negligible, [X]$_0$=1.5 mM; dashed line, marked as #1) and 2.8 (k=1000 M^{-1}s^{-1}, [P]$_0$ = 1 mM, [X]$_0$=1.5 mM; continuous line, marked as #2). Compared to the exponential, pseudo-first order formation of the complex (dashed line), the second-order reaction time course (continuous line) decelerates, reflecting the effect of decreasing ligand concentration. Inset: difference between time courses #2 and #1.

so that new chemical bonds, characteristic of the product, may be formed. The energy required by this process is called the *activation energy* (Ea) and the statistical distribution of the molecules with respect to this energy is described by Boltzmann's law. The relationship between the activation energy and the rate constant describes a distribution identical to that predicted by Boltzmann, as formalized by Arrhenius law:

$$k = A\,e^{-Ea/RT}$$

An important practical aspect of Arrhenius law is that it predicts a positive relationship between the temperature and the kinetic constant. That is, all chemical reactions become faster as the temperature is raised.

2.2 Reversible Reactions: Equilibrium and Kinetics

Because chemical equilibrium is defined as the condition where the concentrations of reactants and products are constant in time, it follows that at equilibrium the rate of the forward reaction equals that of the backward reaction. In the simplest possible case of a protein binding a ligand we have:

$$\left[PX\right]_{eq} k_d = \left[P\right]_{eq}\left[X\right]_{eq} k_a$$
$$K_d = \left[P\right]_{eq}\left[X\right]_{eq} / \left[PX\right]_{eq} = k_d / k_a$$

(eqn. 2.10)

where, as usual, lower case k is used to refer to rate constants, and upper case K to equilibrium constants. The subscript eq refers to the concentration at equilibrium.

The above relationship implies that the equilibrium constant equals the ratio of the appropriate rate constants, provided that the order of the reaction equals its molecularity (i.e., the reaction mechanism does not involve long lived intermediates). In those cases where the order does not equal molecularity, a more complex relationship between the kinetic rate constants and the equilibrium constant may apply (see below).

In the case hypothesized in eqn. 2.10, that hopefully is rather frequent, long-lived reaction intermediates are not populated, and the ligand association reaction is second order, whereas the dissociation reaction of the protein-ligand complex is first order.

If the protein-ligand complex has high affinity and the dissociation rate under the chosen experimental conditions is negligible with respect to the association rate, then upon mixing the unliganded protein with its ligand one will observe a time-dependent signal change related to the complete conversion of P to PX. Under the more interesting and more frequent condition in which the dissociation rate constant is not negligible with respect to the association rate, the total amplitude of the signal change depends on ligand concentration. The apparent rate of the approach to equilibrium depends on both the association and dissociation rate constants, and under the pseudo-first order approximation results:

$$-\delta\left[P\right]/\delta t = \delta\left[PX\right]/\delta t = k_a\left[P\right]\left[X\right] - k_d\left[PX\right] = k_a{}'\left[P\right] - k_d\left[PX\right]$$

In the absence of long lived intermediates $[P]_{tot} = [P] + [PX]$, and we substitute:

$$-\delta\left[P\right]/\delta t = k_a{}'\left[P\right] - k_d\left(\left[P\right]_{tot} - \left[P\right]\right) = \left[P\right]\left(k_a{}' + k_d\right) - k_d\left[P\right]_{tot}$$

which integrates to:

$$\left[P\right]_t - \left[P\right]_{eq} = \left(\left[P\right]_{tot} - \left[P\right]_{eq}\right)e^{-\left(k_a[X]+k_d\right)t} \qquad \text{(eqn. 2.11)}$$

This equation shows that under the pseudo-first order approximation the time course of the approach to equilibrium is exponential. Its apparent rate constant equals the sum of the rate constant of association multiplied by the ligand concentration plus the rate constant of dissociation of the complex. Moreover, the amplitude of the signal change is a function of ligand concentration, because this parameter determines the equilibrium concentrations of P and PX:

$$\text{signal} = f\left\{[X]/\left([X]+k_d/k_a\right)\right\}$$

It is important to stress these relationships because data analysis should always be carried out using the minimum possible number of variable parameters. Thus, the amplitudes of a series of time courses collected at constant $[P]_{tot}$ and variable $[X]_{tot}$ should be fitted using the above relationship.

A series of simulated time courses for reversible binding of a ligand (calculated using eqn. 2.11) is presented in Figure 2.2A, and the corresponding equilibrium isotherm, taken from the amplitudes measured at $t \rightarrow \infty$, in practice $t > 5\,t_{1/2}$, is reported in Figure 2.2B.

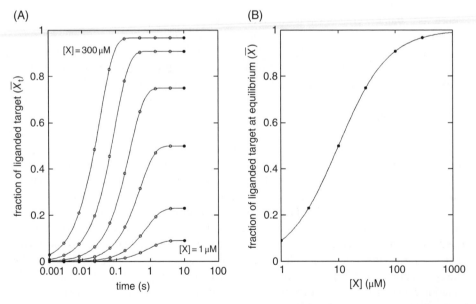

(A)

(B)

Figure 2.2 Panel A: time courses of reversible ligand association simulated using eqn. 2.11. Panel B plot of the end points of the time courses reported in Panel A as a function of ligand concentration. Data used in the simulation are as follows: $k_a = 10^5\,M^{-1}s^{-1}$; $k_d = 1\,s^{-1}$; ligand concentrations 1, 3, 10, 30, 100 and 300 μM. Pseudo-first order was assumed throughout (i.e., [P]<<0.1 μM). The equilibrium dissociation constant equals $X_{1/2}$ and corresponds to $k_d/k_a = 10^{-5}$ M (Panel B). Notice that each time course is exponential (i.e., pseudo-first order), but they become faster and faster as the ligand concentration increases. As a consequence, the time required to approach equilibrium is longer at low ligand concentrations.

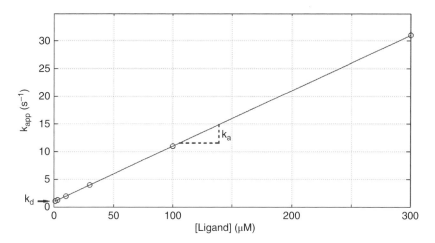

Figure 2.3 Plot of the apparent rate constants from Figure 2.2A as a function of ligand concentration. The rate constants should be obtained from global fitting of experiments like that simulated in Figure 2.2A, rather than by fitting each time course to an exponential as the former procedure is statistically sounder. The plot is a straight line whose slope corresponds to k_a and whose intercept corresponds to k_d.

An important property of the system that we derive from eqn. 2.11 is that the apparent rate constant for the approach to equilibrium under the pseudo-first order approximation depends linearly on [X], with intercept k_d and slope k_a, as shown in Figure 2.3 (calculated for the rate constants and ligand concentrations used in Figure 2.2A). The practical interest of this relationship lies in the fact that it allows the researcher to estimate the k_d from experiments in which the ligand and the target are mixed and the association of the two is followed. This property of the system is useful because it is not always possible to design an experiment in which the time course of dissociation of the ligand can be followed directly, whereas designing an experiment to follow the time course of the association reaction is usually straightforward. As usual, the relationship illustrated in Figure 2.3 should be exploited by directly fitting the experimentally recorded time courses using eqn. 2.2, rather than by linear regression of a set of apparent rate constants estimated independently from each time course, as the former procedure is statistically sounder. *Measuring k_a and k_d and taking their ratio is an alternative way of determining the equilibrium constant, provided that the reaction order coincides with its molecularity.*

2.3   More Complex Kinetic Mechanisms

In biological systems it is quite possible to find reactions having complex mechanisms that nevertheless more or less obey eqn. 2.11. We consider below a reaction scheme that includes one intermediate and may behave in quite different ways, depending on the values of the pertinent rate constants and on whether the intermediate yields some signal or is silent. The prototype system is as follows:

$$P + X \underset{k_2}{\overset{k_1}{\rightleftharpoons}} P-X \underset{k_4}{\overset{k_3}{\rightleftharpoons}} PX$$

In this case the ligand forms an initial, unstable complex P—X with the protein, that may either dissociate or convert to a more stable complex (PX). This is for all purposes a three-state system described by two equilibrium constants and four rate constants, whose relationships (written in the direction of dissociation) are:

$$K_{d,1} = k_2/k_1$$
$$K_{d,2} = k_4/k_3$$

The binding polynomial of this system results:

$$[P]_{tot} = [P]\left(1 + [X]/K_{d,1} + [X]/K_{d,1}K_{d,2}\right) \qquad \text{(eqn. 2.12)}$$

Several possible cases may occur, depending on the relative amplitudes of the rate constants, of which we shall consider only two:

i) The P—X intermediate is unstable and its equilibrium population is negligible. This occurs if the first-order rate constants k_2 and k_3 are both much greater than the first-order rate constant k_4 and the pseudo-first order rate constant $k_1[X]$. A typical (but extreme) example is given by myoglobin, whose ligands must first migrate to the heme pocket to form the unstable *geminate pair*, described in Section 2.5, and then, in a second step, chemically combine with the heme iron (Austin *et al.*, 1975; Gibson *et al.*, 1986).

In this case, the equilibrium constant is actually an apparent one:

$$K_{d,app.} = [P][X]/[PX] = K_1 K_2 = k_2 k_4 / k_1 k_3 \qquad \text{(eqn. 2.13)}$$

The apparent kinetic constants are:

$$k_{a,app.} = k_1 k_3/(k_2 + k_3) \qquad \text{(eqn. 2.13a)}$$

$$k_{d,app.} = k_4 k_2/(k_2 + k_3) \qquad \text{(eqn. 2.13b)}$$

and their ratio yields the apparent equilibrium constant.

We notice that $k_{a,app.}$ is second order because it equals the product of the second-order rate constant k_1 times the dimensionless ratio $k_3 / (k_2+k_3)$. Under pseudo-first order conditions the following relation applies: $k'_{a,app.} = k_1[X]k_3/(k_2 + k_3)$.

ii) Another quite common case is that of a system in which the intermediate complex P—X is populated to a non-negligible extent, but it is indistinguishable from the unbound state (P+X). In this case, the equilibrium fractional ligand saturation, as calculated from the observed signal, depends only on [PX] and $[P]_{tot}$, that is:

$$\bar{X} = [PX]/[P]_{tot} = [X]/K_{d,1}K_{d,2}/\left(1 + [X]/K_{d,1} + [X]/K_{d,1}K_{d,2}\right) \qquad \text{(eqn. 2.14)}$$

This case may initially escape detection, unless one has independent evidence of the existence of the silent species P—X. In particular eqn. 2.14 describes a rectangular hyperbola with asymptote $1/(K_{d,2}+1)$ (instead of 1 as Figures 1.1 and 2.2). It is important to remark that this case will be recognized only if the signal effectively reports that at the asymptote of the binding isotherm $[PX] \neq [P]_{tot}$ (e.g., because of an NMR peak that is expected to disappear but does not). If the signal is normalized to its end point

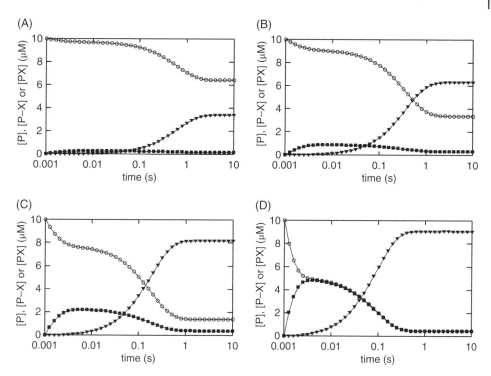

Figure 2.4 Simulated time courses for the reaction scheme $P + X \Leftrightarrow P-X \Leftrightarrow PX$. Conditions, all panels: total protein concentration 10 uM, $k_1=10^6\,M^{-1}s^{-1}$; $k_2=1000\,s^{-1}$; $k_3=20\,s^{-1}$; $k_4=1\,s^{-1}$; $K_{d,1}=10^{-3}\,M$; $K_{d,2}=0.05$. Circles represent the concentration of unliganded protein (P); squares that of the intermediate complex P–X; triangles that of the end complex PX. Total ligand concentrations: 30 µM (Panel A); 100 µM (Panel B); 300 µM (Panel C); and 1 mM (Panel D).

(as is usually done, e.g., when the signal is fluorescence or absorbance) one cannot easily realize that 100% of the transition does not correspond to 100% of PX.

The time courses of ligand association for this system are simulated in Figures 2.4 and 2.5A, and its apparent equilibrium curve together with its best hyperbolic approximation is reported in Figure 2.5B. Unfortunately, the hyperbolic approximation provides an imprecise description.

Evidence of a system in which a silent intermediate is significantly populated is that, when studying the kinetics of complex formation and dissociation, one observes that at low concentrations of the ligand the reaction approximates the second order, whereas at high concentrations of ligand it tends to first order (compare Figure 2.5B, inset with Figure 2.2). This apparently paradoxical effect occurs because at low [X] the *rate-limiting step* is the second-order formation of the initial complex P–X, whereas at high [X] the initial complex forms rapidly, but the reaction is rate limited by the first-order conversion of P–X to PX. Thus, in this case, kinetic analysis may reveal a subtle feature of the reaction mechanism, which might be overlooked in an equilibrium experiment. Clearly, this is one of those cases in which the ratio of the kinetic constants cannot be directly equated to the apparent equilibrium constant.

The interpretation of the rate limiting step requires caution, because in some cases the rate may be determined, or limited, by more than a single reaction step. For example,

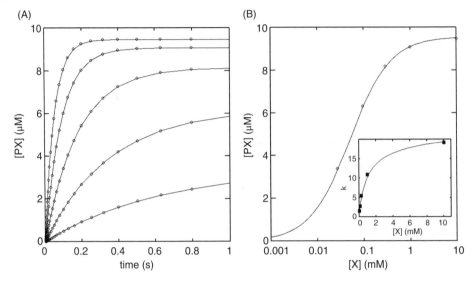

Figure 2.5 Time courses of the approach of equilibrium (Panel A) and equilibrium ligand-binding isotherm (Panel B) of the system simulated in Figure 2.4. Panel A: the time courses of formation of species PX at [X]=0.03, 0.1, 0.3, 1, and 10 mM were fitted to single exponentials; the apparent rate constants obtained were k=1.6, 2.8, 5.5, 11, and 19 s^{-1}. Only the initial part of the simulated time courses is represented in the figure in order to expand the differences. Panel B: the simulated end points of the time courses from Panel A (at 100 s, not shown in the panel) are plotted as a function of the free ligand concentration at the end of the time course. The line is calculated for a single site binding with $K_{d,app} = K_{d,1}K_{d,2} = 5\times10^{-5}$ M and asymptote 9.5 µM. Inset: kinetic constants from panel A as a function of ligand concentration. In the absence of the intermediate this plot should be a straight line with slope k_a and intercept k_d (see Section 2.2 and Figure 2.3); in the present case the order of the reaction decreases from 2 to 1 as [X] increases.

in the mechanisms presented in this section, at low [X] the rate of formation of PX is second order, but the apparent rate constant cannot exceed the limit fixed by eqn. 2.13a, which is obviously lower than k_1.

Other cases are possible (e.g., the intermediate species P−X may give the same signal as PX or a signal different from both those due to P+X and PX), but they are less common and will not be discussed.

2.4 Reactions with Molecularity Higher Than Two

Although there is no theoretical limit to the molecularity of a chemical reaction, reactions of order higher than two are exceedingly rare. The reason why the order of reaction does not exceed two is that a true order 3 reaction would require the simultaneous collision of three molecules of the reactants, an event that is extremely unlikely, except (possibly) at very high concentrations. Thus reactions with molecularity higher than two typically occur as sequences of bi-molecular collisions populating semi-stable intermediates whose fate is to dissociate toward the reactants or to further react toward the products. For example, the reaction $2\,NO + O_2 \rightarrow 2\,NO_2$ proceeds *via* more than a single mechanism, the most efficient involving two bimolecular steps, the first of which populates the unstable intermediate N_2O_2. N_2O_2 may collide with O_2 to yield the product or dissociate to two molecules of NO.

In the study of ligand binding we may encounter reactions with molecularity higher than two in two cases:

i) that of a protein that binds two non-competing ligands: $P + X + Y \rightarrow PXY$;
ii) that of a protein that binds more than one molecule of the same ligand: $P + 2X \rightarrow PX_2$.

In both cases the end product is reached via two second-order steps, and possibly one or more first-order steps. We shall consider the latter reaction under Chapter 6, because they imply ligand:target stoichiometries higher than 1.

2.5 Classical Methods for the Study of Ligand-Binding Kinetics

In this book we shall not be concerned with the technical details of the study of ligand binding kinetics and mechanisms, thus we shall not present an extensive description of the instrumentation. However, it is impossible to describe the kinetics of ligand-binding without at least a rudimentary description of the different experiments that can be carried out. This is particularly relevant if one wants to explore the relationships between equilibrium and kinetic experiments. Our main interest is two-fold:

(i) No reliable equilibrium experiment can be carried out in the absence of at least rudimentary information about the reaction kinetics, because, by definition, the equilibrium state is the asymptote of the reaction time course (Figure 2.1; see also Chapter 3 and Figure 3.3). (ii) In some cases kinetic experiments are easier to carry out than equilibrium ones and the researcher may infer the equilibrium constant from kinetic data.

The most intuitive method for studying the ligand-binding time course is to mix a solution of the protein with a solution of the ligand, at suitable concentrations, and to follow the formation of their complex. Many protein-ligand complexes form quite rapidly, thus their concentration must be followed in real time, rather than by sampling the reaction mixture and submitting the sample to chemical analysis, and for this reason spectroscopic methods are preferred. Reaction half-times ranging between microseconds and seconds at ligand concentrations in the µM to mM range are commonly observed for the association, whereas shorter half-times are uncommon, as they approach or exceed the diffusion limit. This limit can in principle be calculated, provided that the physical dimensions of the ligand, the protein and the binding site are known. Longer half-times are undoubtedly possible and are practically observed, for example, in covalent protein-ligand complexes, whose activation energy may be very high, but they are rarely of physiological interest, except for toxicology (e.g., in the case of irreversible enzyme inhibitors). Another category of potentially interesting slow association rates includes rate-limiting intramolecular conformational changes. Measurement of association rates can point to the existence of such processes.

Because the ligand association reaction must have at least one second-order step, its half-time increases with dilution of the ligand (or the protein, unless under pseudo-first order conditions). Thus, dilution is widely employed to lower the apparent association rate in order to bring it within the time window accessible to the method used to detect the reaction. However, in practice, dilution is limited by the intensity of the signal, and by the affinity of the complex. Quickly binding, low-affinity ligands do not tolerate dilution because the yield of their complexes may become too low to be detected. An interesting example is provided by comparison of the binding of NO and O_2 to myoglobin. Both ligands bind with second-order rate constants in the order of $3-5 \times 10^7 \, M^{-1}s^{-1}$; however, the equilibrium affinities of the two differ by over 100,000-fold. As a consequence, in the case of NO the only limit to dilution of both the protein and the ligand is due to the signal one can reliably measure, and one can carry out the measurement by rapid mixing methods at protein and ligand concentrations in the low µM range. By contrast in the case of oxygen dilution rapidly leads to a condition where binding becomes negligible.

Manual mixing the protein and its ligand in a spectrophotometric cuvette and recording the time course of the reaction is feasible for half-times in the order of some tens of seconds or longer. Rapid mixing methods significantly reduce the dead time of mixing. A typical stopped-flow instrument can easily record the time courses of reactions with half-times as shorter as a few ms.

The stopped-flow apparatus (Gibson and Milnes, 1964; see Figure 2.6) is the most versatile kinetic instrument devised thus far: it uses two driving syringes, filled with the protein and the ligand respectively, to mechanically drive a small (sub-mL) volume of each solution into a mixing chamber and from there to an observing chamber fitted to an absorbance, fluorescence, or CD spectrophotometer. The excess volume is collected into the stopping syringe whose piston triggers the recording apparatus. The time the

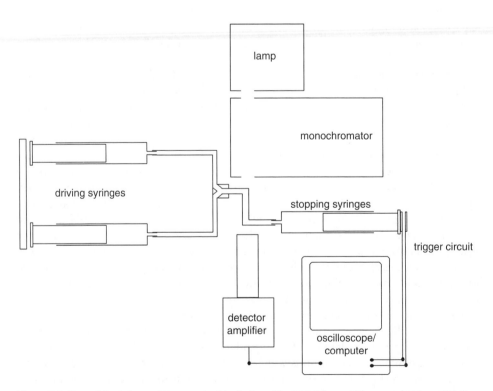

Figure 2.6 The original stopped-flow apparatus designed by Q.H. Gibson (Gibson and Milnes, 1964).

freshly produced reaction mixture spends in the mixing chamber is not accessible to measurement, and constitutes the so-called dead time of the instrument. It is usually of the order of 2–3 ms. The driving syringes, mixing, and observation chamber are kept in a water bath at the desired temperature.

Faster mixing instruments have been devised, for example, by means of continuous flow methods (e.g., Shastry *et al.*, 1998), but none has equaled the versatility of the classical stopped flow.

The rapid-mixing apparatus can also be used to record the time course of dissociation of the protein-ligand complex. This seemingly counter-intuitive measurement can be carried out in several ways:

i) One possibility is to mix the complex with a reagent that chemically destroys the dissociated ligand at a rate faster than the dissociation of the complex. The complex dissociates in the complete absence of free ligand thus the reaction is complete, and the time course is dominated by the kinetic rate constant of dissociation (k_d).

ii) The association time course of a low-affinity ligand contains information on the dissociation reaction that can be taken advantage of by following the reaction over an extended concentration range of the ligand, and extrapolating the dissociation rate constant (see Section 2.2 and Figure 2.3 above). In this case each single time course is governed by an apparent rate constant that is the sum of the association and dissociation ones (eqn. 2.11).

iii) Rapid dilution of the PX complex by mixing with buffer leads to a partial dissociation of the complex. Like for case (ii), the time course in this case is governed by eqn. 2.11, thus the k_d should be extrapolated from a series of measurements at different starting concentrations of the ligand, as in Figure 2.3.

iv) The complex PX can be mixed with a competing ligand and the replacement time course, whose apparent rate constant is a complex function of the association and dissociation rate constants of both ligands, can be followed (see Section 2.7).

v) The complex PX can be mixed with a competing protein, and the ligand exchange between the two proteins can be followed (this is a variant of method iv). These experiments can be carried out also by manual mixing (without a stopped-flow instrument) if their half time is in the order of several seconds or longer; unfortunately, only methods (iv) and (v) allow the researcher to vary the ratio of competing ligand (or protein) and hence to exert a significant control on the half time of the reaction.

Instruments capable of recording significantly faster reactions can be devised if the mixing step can be omitted. This is done by the so-called equilibrium perturbation methods. An equilibrium mixture of the protein and its ligand is set into a spectrophotometric cuvette and an energy pulse is suddenly applied to perturb the equilibrium condition. For example, in the T-jump apparatus the solution is rapidly heated by a current discharge, whereas in a photolysis apparatus the protein-ligand complex is dissociated by a light pulse. The main drawback of perturbation methods is that the extent of the response of the protein-ligand complex is system-specific and highly variable. The response to heat is a function of the ΔH of the reaction, and not all (actually very few) protein-ligand complexes are photosensitive. Moreover, T-jump never does promote 100% dissociation of the protein-ligand complex, and photochemical methods do so only in some particularly favorable cases. In spite of these limits, that may be more relevant in the study of multi-subunit proteins, in suitable biochemical systems a spectacular amount of information can be gathered by applying these methods. We describe below some important results obtained by photochemical methods on monomeric hemoproteins, which may be considered of general relevance.

2.6 Photochemical Kinetic Methods

The essential requirement of the experimental system to be studied by photochemical methods is that the protein-ligand complex must be photosensitive, that is, absorption of a photon must cause its dissociation. The best-studied example is that of hemoproteins because the chemical bond between the hemoprotein and its ligand is a coordination bond between the heme iron and its ligand, thus it represents a true covalent bonding orbital. Absorption of a photon promotes one bonding electron to an anti-bonding orbital and has a finite probability (that may approach 100%) of causing bond breakage and thus ligand release. After an intense but short light pulse the rebinding of the photodissociated ligand can be followed by time resolved spectroscopy. Photochemical methods were initially applied to the study of the CO complexes of hemoglobin and myoglobin by Q.H. Gibson, who used a photographic flash (Gibson, 1959), and later received great advance by the availability of pulsed lasers (Austin *et al.*, 1975).

The reason photochemical methods are applicable to very few protein-ligand complexes is that most bonding interactions are unsuitable. Most complexes are due to weak bonds (e.g., hydrogen bonds or hydrophobic interactions), in which no bonding orbital is formed, or to covalent bonds that can dissipate excitation energy via thermal decay, fluorescence, or phosphorescence without bond breakage. Even among hemoproteins, only the complexes of ferrous heme iron are photolabile, whereas those of ferric iron are not photosensitive.

The time courses of CO rebinding to myoglobin recorded after photolysis by a photographic flash are essentially superimposable to those obtained by rapid mixing in the stopped flow. Much different, and more interesting, results are obtained by using very short light flashes by pulsed lasers, possibly at low temperature or high viscosity.

In 1973–1975, Frauenfelder and co-workers, using a photolysis instrument setup at very low temperatures, followed CO rebinding from inside the myoglobin protein matrix, that is, before the photodissociated ligand could diffuse to the solvent. These authors demonstrated that the protein has access to multiple structural substates that rebind CO with different rates (Austin *et al.*, 1975). The broader significance of this result was its suggestion that such substates exist independently of the presence of the ligand, one of the first clues to the extensive dynamics of protein structures.

The development of ns and sub-ns pulsed lasers allowed several researchers to study the rebinding of the photodissociated ligand from inside the protein matrix at room temperature. The non-covalent protein-ligand complex has been studied extensively in ferrous hemoproteins, where it is called the *geminate pair* or *geminate couple*. Geminate pairs are very short-lived, because the ligand either rapidly rebinds or escapes to the solvent. Thus, they can be significantly populated only if the covalent hemoprotein-ligand bond is broken in a significant fraction of the molecules at a rate faster than the sum of those of ligand rebinding and escape. Short but intense laser pulses are the only method thus far devised that conveys the required amount of energy to the protein-ligand complex in a sufficiently short time.

There are at least two reasons why geminate pairs are so relevant in the study of protein-ligand complexes. The short-lived geminate pairs of hemoproteins and their ligands probe the dynamics of ligand diffusion inside the protein matrix and the intrinsic reactivity of the iron for the chosen ligand (Gibson *et al.*, 1986). Moreover, the dynamics of ligand diffusion occurs over the same time regime as protein structural relaxations, hence rebinding or escape of the ligand probes protein dynamics (Austin *et al.*, 1975; Agmon and Hopfield, 1983).

Interpretation of the very early events that follow the photochemical dissociation of the ligand is quite complex, because light pulses of different wavelength and duration yield different photochemical intermediates, due to the fact that absorption of one photon, besides breaking the Fe-ligand bond, causes significant local heating of the metal and the heme. The photochemical event itself, that is, absorption of a photon and breakage of the iron-ligand bond, may be considered instantaneous on the time scale of other protein processes. This event yields the first photoproduct: a hemoprotein containing an excited (i.e., hot) heme, with the photodissociated ligand in close proximity to the heme iron but unbound. This type of photoproduct is called the proximate geminate pair and accounts for the unliganded state of the hemoprotein at the end of picosecond or femtosecond laser pulses. The photodissociated ligand moving within the heme pocket may rebind very rapidly or may move farther from the heme, in which case rebinding is delayed and escape to the solvent is possible. Ligand rebinding within the proximate geminate pair is a monomolecular process, in the sense that it does not depend on the ligand concentration in the bulk, but is non-exponential, and the reaction's rate decelerates as ligand binding proceeds. At room temperature, the ligand rebinding time course may be described as the sum of two exponential decays, but at lower temperatures two exponentials may not be sufficient.

There are at least three different, non mutually exclusive, explanations of the non-exponential rebinding of the geminate pair in the ps time regime. (i) The liganded protein may be present in different conformational isomers prior to photolysis. Differences may be as subtle as the rotational isomers of single amino-acid side chains. Over the ps and sub-ps time regimes these isomers have no time to interconvert and each rebinds the ligand at its own rate (Petrich *et al.*, 1988). This explanation was preferred by Frauenfelder and co-workers and is consistent with the temperature dependence of the geminate processes (Austin *et al.*, 1975). (ii) The light pulses used for photolysis cause a significant heat transfer and the very early geminate rebinding processes occur on a thermally excited heme, overlapping the heat transfer to the protein matrix. Cooling of the photoexcited heme is a non-exponential process with half time in the order of 30 ps (Henry *et al.*, 1986). Rebinding-while-relaxing yields highly non-exponential processes that Agmon and Hopfield [1983] described as a "changing barrier." (iii) Rebinding competes with random migration of the photodissociated ligand inside the heme pocket. If the ligand moves away from the heme iron, while still remaining in the pocket, further geminate rebinding is slowed down, resulting in a non-exponential process. This explanation was preferred by Q.H. Gibson, based on experiments carried out on site directed myoglobin mutants (Carlson *et al.*, 1994).

In hemoglobin and myoglobin, the rebinding of the proximate geminate pair is complete approximately 1 ns after the photolysis pulse. The fraction of the photolyzed ligand that rebinds over this time window varies greatly: it is close to 100% for NO, significant for O_2, and nil for CO. The photodissociated ligand that has not recombined within 1 ns is still trapped inside the protein matrix, but has diffused away from the heme iron, forming a second geminate pair that either rebinds exponentially in a time regime of 200-500 ns or diffuses to the solvent. Also in this case, the yield of geminate rebinding depends on the ligand: NO and O_2 both present recombination with sperm whale myoglobin in this time window, whereas CO rebinds to a very minor extent (Gibson *et al.*, 1986). Because this time window does not reflect fast protein dynamics, at least if the experiment is carried out at room temperature and low viscosity, it has received less attention.

The study of ligand rebinding from inside the protein matrix adds some relevant information on the overall process of ligand association. Indeed we can draw a global picture of the ligand association processes, applicable to both rapid mixing and photochemical experiments, with one caveat. Because the cooling of the heme occurs over a few tens of ps, the correspondence of photochemical and rapid mixing experiments is strict only for laser pulses longer than some hundreds of ps. The kinetic mechanism for the binding of a ligand to a hemoprotein occurs via at least two steps: the initial second-order diffusion of the ligand to inside the protein matrix, to form the geminate pair, followed by the first-order rearrangement of the pair to its equilibrium structure, with formation of the Fe-ligand bond. This reaction mechanism is a variant of the two-step mechanism described in Section 2.3, and its peculiarity is that the rate constants k_2 and k_3 may be very large, and the population of the intermediate species P−X at equilibrium is indistinguishable from zero. The association reaction behaves as a perfect second-order reaction, because the intermediate P-X does not accumulate, and the dissociation reaction as a perfect first-order one, but the observed rate constants are apparent constants, whose relationships with the real rate constants of the process are as eqns. 2.13a and 2.13b.

It is very plausible that this description may be generalized at least to those cases in which the protein-ligand bond is covalent, for example, to the case of irreversible enzyme inhibitors, hemoproteins having the peculiarity that their photosensitivity allows the geminate pair to be populated and investigated, rather than hypothesized.

Photochemical methods may be applied to a larger number of biological systems if one has access to an artificial photosensitive ligand. Photosensitive ligands (often called caged ligands) are small molecules or ions bound or coupled to an inert molecule that prevents their binding to the target protein. The bond between the ligand and the inert molecule is photosensitive, thus one can easily prepare an equilibrium mixture of the caged ligand and the unliganded protein, and promote their binding by photochemically breaking that bond (Kramer *et al.*, 2005). The experiment is conceptually similar to a rapid mixing experiment, except that the time required to photochemically break the bond between the ligand and its cage is much shorter than that required to mix the two separate solutions of the protein and the ligand. Thus, rapidly binding ligands greatly benefit of this approach, provided that they can be coupled to an inert molecule by means of a photosensitive chemical bond. This approach has not the flexibility of the preceding one, because: (i) it does not populate geminate pairs, thus it does not allow the researcher to explore the protein dynamics; and (ii) in the case of multi-subunit proteins it does not allow the researcher to vary the post-flash ligand saturation of the protein.

Finally we may mention the application of photochemical methods to naturally photosensitive biological systems such as flavin-based photosensors (Losi, 2007); or photosynthetic centers (Mamedov *et al.*, 2015). These applications may not be directly related to the problem of ligand binding because the light flash is used to populate an excited state of the protein chromophore, whose decay to the ground state may primarily involve structural changes, electron transfer, or other events.

2.7 The Kinetics of Replacement Reactions

Ligand replacement experiments under equilibrium conditions have been described in Section 1.8 and have been demonstrated to be useful for the study of high-affinity ligands and for protein-ligand couples that yield poor signals. When we are interested in the reaction rate constant(s) and mechanism, ligand replacement experiments are even more precious, because, besides the advantages listed for equilibrium experiments, they provide an indirect tool to determine the rate constant of the dissociation reaction, whose direct study, as explained in Section 2.5, may be difficult. The time course of the approach to equilibrium for a ligand replacement reaction depends on four rate constants and the concentrations of two ligands. Thus it does not lend itself to obvious simplifications, unless some special conditions are met. To derive the fundamental relationships for this type of experiment, we follow the exhaustive treatment reported by Antonini and Brunori (1971).

The experiment is carried out by mixing a solution of the second ligand (X) with a solution of the protein of interest saturated with the first ligand (Y). The reaction time course will asymptotically approach the equilibrium condition of the complexes PY and PX (see Section 1.8).

We schematize the reaction as follows:

$$PY + X \Leftrightarrow P + Y + X \Leftrightarrow PX + Y$$

The pertinent kinetic differential equations are:

$$\delta[PY]/\delta t = k_{a,Y}[Y][P] - k_{d,Y}[PY] \qquad \text{(eqn. 2.15)}$$

$$\delta[PX]/\delta t = k_{a,X}[X][P] - k_{d,X}[PX] \qquad \text{(eqn. 2.16)}$$

Equations 2.15 and 2.16 can be numerically integrated to fit ligand replacement experiments carried out under any experimental condition. This approach is always applicable, and modern computers are more than sufficient to solve the problem.

Even though analyzing one's kinetic data by means of numerical integration of the rate equations yields satisfactory results, this procedure does not enable the researcher to fully rationalize the experiment, because one cannot easily imagine or predict the results of such a procedure. Thus, it is highly advisable, at least for pedagogical purposes, to analyze how this system behaves if experimental conditions are chosen where the two second-order processes behave as pseudo-first order ones, a condition that makes it possible to analytically integrate eqns. 2.15 and 2.16. A description of the system under these conditions is rewarding because of the insight it provides to the reaction mechanism, even though one might still decide to use numerical integration for data analysis.

Because the equilibrium condition depends on the ratios $K_{d,Y}/K_{d,X}$ and $[Y]/[X]$ (see eqn. 1.16), the experiment design does not depend on the absolute concentrations of the two ligands, but only on their ratio, and it is advantageous to have $[Y]>>K_{d,Y}$, $[Y]_{tot}>>[P]_{tot}$ and $[X]>>K_{d,X}$, $[X]_{tot}>>[P]_{tot}$ because these conditions allow two important simplifications, necessary to the analytical integration of eqns. 2.15 and 2.16: (i) The concentration of both ligands may be assumed to be approximately constant during the reaction time-course, and both combination reactions will be pseudo-first order. (ii) The protein will be completely ligand saturated, that is, the concentration of the unliganded state P will be negligible throughout the reaction time course.

If the population of the unliganded intermediate state is negligible with respect to either liganded state, the assumption that any consumption of PY is exactly matched by an equal production of PX (or vice versa) is justified.

$$-\delta[PY]/\delta t = \delta[PX]/\delta t$$

and

$$k_{d,Y}[PY] - k_{a,Y}[Y][P] = k_{a,X}[X][P] - k_{d,X}[PX]$$

which leads to:

$$[P] = \left\{ k_{d,Y}[PY] + k_{d,X}\left([P]_{tot} - [PY]\right) \right\} / \left(k_{a,Y}[Y] + k_{a,X}[X] \right) \qquad \text{(eqn. 2.17)}$$

Substituting eqn. 2.17 into 2.15, one obtains:

$$-\frac{\delta[PY]}{\delta t} = \frac{[PY]\left(k_{a,Y}[Y]k_{d,X} + k_{a,X}[X]k_{d,Y}\right)}{\left(k_{a,Y}[Y] + k_{a,X}[X]\right)} - \frac{[P]_{tot}\,k_{a,Y}[Y]k_{d,X}}{\left(k_{a,Y}[Y] + k_{a,X}[X]\right)} \qquad \text{(eqn. 2.18)}$$

We define the apparent rate constant of the process as:

$$k_{app} = \left(k_{a,Y}[Y]k_{d,X} + k_{a,X}[X]k_{d,Y}\right) / \left(k_{a,Y}[Y] + k_{a,X}[X]\right) \qquad \text{(eqn. 2.19)}$$

And rewrite eqn. 2.18 as

$$-\delta[PY]/\delta t = [PY]k_{app} + \text{constant} \qquad \text{(eqn. 2.20)}$$

Notice that k_{app} (as well as the constant term that occurs in eqn. 2.20) is a complex function of both ligand concentrations and that its definition requires that pseudo-first order conditions apply to both ligands.

Eqn. 2.20 is analogous to the rate equation of reversible binding, and integrates to a form very similar to eqn. 2.11:

$$\left([PY]_t - [PY]_{eq}\right) = \left([P]_{tot} - [PY]_{eq}\right)e^{-k_{app}\,t} \qquad \text{(eqn. 2.21)}$$

Eqn. 2.21 shows that the approach to equilibrium of the replacement of ligand Y by ligand X, under condition of pseudo-first order for both ligands, follows an exponential time course.

If the apparent rate constant k_{app} of the ligand replacement reaction is measured at constant concentration of either ligand while systematically varying the other, it exhibits a hyperbolic dependence on the concentration of the variable ligand (or a sigmoidal dependence on its logarithm), with asymptotes equaling the dissociation rate constants of the variable and fixed ligand, respectively. For example, if [Y] is kept constant, and [X] is varied, k_{app} will depend hyperbolically on [X], and will approach $k_{d,X}$ at low values of [X] and $k_{d,Y}$ at high values of [X]. Moreover, the midpoint of the curve, $k_{app} = (k_{d,X} + k_{d,Y})/2$, corresponds to the condition where the ligand concentrations are so arranged to yield $k_{a,Y}[Y] = k_{a,X}[X]$ (from eqn. 2.19).

Under pseudo-first order experimental conditions, the time course of ligand replacement is mainly governed by the dissociation rate constants, and the association rate constants play a comparatively minor role, consistent with the fact that the unliganded protein, that is the reactant of the association reaction, is populated to a negligible extent. Indeed, we may decompose the apparent rate constant of replacement in the sum of two perfectly symmetric terms: $k_{d,X}k_{a,Y}[Y]/(k_{a,Y}[Y]+k_{a,X}[X])$ and $k_{d,Y}k_{a,X}[X]/(k_{a,Y}[Y]+k_{a,X}[X])$. Each of these contains the rate constant of the dissociation of either ligand (i.e., its k_d) times the probability that the vacant binding site is occupied by the other ligand (via the ratio between either pseudo-first order k_a and their sum). Thus the association rate constants: (i) only appear in a fraction in which they are divided by a sum of terms that includes themselves, and contribute to the k_{app} *via* a correction fraction ranging between 0 and 1, applied to the dissociation rate constants; and (ii) govern the dependence of the k_{app} on the variable ligand concentration, but not its asymptotes.

Special experimental conditions may be devised to linearize the dependence of k_{app} (or its inverse) on the ratio [Y]/[X] (see Figure 2.7), but these require assumptions on the relative magnitudes of the rate constants $k_{d,Y}$, $k_{a,Y}$, $k_{d,X}$ and $k_{a,X}$ that may not be warranted. We take this chance to reinforce our suggestion that in the era of personal computers and beyond there is no real need to linearize a function. Actually, the statistically soundest procedure to determine the four rate constants that determine the k_{app} is to globally fit a series of experiments carried out at different concentration of either or both ligands using eqn. 2.19 to construct the k_{app} for each data set, and eqn. 2.21 for calculating the squared residuals between the experimental data and the calculated time courses. An analogous global procedure may be applied if one chooses instead to simulate the experimentally recorded time courses by numerical integration of eqns. 2.15 and 2.16, and to calculate the squared residuals using the simulated data (see the appendix to this chapter).

A case that may be quite puzzling is that of *association reactions whose dependence on the ligand concentration resembles that of a replacement.* This case is not infrequent and requires careful analysis; it is often due to the unanticipated *replacement of an internal ligand.* A typical example is provided by the association kinetics of ligands of ferric hemoglobin and myoglobin: in this case the "unliganded" protein is not at all unliganded: it contains a water molecule or hydroxyl ion coordinated to the heme iron. When an external ligand is added, the association kinetics may be rate-limited by the dissociation of the internal ligand, and thus behaves as a replacement, rather than as a bimolecular association. A case of special interest is provided by those proteins that may or may not have the internal ligand, depending on the experimental conditions. In these proteins the mechanism of binding of the

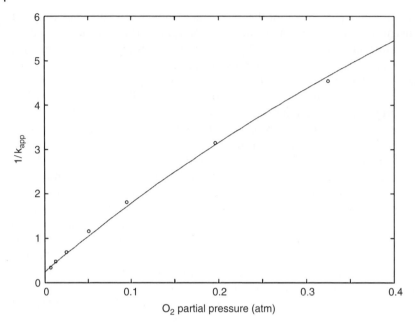

Figure 2.7 Rate of replacement of O_2 by CO in myoglobin (redrawn after Antonini and Brunori, 1971). In this series of experiments MbO_2 (equilibrated with several concentrations of O_2) was rapidly mixed with a solution of CO (at 0.5 mM constant concentration); the concentrations of both gases were largely in excess with respect to that of the protein, thus they may be assumed as constant throughout the reaction time course. As expected, the limiting value for the kinetic constant of the ligand replacement reaction extrapolated to zero concentration of O_2 equals the rate constant of O_2 dissociation. Data points are from Antonini and Brunori (1971); line was calculated from eqn. 2.19, using the following rate constants (also from Antonini and Brunori, 1971): $k_{a,O2} = 10^4\,atm^{-1}s^{-1}$; $k_{d,O2} = 4\,s^{-1}$; $k_{a,CO} = 3 \times 10^5\,M^{-1}s^{-1}$; $k_{d,CO} = 0.04\,s^{-1}$. *Source*: Adapted from Antonini and Brunori (1971).

external ligand switches from a bimolecular, second-order association reaction to a pseudo-first order replacement depending on the experimental conditions (Giacometti *et al.*, 1975).

The internal ligand may be provided by any component of the solution, or by amino-acid residues that may partially occupy the ligand binding pocket. In the latter case the internal ligand behaves as if its concentration were fixed, and binding of the external ligand requires a local conformational rearrangement.

References

Agmon N. and Hopfield J.J. (1983) CO binding to heme proteins. J Chem Phys, 79: 2042–2053.

Antonini E. and Brunori M. (1971) *Hemoglobin and Myoglobin in Their Reactions with Ligands*. North Holland, Amsterdam.

Austin R.H., Beeson K.W., Eisenstein L., Frauenfelder H., and Gunsalus I.C. (1975) Dynamics of ligand binding to myoglobin. Biochemistry, 14(24): 5355–5373.

Carlson M.L., Regan R., Elber R., Li H., Phillips G.N. Jr, Olson J.S., and Gibson Q.H. (1994) Nitric oxide recombination to double mutants of myoglobin: role of ligand diffusion in a fluctuating heme pocket. Biochemistry, 33: 10597–10606.

Cornish-Bowden A. (1995) *Fundamentals of Enzyme Kinetics*. London: Portland Press.

Giacometti G.M., Da Ros A., Antonini E., and Brunori M. (1975) Equilibrium and kinetics of the reaction of Aplysia myoglobin with azide. Biochemistry, 14: 1584–1588.

Gibson Q.H. (1959) The photochemical formation of a quickly reacting form of haemoglobin. Biochem J, 71: 293–303.

Gibson Q.H. and Milnes L. (1964) Apparatus for rapid and sensitive spectrophotometry. Biochem J, 91: 161–171.

Gibson Q.H., Olson J.S., McKinnie R.E., and Rohlfs R.J. (1986) A kinetic description of ligand binding to sperm whale myoglobin. J Biol Chem, 261: 10228–10239.

Henry E.R., Eaton W.A., and Hochrasser R.M. (1986) Molecular dynamics simulations of cooling in laser-excited heme proteins. Proc Natl Acad Sci USA, 83: 8982–8986.

Kramer R.H., Chambers J.J., and Trauner D. (2005) Photochemical tools for remote control of ion channels in excitable cells. Nat Chem Biol, 1: 360–365.

Losi A. (2007) Flavin-based blue-light photosensors: a photobiophysics update. Photochem Photobiol, 83: 1283–1300.

Mamedov M., Govindjee, Nadtochenko V., and Semenov A. (2015) Primary electron transfer processes in photosynthetic reaction centers from oxygenic organisms. Photosynth Res, 125: 51–63.

Petrich J.W., Poyart C., and Martin J.L. (1988) Photophysics and reactivity of heme proteins: a femtosecond absorption study of hemoglobin, myoglobin, and protoheme. Biochemistry, 27: 4049–4060.

Shastry M.C.R., Luck S.D., and Roder H. (1998) A continuous-flow capillary mixing method to monitor reactions on the microsecond time scale. Biophys J, 74: 2714–2721.

Appendix to Chapter 2: Principles of Data Analysis

The statistically soundest procedure to analyze one's experimental data is to globally fit them with the desired equation by means of a robust least squares reduction routine. We take as an example the kinetics of ligand replacement described in Section 2.7 because its complexity allows us to consider different possible approaches, and to illustrate the complementary roles of global fitting and linearization.

The researcher will have collected an extensive series of measurements (time courses of ligand replacement), carried out at several concentrations of the variable ligand. Each data set is a series of values of the signal that monitor $\bar{X}_t$ as a function of the variables of the system ([X], [Y] and time after mixing, t). It is essential that the data set explores a significant range of the involved variables, which one may imagine as a grid, in which t must cover at least five half-times for each time course. The definition of useful ranges of [X] and [Y] should fulfill the following conditions: (i) [X] and [Y] should never be less than 10-times their respective equilibrium dissociation constant to ensure that the fraction of unliganded protein is negligible; (ii) the concentration of both ligands should be significantly higher than that of the protein in order that the association reactions are pseudo-first order; and (iii) the ratio between the two ligands should vary over an interval sufficient to explore a significant change of k_{app}.

The analysis of the extended data set requires the iteration of two steps. First, one should possess an equation capable of predicting the signal using the pertinent variables: that is, an algebraic function with the general form:

$$\bar{X}_{calc,t} = f\left([X],[Y],t,\text{parameters}\right)$$
$$\text{signal}_{calc} = f\left(\bar{X}_{calc,t}\right)$$

Often the signal is directly proportional to $\bar{X}_t$; if this is the case the latter function reduces to:

$$\text{signal}_{calc} = \text{signal}_{calc,0} + \bar{X}_{calc,t} \times \Delta_{calc}$$

where $\text{signal}_{calc,0}$ represents the signal one records at time=0, and Δ_{calc} the change recorded between time=0 and time $\rightarrow \infty$.

If the experimental conditions of the different experiments are such that the total signal change is not constant, Δ_{calc} should be replaced by the function:

$$\Delta_{calc} = \Delta_{tot} \times [X]/\left([X]+[Y]k_{d,X} k_{a,Y}/k_{d,Y}k_{a,X}\right)$$

that we derive from eqn. 1.15.

In the case of ligand replacement, parameters are $k_{a,X}$, $k_{d,X}$, $k_{a,Y}$ and $k_{d,Y}$, signal_0 and Δ_{tot}. The function may be provided by eqns. 2.15 and 2.16 or by eqns. 2.19 and 2.21, if applicable. The parameters are to be initially guessed based on the available knowledge. If the integrated equation eqn. 2.21 is used, it directly calculates $\bar{X}_{calc,t}$. If, on the contrary, the differential kinetic equations are resorted to, these do not directly calculate $\bar{X}_{calc,t}$; rather, these are used to stepwise reconstruct the time course of the reaction, from which the points at the desired times can be extracted.

The second step is to feed the couples $\bar{X}_t - \bar{X}_{calc,t}$, or, better, $\text{signal-signal}_{calc}$ to a computer routine that calculates the sum of the squared residuals $\Sigma\left(\bar{X}_t - \bar{X}_{calc,t}\right)^2$ or $\Sigma(\text{signal-signal}_{calc})^2$ and elaborates a new guess for the parameters, expected to reduce the sum of the squared residuals. At this point the procedure is iterated and a new set of $\bar{X}_{calc,t}$ or signal_{calc} is calculated using the newly guessed parameters, and again used to estimate the sum of the squared residuals and to further refine the parameters. The procedure is repeated until the reduction of the sum of the squared residuals over two successive iterations becomes negligible.

Given that the above procedure is entirely delegated to a computer and only requires the function to calculate $\bar{X}_{calc,t}$ and the least squares minimization routine, one may wonder which role we attribute to the extended analysis of the system (i.e., to eqns. 2.17 through 2.21), and to the recommendations on experimental design. Indeed eqns. 2.15 and 2.16 are perfectly sufficient for data analysis (by numerical integration), and resort to pseudo-first order conditions is unnecessary, as is consideration of the dependence of k_{app} on the concentration of the variable ligand. The reason to devote our attention to these matters is that least-squares minimization of a numerically integrated differential kinetic equation is a procedure humans have difficulties to visualize. Thus in the absence of these considerations, one would not have any grasp on the behavior of the system and would be at a loss to guess its parameters and to realize whether the equations he/she is using are adequate or not. Thus, even though we discourage the use of algebraic transformations, and recommend global-fitting procedures applied to the raw experimental data, we recognize that the algebraic elaboration of the equations used is important to conceptually explore the behavior of the system, and to design the relevant experiments.

3

Practical Considerations and Commonly Encountered Problems

In this chapter we shall be concerned with the sample preparation and the possible practical difficulties one may encounter while measuring the equilibrium constant of a monomeric protein binding its ligand with 1:1 stoichiometry. In a sense, this chapter deals with common errors and common problems that have often been dealt with in the form of recommendations in Chapter 1. Even the most skilled researcher may encounter systems that behave in an unexpected manner and yield inconsistent results, thus an analysis of errors and problems seems worthwhile. Recognizing the actual problem (or error) is a bottom-up process, carried out in the opposite direction from that followed in Chapter 1. We believe that a bottom-up treatment, as the one given in this chapter, with figures to illustrate common cases may be helpful to the PhD student or to the researcher, and may save some of his/her time.

3.1 Design of the Experiment: The Free Ligand Concentration

The most obvious, but not the only, way to measure the affinity of a protein for its ligand(s), that is, its K_d, requires the researcher to prepare a suitable sample of the protein of interest in the desired buffer, to add aliquots of the ligand from a concentrated solution, and to determine the amount of complex formed after each addition of ligand. It is essential to allow time enough, at constant temperature, for the protein-ligand mixture to equilibrate. *If no independent information is available to establish the length of time required to reach equilibrium, then the only way to be sure it has been reached is to repeat the entire experiment using a longer equilibration time, and determine if the data are reproducible.* Better still, one may follow the time course of ligand association at selected ligand concentrations, and determine its $t_{1/2}$. Equilibration is reasonably approached after an incubation at least five times longer than the $t_{1/2}$ (see Figure 2.2). It is also essential that the ligand concentration is varied over a range wide enough to cover fractional saturations between, say, 0.05 and 0.95. Meeting these requirements may pose several different problems that we want to examine in detail.

We will assume that the system described in this chapter conforms to the equations developed in the preceding ones, and it is important to stress once more that the ligand concentration that appears in the equilibrium expression refers to the *free* ligand, not to

Reversible Ligand Binding: Theory and Experiment, First Edition. Andrea Bellelli and Jannette Carey.
© 2018 John Wiley & Sons Ltd. Published 2018 by John Wiley & Sons Ltd.

the total ligand present in solution after each addition, the relationship between the two being, in the case of 1:1 binding: $[X]_{tot} = [X] + [PX]$.

To design the experiment one must meet two requirements, which unfortunately may be conflicting:

i) one must explore a range of free ligand concentration extending over at least two to three logarithmic units, centered on K_d; and

ii) one needs to use a target concentration that provides a signal readable by the chosen detection method (e.g., if the signal is absorbance, the target concentration should be such that the absorbance difference between P+X and PX, at the concentration used, is in the order of 0.1 to 1 Abs units).

Usually it is easier to have the higher molecular weight species (the macromolecule) at fixed concentration and to vary the concentration of the lower molecular weight species.

The lower the target concentration, the easier to meet requirement (i). The reason for this is that one usually, if not always, knows $[X]_{tot}$, and the lower the concentration of $[P]_{tot}$ the less the correction to calculate $[X]$ from $[X]_{tot}$. Ideally, one would like to have $[X] \approx [X]_{tot}$, a condition that requires $[P]_{tot} << [X]_{tot}$. Because the range of ligand concentrations to explore is centered on K_d, what one wants in practice is $[P]_{tot} << K_d$. Unfortunately this requirement may clash with requirement (ii) because lowering $[P]_{tot}$ diminishes the observable signal change. Moreover, in the experimental design one has to guess K_d, for example, from literature data on similar proteins. It is not uncommon to find that the target is completely titrated upon the first addition of ligand, because one overestimated K_d. In these cases one lowers his/her guess of K_d by a factor of at least 100-fold and re-designs the experiment.

If no information is available to formulate a guess on K_d, a preliminary test over an extended ligand concentration range should be carried out. It is usually adequate to measure the fractional ligand saturation in 10-fold steps of ligand concentration, for example, at $[X] = 100$ nM, $1\,\mu M$, $10\,\mu M$, $100\,\mu M$, 1 mM, 10 mM. Having a rough estimate of the K_d, one can design the experiment taking advantage of the relation:

$$[X] = K_d \, \bar{X} / (1 - \bar{X})$$

that derives from eqn. 1.6. This allows one to calculate that:

i) to achieve $\bar{X} = 0.05$ one needs $[X] = 0.053\,K_d$

ii) to achieve $\bar{X} = 0.95$ one needs $[X] = 19\,K_d$

The range of ligand concentration to be explored is thus set to the interval of approx. $K_d/20 < X < 20\,K_d$. Since the plot $\bar{X}$ versus log $[X]$ is a symmetric sigmoidal curve, it is reasonable to express the ligand concentration range as a logarithmic interval divided into some six to ten approximately equal segments. For example, one may want to measure $\bar{X}$ at the following values of free ligand concentration: $0.05\,K_d$, $0.15\,K_d$, $0.5\,K_d$, $1.5\,K_d$, $5\,K_d$, $15\,K_d$, or: $0.03\,K_d$, $0.1\,K_d$, $0.3\,K_d$, K_d, $3\,K_d$, $10\,K_d$, $30\,K_d$. Notice that it is required neither that the points are evenly spaced over the logarithmic scale, nor that the point $[X] = K_d$ is included in the series.

Under the favorable condition $[X] \approx [X]_{tot}$, the above series allows the experimenter to calculate with reasonable approximation how to prepare each experimental point.

If this condition is not met, the amount of bound ligand ([PX]) should be calculated from the guessed value of K_d and added to the ligand concentrations, but an experimental condition in which [PX]>[X] is subjected to large errors and should be avoided.

The series given above will tolerate relatively large errors in the initial guess of K_d; however, if K_d was severely underestimated or overestimated, the data set will lack experimental points at one end and can be used as a preliminary test to design a new experiment. Depending on the measurement method chosen, it may also be possible to add more points at either end of the range of ligand concentration during the course of the experiment, in order to compensate for errors in the initial guess of K_d.

Samples corresponding to each experimental point may be prepared in several ways that are variants of two basic experimental designs. One may either prepare a single sample of the protein to which aliquots of the stock ligand solution are added sequentially, or one may prepare as many samples as ligand concentrations to explore.

The single protein sample method is often used because it requires less protein. Moreover, if special (e.g., gas tight) glassware or other equipment is needed it may be the only practical option because preparation of multiple samples would require multiples of the special equipment. To carry out the experiment, the researcher prepares a sample of the protein dissolved in the desired buffer at the appropriate concentration (see Section 3.2) and measures the signal. The signal is chosen in order to give information about the concentration of the unliganded protein P (which at this stage equals $[P]_{tot}$), or the complex PX (which at this stage is zero). One then adds the amount of ligand required to obtain the final concentration chosen for the first point, incubates the sample at the desired temperature and waits for the mixture to reach the equilibrium condition before repeating the measurement of the signal. The procedure is repeated until all ligand concentrations have been explored. Obviously, one must calculate the amount of ligand to be added so that the sum of all preceding additions equals the concentration desired. Depending on its physical state, the ligand may be added from a concentrated solution, or as a solid, or in the gas phase. If added from a concentrated solution, each ligand addition dilutes the sample and this affects the signal in addition to the effect of binding. If necessary the signal may be corrected, or the target, at the same concentration as that in the experimental sample, can be added to the stock ligand solution.

Unfortunately, protein and glassware economy constitute all the advantages of the single sample method, to be weighed against its disadvantages: (i) It is a destructive method that uses up each sample in order to prepare the next; if for any reason one wants to repeat one reading, the only way is to repeat the entire experiment. (ii) The time of incubation required to reach equilibrium must be known in advance. (iii) If an error is made or a problem occurs at some point (e.g., protein precipitation) all successive points will be affected, and in severe cases the entire experiment must be discarded and repeated. (iv) The method introduces the extraneous variable time since starting the experiment, which is correlated with the independent variable (ligand concentration) in such a way that an effect of the former (e.g., protein precipitation, slow drifts of the instrument) may be erroneously attributed to the latter. (v) Each successive addition of the ligand dilutes the protein sample, unless special precautions are taken. (vi) If the stock ligand solution contains some substance that is extraneous to the protein sample (e.g., DMSO or NaOH added to increase the solubility of the ligand) the experiment will be carried out in the presence of a variable concentration of this substance, proportional to that of the total ligand, which may affect binding.

The multiple protein samples method consumes more protein, but has none of the disadvantages of the other method. To carry out the experiment, the researcher prepares as many samples as experimental points desired in the experiment. Each sample contains the desired buffer, the protein, the desired amount of the ligand, and water (or the same buffer used to dissolve the ligand) to the final volume. Proteins typically have more restricted buffer requirements than do their ligands, thus it is preferable to use the protein buffer to dissolve the ligand.

Accurate sample preparation ensures that the concentration of the protein and all other chemical components except for the ligand is the same in all samples. For example, if the ligand is dissolved in a water/DMSO mixture, one can add DMSO to all samples, taking care that the final concentration, which includes that added with the ligand, is the same in all samples and equals that of the sample at the highest ligand concentration. The samples are then incubated at the desired temperature for the time necessary to reach equilibrium, and the concentration of the complex PX in each sample is assessed using the chosen method. Because the procedure is non-destructive, the researcher can repeat a reading if necessary, and it is good practice to actually repeat them all at least once after allowing a second time interval for further equilibration. Doing so will ensure that equilibrium had been reached, or reveal that the incubation time was not long enough. Should anything go wrong in one sample, the corresponding point could be discarded without affecting all the others. A further advantage of this method is that the incubation time is rigorously the same for all samples. Thus no spurious independent variables are added to the experiment, and one has to wait for completion of the binding process only once, rather than after each addition of ligand, a significant saving of time.

No matter which method is used, if one realizes that the highest concentration of ligand tested is too low, additional points can be added at higher concentrations of the ligand. By contrast if one realizes that the lowest ligand concentration explored was too high, only the multiple samples method allows the researcher to add points at lower ligand concentrations.

3.2 The Signal and the Concentration of the Target

As stated above, a particularly favorable condition for measuring K_d is $[P]_{tot} << K_d$, because it implies $[X] \approx [X]_{tot}$. Meeting this condition or not is a matter of the physico-chemical properties of the protein and of the method chosen to determine the concentration of the protein-ligand complex PX. To measure the concentration of PX, one needs a method or procedure that is able to discriminate a mixture of P and X from an equimolar amount of the complex PX. This measurement uses the *signal*, a physical property of the system that differs for PX and an equimolar amount of P+X.

Each experimental method for detecting the signal has an optimal range of sensitivity and it is not advisable to use the method beyond this range. Absorbance spectroscopy provides a clear-cut example. One can use this method if the molar extinction coefficient of the protein plus that of the free ligand (P+X) is different from that of the complex (PX). If this condition is met, then the sensitivity of the method depends on the difference between the extinction coefficients at the chosen wavelength, $(\varepsilon\lambda_{,P} + \varepsilon\lambda_{,X})$ versus $\varepsilon\lambda_{,PX}$. A very favorable case is that of ferrous myoglobin or hemoglobin and their ligands, where the ligands usually have no significant absorbance in the visible or the Soret regions of the spectrum, whereas the protein has different extinction coefficients in states P and PX (e.g., Mb and MbO_2). The differences in the extinction coefficients range between ~70,000 $M^{-1}cm^{-1}$ in the Soret region of the spectrum to ~4,000 $M^{-1}cm^{-1}$ in the visible region and ~500 $M^{-1}cm^{-1}$ in the far red/NIR. A good spectrophotometer can reliably record an absorbance change of 0.25 units with 2% precision (0.005 units). This fact sets the optimal range of protein concentration between approximately 3 and 500 μM. Other proteins contain much less-fortunate chromophores than the heme and it is safe to assume that the typical range of protein concentration required for absorbance measurements lies between 20 and 200 μM. The case of a chromogenic ligand whose absorbance changes upon binding to the protein might in principle be different and there are no typical extinction coefficients for such cases. However, compounds having extinction coefficients greater than those of the heme are uncommon, and thus cases that lie outside the concentration range defined for myoglobin are rare.

The use of chromogenic ligands is subject to further problems and restrictions that depend on the experimental design. Indeed, the experiment requires that either the concentration of the chromogenic ligand or that of the protein is varied, and difficulties may arise in either case. If the ligand concentration is varied this may lead to an increase of the absolute absorbance without increase of the difference signal, as the fraction of the bound ligand ($[PX/[X]_{tot}$) decreases. This condition is unfavorable and can be avoided by designing an experiment in which the ligand concentration is kept constant and that of the protein is varied. This, however, may conflict with the solubility or availability of the protein, given that the concentration of the varied partner must extend to 20 K_d.

Our reasoning tells us that, unless some very favorable conditions occur (e.g., the ligand is a gas or can be partitioned in an immiscible solvent), absorbance spectroscopy meets the ideal condition $[P]_{tot}<<K_d$ if K_d is in the order of tens or hundreds of μM or higher. In other words, absorbance spectroscopy is a suitable method only for systems that exhibit relatively weak binding, unless one adopts some stratagem to overcome this limit.

If the K_d of the protein-ligand complex is lower than the limit set by the sensitivity of the chosen method, the condition $[P]_{tot}<<K_d$ cannot be met. Several non-mutually exclusive solutions may be adopted, of which we consider below the most obvious. If these solutions fail, drastic changes in the design of the experiment intended to reduce the affinity of the complex are required, whose discussion is detailed in Section 3.7. We consider here only those approaches that do not require extensive redesign of the experiment.

The simplest solution for insufficient sensitivity of the chosen detection method is to select a different experimental method with higher sensitivity. For example, fluorescence usually affords at least a hundred-fold increase in sensitivity over absorbance, provided that the protein-ligand complex generates a suitable fluorescence signal. A very sensitive and robust method is to use a radiolabelled ligand (e.g., Maguire *et al.*, 2012). This method typically requires one to physically separate the bound and free ligand. There are many methods for doing so including precipitation of the protein, equilibrium dialysis, gel filtration, and filter-binding (e.g., Kariv *et al.*, 2001). However, with the exception of equilibrium dialysis, all separation methods have the potential to perturb the equilibrium whose K_d is being measured, by promoting the dissociation (or in some cases the association) of the complex.

Another possible solution is to give up the ideal condition and carry out the measurements at protein concentrations comparable to K_d. If one chooses to do so, one cannot assume $[X] \approx [X]_{tot}$, and if only $[X]_{tot}$ is known, one should measure $[PX]$ from the signal, and should use this value to calculate $[X] = [X]_{tot} - [PX]$. This strategy, however is error prone since it propagates the measurement errors on $[PX]$ to $[X]$, and is to be avoided if $[P]_{tot} > K_d$. A more robust procedure is to use eqn. 1.4 instead of eqn. 1.3. In either case, this approach allows one to gain one order of magnitude in the protein concentration but hardly more than that.

Finally, one obtains a very significant gain if one couples the above strategy to the direct measurement of the free ligand in solution, for example, by means of an appropriate electrode, if the ligand is amenable to this type of measurement, or by equilibrium dialysis. In this case $[X]$ can be significantly less that $[X]_{tot}$ because one does not use the error-affected $[PX]$ to calculate $[X]$, and the error on $[X]$ is as small as it would be in the ideal case.

A further point that needs consideration is how uniquely the chosen method reports on the binding process. Indeed the signal might report on some event indirectly associated with ligation, for example, a structural change occurring in the protein, rather than binding itself. Commonly employed methods range from extremely specific measurements, which include quantitative chemical analysis or radioactivity, to less-specific ones, for example, absorbance or fluorescence spectroscopy. This point must be assessed by experiment and takes great advantage from the condition $[P]_{tot} >> K_d$, which implies $[X]_{tot} \approx [PX]$. The rationale of the experiment is that if the signal is directly proportional to the concentration of the protein-ligand complex, then under conditions in which $[X]_{tot} \approx [PX]$, the signal is expected to be also directly proportional to $[X]_{tot}$, which is usually known with great precision. Conversely, a non-linear relationship between the signal and $[X]_{tot}$ is proof of a more complex relationship between the signal and $[PX]$. An example of this type of experiment is provided by the test of the linearity of the absorbance signal due to CO binding to hemoglobin (Figure 3.1).

There is another aspect of the relationship between binding and signal worthy of our attention, that is, whether interpretation of the signal is *absolute* or *relative*. The signal is always an absolute physicochemical property of one or more components of the mixture, be it PX, P, or X. However, in some cases interpretation of the signal may require information that the researcher does not know, and the relationship between the signal and the concentration of the pertinent chemical species is included in the analysis of the binding isotherm itself. Direct chemical determination and radioactively labeled ligands are examples of absolute signals: every single determination quantitates the molar amount or concentration of PX. Fluorescence is a relative signal, and the concentrations of bound and free target are usually calculated as a function of the fraction of the total fluorescence change recorded in the titration experiment: the fraction of the total signal change equals the fraction of ligand saturation ($\overline{X}$) or desaturation $(1 - \overline{X})$. If, for any reason, the total signal change is not known or is not recorded in the experiment, one relies on fitting procedures to relate the observed signal to $\overline{X}$ or $[PX]$.

This is not an intrinsic property of the signal itself, but a consequence of our ignorance of relevant information. For example, absorbance provides absolute determination of PX if one knows exactly the pertinent extinction coefficients; these, however, can be determined only for molecular species that can be obtained in pure form. The extinction coefficients of ligation intermediates that do not attain 100% population are usually

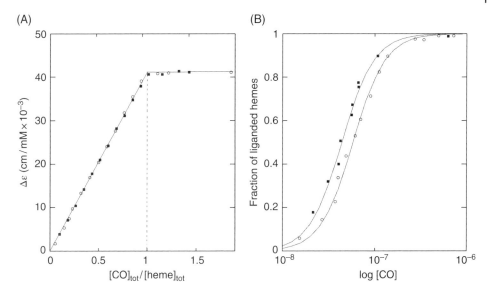

Figure 3.1 Binding isotherms of ferrous human hemoglobin and CO, a high-affinity ligand. Panel A: at high hemoglobin concentration (6.4 to 6.9 uM on a heme basis, at pH=7 (open circles) and 9.2 (closed squares)) the binding isotherm is a stoichiometric titration and $[CO]_{free} << [CO]_{tot}$. The linear relationship between absorbance and binding is proven by the signal being a straight line, that breaks at the equivalence point $[CO]_{tot} = [heme]_{tot}$. Panel B: at low hemoglobin concentration (45 nM (closed squares) and 234 nM (open circles), both at pH=7) $[CO]_{free}$ is large enough to be calculated with precision from $[CO]_{tot} - \bar{X}[heme]_{tot}$, and a hyperbolic equilibrium curve is obtained. The concentration-dependent difference in affinity is a consequence of ligand linked dissociation (see Chapter 5). Source: Adapted from Anderson and Antonini (1968).

determined by calculations that rely on estimates of their relative population, and are thus model dependent. Even in the most straightforward cases, approaching 100% ligand saturation requires quite high free ligand concentrations: for example, at $[X] = 50K_d$, $\bar{X} = 0.98$, still appreciably different from unity (see eqn. 1.6).

3.3 Test of the Reversibility of the Reaction

For each experimental point of a ligand-binding experiment the fractional ligand saturation is independent of the manner in which the equilibrium condition is reached. The same extent of ligand binding must be attained by mixing the unliganded protein with the appropriate amount of the ligand; or by removing the bound ligand (not always an easy task) until its concentration reaches the desired value; or by diluting a ligand/protein mixture until the ligand concentration reaches the appropriate value. Preparing at least some of the samples in two different ways, whenever possible, provides an important experimental test of reversibility and of whether the system has effectively reached its true equilibrium condition.

An interesting example is provided by the gas current method devised by K. Imai to record the oxygen-binding isotherms of hemoglobin. The instrument is an optical cell in which a hemoglobin solution is equilibrated by constant stirring with a changing atmosphere of oxygen, nitrogen, or a mixture of the two. The free oxygen concentration in solution is measured using an oxygen electrode, and the fractional oxygen saturation of hemoglobin is measured by absorbance spectroscopy. The instrument allows the researcher to record an oxygenation isotherm, starting from deoxygenated hemoglobin in pure nitrogen and flushing the cell with a nitrogen/oxygen mixture, or a deoxygenation isotherm, starting from oxygenated hemoglobin and flushing the cell with pure nitrogen. The two experiments can be run sequentially on the same hemoglobin sample to directly demonstrate the reversibility of binding (Imai *et al.*, 1970; Imai, 1983).

In some cases, however, there are obvious difficulties in running the experiment in both the ligation and de-ligation directions. For example, the ligand may be removable only by dialysis and the extent of de-ligation cannot be checked in real time. In such cases it is essential that the experimenter tries to reproduce the initial, unliganded state of the protein, in order to demonstrate the reversibility of binding. This operation demands that at least the following two points be checked:

i) it must be possible to remove the bound ligand (e.g., by dialysis) to restore the unliganded state of the protein, and
ii) the unliganded state thus restored should be able to bind the ligand again and to yield a superimposable ligand binding isotherm.

Ideally, to fully demonstrate thermodynamic reversibility, one should also be able to recover the chemically unchanged ligand after dissociation from its complex with the protein, but this is not always possible. However, the theoretical requirement remains that the reaction is reversible only if the dissociation of the protein-ligand complex yields the native, original states of both the ligand and the protein (see the discussion of Cys ligands in Section 1.11).

3.4 Frequent Abuses of the Concept of $X_{1/2}$

$X_{1/2}$, defined in Section 1.5, is an important empirical parameter of the ligand binding reaction, and its dependence on other variables may provide important clues on the binding process. For example, we remarked in Section 1.9 that dependence of $X_{1/2}$ on the concentration of a second ligand Y linked to it is characteristic of the type of linkage between X and Y. Thus, a plot of $X_{1/2}$ as a function of the concentrations of other molecules may yield crucial information on the protein we are studying.

However, one must be aware of the potential problems implicit in the use of an apparent parameter such as $X_{1/2}$ to replace better defined thermodynamic parameters. $X_{1/2}$ is often misunderstood as a parameter in itself, unrelated to the K_d, by analogy with empirical concepts such as LD_{50} (the dose of a poison that is lethal to 50% of the cells or experimental animals) or the ED_{50} (the dose of the drug which achieves 50% of the maximal effect). Unpredictable, irreproducible, and sometimes meaningless results can be obtained if $X_{1/2}$ is used in the absence of the stringent controls usually adopted in the measurement of ligand binding isotherms (e.g., check of reversibility, and of the incubation time allowed to reach equilibrium). Two frequent abuses of the concept of $X_{1/2}$ are: (i) application to irreversible ligands ($X_{1/2}$ of an irreversible ligand is zero, by definition), and (ii) application to reaction schemes that do not conform to eqn. 1.1 (see Sections 1.10 and 1.11).

The literature abounds with reports of ligands (e.g., enzyme inhibitors, in which case $X_{1/2}$ is called the inhibitory concentration 50, IC_{50}) that are at the same time declared irreversible and assigned a finite $X_{1/2}$ (or IC_{50}). A common reason for this type of error is that the incubation time is insufficient to allow the mixture to reach equilibrium. In such cases $X_{1/2}$ inadvertently becomes a kinetic rather than a thermodynamic parameter. *The effect of insufficient incubation time is to overestimate $X_{1/2}$.* As stated in the preceding chapters, the researcher needs to determine carefully the incubation time necessary for the reaction mixture to reach the equilibrium condition. The incubation time should be equal to or greater than five-fold the half time of the reaction. The half time of the binding reaction depends on the ligand concentration, and is greatest at the lowest concentrations of the ligand (see Figures 2.2 and 3.2); under pseudo-first order conditions it corresponds to $t_{1/2} = \ln(2)/(k_a[X] + k_d)$ (see eqn. 2.11 and Section 2.2). If too short an incubation time is selected, one will overestimate the equilibrium dissociation constant of the protein-ligand complex, sometimes quite severely. In order to demonstrate this point Figure 3.2 reports a simulation for the limiting condition of a ligand that binds irreversibly (i.e., $K_d = 0$), whose affinity has been measured after an insufficient incubation time (the reader may compare Figure 3.2 with Figure 2.2).

The time courses in Figure 3.2A were simulated under the assumption of irreversible pseudo-first order association rate; hence they are exponentials and obey the law:

$$\bar{X}_t = 1 - e^{-k[X]t}$$

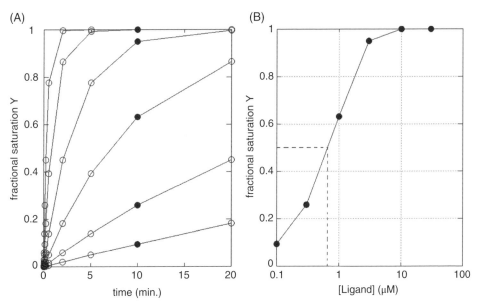

Figure 3.2 Panel A: Simulated time courses of irreversible ligand binding for a protein having infinite affinity and $k_a = 10^5\,M^{-1}min^{-1}$; ligand concentrations 0.1, 0.3, 1, 3 10, 30 μM. All time courses tend asymptotically to 1, with different rates. Panel B: Fractional saturation measured after 10 min. incubation (the points correspond to the full circles of Panel A). The true $X_{1/2}$ in this case is zero; however, given that the incubation time is insufficient to allow the reaction mixture to reach its equilibrium condition, a curve resembling equilibrium binding with non-zero apparent $X_{1/2}$ is obtained (see text).

After the chosen incubation time (10 min. in this example), each sample has reached a degree of saturation that depends exponentially on the ligand concentration; hence the "binding" curve reported in Figure 3.2B is an exponential rather than a hyperbola (both would be distorted, in any case, by the logarithmic scale chosen for the abscissa). Unfortunately, an exponential and a hyperbola may to some extent resemble each other, and the former may be misinterpreted for the latter, more so if we allow for the experimental error. The apparent "$X_{1/2}$" graphically determined in Figure 3.2B represents *the free ligand concentration required in order that the arbitrarily chosen incubation time is the $t_{1/2}$ of the reaction time course, and has nothing to do with a true $X_{1/2}$ or with K_d.*

We suspect that an error of this type is at the basis of the numerous instances in the scientific literature where a ligand is declared irreversible (i.e., it cannot be dialysed away) and is assigned a non-zero, and often quite high K_d. A good recipe to avoid this error, or at least to make it unlikely, is to adopt the multiple protein samples method (see Section 3.1) and to read each sample several times to ascertain that all of them have finally reached the equilibrium condition (i.e., that the signal is constant). Notice that if the incubation time is insufficient, but binding is reversible, $X_{1/2}$ will be overestimated and the error will be more difficult to detect because a finite $X_{1/2}$ is consistent with reversibility by dialysis.

Unfortunately, there is no easy way to speed up a slow binding reaction, except carrying out the experiment at a higher temperature, and there is no alternative to wait as long as needed by the system to reach its equilibrium condition. This is another good reason to study the kinetics of ligand binding as well as the equilibrium isotherms.

3.5 Two Common Problems: Protein Precipitation and Baseline Shifts

Proteins may undergo denaturation under the experimental conditions chosen. If the single sample method has been used, the researcher will observe a gradual change of the signal during the experiment. The direction of the time-dependent signal change due to protein precipitation during ligand binding cannot be defined *a priori*. For example, if absorbance spectroscopy is used, the signal may increase because of turbidity or it may decrease because of the loss of ligand binding sites. The maximum slope of the $\Delta \bar{X}$ versus $\log([X])$ plot will usually be decreased. In some cases it will be possible to estimate the amount of protein that is being lost and to correct the estimated fractional ligand saturation accordingly. However, correcting for the loss of more than a small fraction (>5%) of the total protein usually leads to unreliable results. Repetition of the experiment under conditions that increase the stability of the protein is strongly advisable. A typical example is given by experiments designed to measure K_d of a ligand at the body temperature of mammals. If the protein proves unstable at this temperature, it is highly recommended to record ligand binding isotherms at several lower temperature and to extrapolate the K_d at physiological temperature from the ΔH (Section 1.7); the value may then be compared with the one obtained directly, if the sample precipitation allowed the measurement, imprecise as it may have been.

Baseline shifts may be due to several causes, including protein precipitation (with or without associated turbidity). Other causes may depend on instrumental artifacts (e.g., power supply instability) or incomplete control of temperature. Unfortunately, the correction of baseline shifts is difficult, at best, and in severe cases it is advisable to discard the data and to repeat the experiment. A special case of baseline drift is due to insufficient incubation time to reach the equilibrium, especially if the single protein sample method has been adopted. In this case the ligand introduced in the sample with the initial additions continues to bind during the successive ones. It is very important to identify this type of drift because the only possible correction is to increase the incubation time. This is easy to do if the researcher has used the multiple samples method, in which case one needs only to repeat the readings on each sample until they become constant. If the single sample method had been adopted, the whole experiment must be repeated.

3.6 Low-Affinity Ligands

Strictly speaking, no such things as low- or high-affinity ligands exist: the affinity of any ligand is defined by its K_d. Thus, the terms have only relative meaning, and require definition. From the point of view of physiology, low-affinity applies to systems that achieve low or negligible saturation at the ligand concentrations present in the natural environment (these systems may require cofactors or other conditions that increase their affinity); high affinity applies to those systems that do not appreciably dissociate at the ligand concentrations present in the environment. In the biochemistry laboratory, low- and high-affinity systems are those that require a range of concentrations difficult to explore. Thus, we define low- and high-affinity systems in relation to specific practical difficulties we may encounter in their study, and the definition is conventional, related to the available experimental methods, rather than to intrinsic characteristics of the system.

Determination of the equilibrium constant of a ligand may pose a problem in the case of low affinity if K_d is so high as to require ligand concentrations that exceed its solubility or that alter the properties of the solution (e.g., ionic strength in the case of ionic ligands). Within this context, high affinity is not the opposite of low affinity. Rather, high affinity can be said to occur when K_d is low with respect to the protein concentration required to detect and measure binding. Thus, at least in theory, it is not impossible that a protein-ligand system presents at the same time the problems of low and high affinity. If one encounters a case of this type, one should critically reconsider if a more sensitive signal can be selected, that allows him/her to use a lower protein concentration.

Low-affinity ligands are common, but they have physiological relevance only under very specific physiological conditions. An interesting example is provided by those fish hemoglobins (Hbs) that present the Root effect (Root 1931; Brunori 1975), whose functional role is to pump oxygen in the swim bladder, and possibly in the choroid rete of the eye, against high hydrostatic pressure. In this case very low affinity is exploited for a precise physiological function. At the physiological pH of the gills, Root effect Hbs have the usual O_2 affinity (e.g., the O_2 partial pressure required to half saturate trout HbIV is $P_{1/2} = 25$ mmHg at pH 7.15 and T=14 °C) and easily upload the gas. In all tissues of the fish, where the pH variations are small, Root effect Hbs release oxygen as required by the partial pressure of the gas in the peri-capillary tissues. In the swim bladder, however, a proton pump significantly acidifies the blood's pH, causing a dramatic decrease of O_2 affinity (the Root effect), and the $P_{1/2}$ of Hb increases to over one atmosphere (e.g., $P_{1/2}$ = 900 mmHg for trout HbIV at pH=6.1 and T=14 °C)(Brunori *et al.*, 1978). This causes the gas to be released almost irrespective of its partial pressure in the swim bladder, and fills the organ against the hydrostatic pressure of the surrounding water to regulate the fish's buoyancy.

Other examples of physiologically relevant low-affinity binding are provided by the many enzymes whose Michaelis-Menten constant is high (possibly because of the catalytic efficiency): for example, the K_M of MAO B (monoamine oxidase B) for oxygen is as high as 240 µM and exceeds the gas concentration of an air equilibrated solution (Edmonson *et al.*, 2009). Interestingly, the evolutionarily related enzyme MAO A has $K_{M\,O2}$=12 µM.

The practical problem posed by low-affinity complexes is that it is difficult or in some cases impossible to reach ligand concentrations high enough to fully explore an acceptably large interval of fractional saturation (Figure 3.3). One possible method to overcome this problem is to take advantage of the ΔH of the reaction and to carry out the measurements at a temperature which, though unphysiological or uninteresting, causes an increase in ligand affinity for the macromolecule. The equilibrium constant at the desired temperature can then be calculated using the van t'Hoff equation, if it is known that the temperature dependence of the heat capacity, ΔCp, for the reaction is

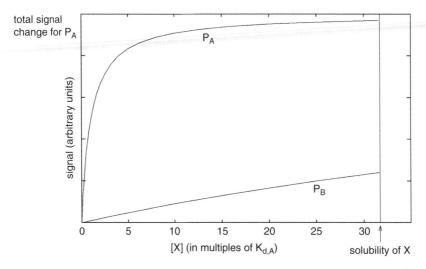

Figure 3.3 Comparison of the binding isotherms of two proteins that bind the same ligand with 100-fold different affinity.

Protein A (P_A) binds ligand X with dissociation constant $K_{d,A}$, much lower that the solubility of X. The binding isotherm of A reasonably approaches saturation. The expected total signal change can be reasonably extrapolated from the recorded curve, and the fractional saturation of protein A, $\bar{X}_A$, for any given value of [X] equals the fraction of the total signal change (the binding curve is simulated under the assumption of a linear relationship between $\bar{X}$ and signal change). Protein B (P_B) binds ligand X with dissociation constant $K_{d,b}$, higher than the solubility of the ligand. Only a small portion of the binding isotherm can be recorded, not sufficient to extrapolate the total expected signal change (in this simulation the total signal change for the two proteins is the same and $K_{d,B}$=100 $K_{d,A}$). The data approximate a straight line and are insufficient to determine $K_{d,B}$, unless additional information is available (see text).

negligible, or if ΔCp can be measured and taken into account. Better still, the experimental design may cover an extended temperature range that includes uninteresting temperatures at which a full ligand-binding isotherm is obtained, as well as the temperature of interest, where the experiment will be incomplete and will lack points in the high saturation range. Global fitting procedures will allow the researcher to take full advantage of the information recorded. Unfortunately, the ΔH of most biological systems can be relatively small, and the practically useful temperature range is limited between the freezing point of the solvent and the resistance of the protein to thermal denaturation. Thus, this solution may not always work.

If the signal recorded for the partial titration of a low-affinity ligand is absolute (i.e., it allows the researcher to calculate $\bar{X}$ without requiring an estimate of the total signal change), or if the total signal change expected is known, at least approximately, by other means (e.g., by analogy with a similar protein having higher affinity), one may still estimate the K_d of a low-affinity ligand whose titration is incomplete, under the assumption that the reaction conforms to eqns. 1.1 to 1.3. Indeed, eqn. 1.6 under the experimental condition $[X] \ll K_d$ simplifies to:

$$\bar{X} \approx [X]/K_d$$

Thus, the initial portion of the hyperbola, or, better, its tangent, is a straight line with slope $1/K_d$. The estimate of K_d as the reciprocal of the slope of the tangent to the first part of the binding hyperbola is a last resort and may yield gross errors because it rests on two implicit assumptions that may both fail: (i) that the binding isotherm is actually a hyperbola (presence of isoforms, and negative or positive cooperativity are only the most obvious examples of conditions yielding non hyperbolic binding isotherms); and (ii) that the estimate of the total signal change used (if one was used) is reliable.

3.7   High-Affinity Ligands

High-affinity ligands pose two types of problems to the determination of equilibrium constants: (i) the free ligand concentration may be too low to be reliably measured or calculated; and (ii) depending on the protein-ligand system and the type of experiment one has designed, approaching the equilibrium condition may be slow, both because of the effect of the low ligand concentration on the association reaction and the (possibly) low kinetic constant of the dissociation reaction (eqn. 2.11).

High affinity poses a severe challenge to the experimental determination of the equilibrium constant when the ligand's K_d is significantly lower than the total protein concentration required to detect a signal related to binding. Depending on the experimental method chosen, the protein concentration required to generate the signal may range from nanomolar to millimolar concentrations, thus high affinity, like low affinity, is a relative concept. To give an example, the K_d of biotin for hen egg avidin under typical experimental conditions is in the femtomolar range (i.e., $K_d \sim 10^{-15}$ M), very low even for highly sensitive detection methods.

The difficulty implicit in the measurement of K_d for high-affinity ligands derives from the combination of conflicting requirements: (i) the K_d dictates the range of ligand concentrations to be explored (i.e. 0.05 K_d < [X] < 20 K_d); (ii) it is desirable that $[X]_{tot} >> [P]_{tot}$ (see Section 3.2), which, combined with requirement (i), amounts to $K_d >> [P]_{tot}$; (iii) $[P]_{tot}$ is determined by the sensitivity of the method chosen for the measurement (Section 3.2). The essential conundrum is the possible conflict between requirements (ii) and (iii): that is, in the case of high-affinity ligands the protein concentration required for determination of K_d is lower than that required by the instrument sensitivity. It is important to stress that calculating the free ligand concentration by subtracting the bound from the total ligand yields reliable results only when the free and bound ligand concentrations are of comparable magnitude. This subtraction should not be attempted if [PX] exceeds [X], because if a small number is calculated as the difference between two much larger ones the expected error is very large.

Figure 3.4 explores the effect of protein concentration on the recorded binding curves, covering the conditions $[P]_{tot} << K_d$ to $[P]_{tot} >> K_d$.

In order to determine the K_d for high-affinity ligands one has a number of possible alternatives, some of which have been considered under Section 3.2: adopting a more sensitive method (e.g., fluorescence instead of absorbance spectroscopy), or direct measurement of the free-ligand concentration (e.g., by using equilibrium dialysis, or a suitable electrode).

More radical alternatives involve the adoption of an experimental design to reduce the ligand affinity, or the separate measurement of the rate constants for ligand association and dissociation, under non-equilibrium conditions.

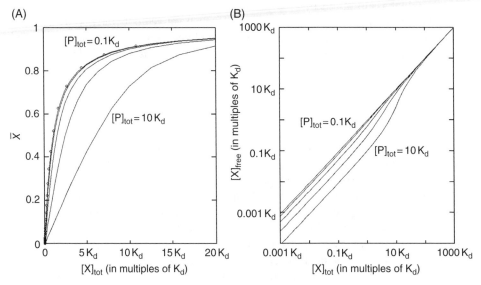

Figure 3.4 Effect of protein concentration on the shape of the ligand binding isotherms. Panel A: fractional saturation ($\bar{X}$) as a function of total ligand concentration, for five different protein concentrations: $[P]_{tot}$ = 0.1 K_d, 0.3 K_d, K_d, 3 K_d, and 10 K_d. Circles report the calculated values of $\bar{X}$ for the highest protein concentration (10 K_d) plotted against [X] instead of $[X]_{tot}$; obviously in a simulation the experimental error is zero, and this allows us to demonstrate that increasing the protein concentration above K_d has no real effect on ligand affinity. Panel B: [X] versus $[X]_{tot}$ as a function of $[P]_{tot}$. Note that when $[P]_{tot} > K_d$ the free ligand concentration ([X] or $[X]_{free}$) is a small fraction of $[X]_{tot}$, hence its calculation is subjected to large errors (Section 3.2).

The simplest method to reduce the ligand affinity is to set up a replacement experiment in which the high-affinity ligand competes with another ligand. This method reduces the apparent affinity of the protein for its ligand (see Figure 1.5) and works fine with monomeric proteins, but has limitations in the case of multimeric ones, as discussed in Chapter 4. Moreover, depending on the properties of the chosen ligands, the approach to equilibrium may be very slow. In some cases it is possible to lower the ligand affinity by means of physical methods and extrapolate the affinity to physiological conditions. For example one may measure the affinity at higher or lower temperature, depending on the sign of ΔH for the reaction; or one may measure the affinity under variable illumination intensity in the case of photolabile complexes, and then extrapolate the value of K_d in the dark (Brunori *et al.*, 1972). A variant of this approach that is useful in certain cases is to alter affinity using a chemical additive, then determine the relationship between K_d and additive concentration to allow extrapolation. An example of this kind is the steep salt concentration dependence of many protein-DNA interactions (Record *et al.*, 1978). Altering pH is another potentially useful means to shift K_d into a measurable range.

Alternatively, one may measure the kinetic rate constants for binding and release of the ligand, whose ratio equals the equilibrium constant, provided that the reaction mechanism does not involve long-lived intermediates, and the rate constants are in a measurable range. This method is quite powerful, provided that the researcher has a way of inducing the dissociation of the protein-ligand complex. For example, Moore and Gibson (1976) were able to promote the dissociation of the Hb-NO complex in the presence of excess CO (which competes with the released NO) and dithionite (which reacts with free NO). This information, together with the previously measured rate constant for NO association (Cassoly and Gibson, 1975) allowed the authors to estimate K_d values in the order of 10^{-11} M. V. Sharma and co-workers (Sharma *et al.*, 1976) were able to promote dissociation of the HbCO complex by rapid mixing with excess microperoxidase, whose CO affinity is higher than that of Hb. These examples point out an important difficulty of the kinetic method, that the above authors cleverly solved, that is, the difficulty in inducing and following in real time the dissociation of the complex. Indeed, measurement of the combination rate constant is usually straightforward and accessible by stopped flow. Because the ligand has high affinity, one can dilute both the protein and the ligand down to the sensitivity limit of the instrument to slow down the reaction. On the other hand, measurement of the rate constant for dissociation of the complex requires that the researcher is able to achieve the rapid and irreversible removal of the released ligand from the solution, in order to allow the dissociation reaction to proceed. Indeed, in the case of high-affinity ligands it may be impossible to extrapolate k_d from the k_{app} of association time courses, as in Figure 2.3, because it may be indistinguishable from zero.

High-affinity ligands are precious tools to explore properties of the protein other than the equilibrium constant including the linearity of the signal (Section 3.2) and the stoichiometry of the protein:ligand reaction (see below).

3.8   Determination of Binding Stoichiometry

The protein-ligand complex has the general formula P_nX_m. Determining the binding stoichiometry equates to determining n and m, the stoichiometric indices, of the fully saturated complex. Chapters 1 through 3 have assumed complexes having, or behaving as if n=1 and m=1. In this section we describe strategies to determine n and m; these strategies apply to any complex, and thus to any ratio n:m.

It is common in biology that the ligand X is monomeric in solution, and does not change its aggregation state upon binding to the protein; by contrast the protein may often be an oligomer of identical or at least functionally equivalent subunits, whose actual aggregation state in solution is unknown. As a consequence, a typical case may be schematized as:

$$P_n + mX \Leftrightarrow P_nX_m$$

Even though we may often know exactly the molecular weight of P and X (e.g., we may know the amino acid sequence of P from its cloned mRNA), the aggregation state in solution must be determined by experiment. In this section we shall consider first the experimental problems one can most often encounter, and then a classical example, which, in our opinion, has great pedagogical value.

If one can prepare a solution containing only the fully saturated complex P_nX_m, in the absence of free P or free X (as occurs with high-affinity ligands), the molar ratio of the complex would simply be the ratio between the molar amounts of X and P (or *vice versa*), expressed by a fraction reduced to its lowest terms. Often, but not always, either n or m is unity. The actual stoichiometric ratio either coincides with the molar ratio or is its multiple by a factor depending on the aggregation state of the macromolecule in solution. Some typical examples of stoichiometric ratios are as follows: (i) oxygen:hemoglobin: 4:4 (molar ratio 1:1); (ii) 2,3 diphosphoglycerate:hemoglobin 1:4; (iii) calcium:calmodulin: 4:1.

The experimental condition $[P]_{tot} >> K_d$, which characterizes high-affinity ligands, is ideal to determine the molar ratio. This is because it entails $[X] << [X]_{tot}$, thus, if one titrates the protein with successive additions of the ligand, the experimental points are fitted by two straight lines and equivalence is reached at their crossing, where $[PX] \approx [X]_{tot}$ and $[PX] \approx [P]_{tot}$ (see Figure 3.1A). The molar ratio at the point of equivalence is thus the ratio between $[X]_{tot}$ and $[P]_{tot}$, reduced to its lowest integer terms.

Determination of the molar ratio requires: (i) an experimental condition where the above conditions are realized; and (ii) the determination of the total concentrations of the protein and the ligand at the point of equivalence. In many cases the total concentration of the ligand is known, whereas that of the protein is not and must be measured with any of the several methods available. If the molecular weight of the protein is known, calculating the molar ratio is straightforward; otherwise one may determine the minimum composition of the complex on a moles of ligand per gram of protein basis, and convert this to the true molar ratio after having determined the MW of the protein. The most precise method to determine the concentration of an unknown (pure) protein sample is by dry weight. If the protein concentration is too low for dry weight determination, absorbance is resorted to (Gill and von Hippel, 1989).

A more difficult step is the conversion of the molar ratio (or mole by weight ratio) to a true stoichiometric ratio. This requires the determination of the actual molecular weight of the protein in solution, that is, of the aggregation state (if any) of the protein monomers. For example, at roughly the same molecular weight *per* monomer, myoglobin is monomeric in solution, vertebrate hemoglobins are tetrameric, and non-vertebrate hemoglobins present association states ranging from monomers and dimers up to 48-mers and in some cases 144-mers. Several experimental methods may be applied to determine the aggregation state of the protein under the chosen experimental conditions, for example, analytical ultracentrifugation, osmometry, gel filtration, and so on.

Even though the experimental approach described above may appear complicated, the procedure is straightforward, and can be better illustrated by the classical example of the determination of the oxygen binding stoichiometry of hemoglobin, each step of which can be historically traced to a specific experiment.

Hemoglobin is much more soluble than oxygen, and one may easily prepare solutions in which the protein has concentrations significantly higher than the K_d. By denaturation of a concentrated solution of hemoglobin, Hufner in 1884 was able to determine the amount of oxygen released, which greatly exceeded that of the freely dissolved gas. Moreover, by desiccation of the protein solution, the same author could determine the amount of protein, by weight. Hufner results demonstrated to the oxygen:Hb ratio is 1.34 mL/gram (with the gas measured under standard conditions). The state equation of perfect gases allowed him to convert the above ratio to 0.06×10^{-3} moles O_2/gram Hb.

The next step involved the determination of the minimum molecular weight of the macromolecule. This determination was carried out by Zinoffsky, who in 1885 was able to determine the iron content of Hb at 0.335%. This estimate leads to a minimum molecular weight of 16700 Da and, coupled to the determination of 0.06×10^{-3} moles O_2/gram Hb, yields the O_2:Hb molar ratio:

$$0.06 \times 10^{-3} \, mol \, / \, g \times 16700 \, g/mol = 1 : 1 \, mol/mol.$$

The molar ratio calculated for the minimum molecular weight is not yet a true stoichiometry: to obtain this parameter one should use in the above calculation the actual molecular weight in solution, or one should multiply the molar ratio by the number of subunits that constitute the macromolecule in solution. The determination was carried out by Adair in 1925, who used precise osmometry measurements to determine that the molecular weight of Hb in solution is 65000 Da, corresponding to an aggregate of four 16700 Da subunits. This finally leads to the correct reaction stoichiometry of 4 O_2/Hb tetramer.

The reader should notice that the experimental conditions which are ideal for the determination of K_d (i.e., $[X] \approx [X]_{tot}$, $[X] >> [PX]$) are opposite to those which are required to determine the reaction stoichiometry (i.e., $[PX] \approx [X]_{tot}$, $[X] << [PX]$).

3.9 Ligands Occupying a Thermodynamic Phase Different from the Protein

In some cases the ligands may occupy a thermodynamic phase other than that of the protein: for example, respiratory proteins and enzymes react with gases, and steroid ligands partition between fat tissues and water-soluble proteins. Ligand binding in this case occurs via a phase exchange, and the system has a huge reserve of the ligand in the gas (or fat) phase, an obvious advantage in many experiments. The reaction of a gaseous ligand may be represented as follows:

$$X_g \Leftrightarrow X_{aq}$$
$$P_{aq} + X_{aq} \Leftrightarrow PX_{aq}$$

The state X_{aq} may be imagined as a scarcely populated thermodynamic intermediate, in equilibrium with the much more populated gaseous phase, according to Henry's law (more complex relationships apply for gases that do not obey Henry's law); thus, provided that the phase equilibrium has been reached, we may simplify the above equations to:

$$P_{aq} + X_g \Leftrightarrow PX_{aq}$$

At the price of a somewhat more complicated experimental setup, which requires gas-tight vessels and the instrumentation required to manipulate gases, the measurement of K_d of gaseous ligands in the gas phase, rather than in solution, offers the advantage that high affinity is usually not a problem, except in really extreme cases; indeed, in this case the protein in solution must compete with the tendency of the ligand to escape to the gas phase, and the concentration of the physically dissolved gas, $[X_{aq}]$ is irrelevant to the determination of K_d. What the researcher records is $\bar{X}$ and gas partial pressures; high-affinity ligands take advantage of larger vessels, which allow a greater volume to the gas phase (Figure 3.5).

The case of poorly soluble ligands that form a solid phase on the bottom of the solution is more awkward than that of gases because $[X_{aq}]$ can never exceed the limit of solubility, and because reaching the phase equilibrium may be agonizingly slow. A way to overcome this problem is to measure the partition of the ligand between the protein of interest and a water-soluble complex of the same ligand.

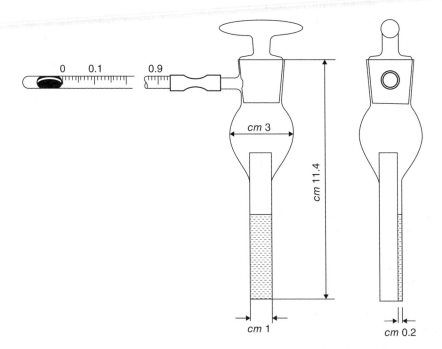

Figure 3.5 A tonometer sealed to a spectrophotometric cuvette, for the determination of the binding isotherms of gaseous ligands. *Source*: Fanelli and Antonini (1958). Reproduced with permission of Elsevier.

3.10   Mixtures of Isoforms

It is a common occurrence that the same organism produces more than a single variety of the same protein, because of heterozygosity or gene duplication, leading to allelic variants. For example, the blood of many vertebrates contains several types of hemoglobin, and their muscles several types of myoglobins. Multiple variants of the same proteins are called isoforms and may differ from each other by as little as a single amino acid residue. Proteins may be purified as mixtures of isoforms, and the researcher may not be aware that his/her preparation is not chemically pure. Occasionally, the stability or other functional properties of an allelic variant may be so grossly altered as to cause a disease, in which case the isoform is called a pathological variant. This type of protein variant is evolutionarily counter-selected, and thus relatively uncommon, except for very special cases like the malaria-protecting pathological hemoglobins (Lopez *et al.,* 2010).

If the researcher is unaware of the presence of isoforms in his/her protein preparation, unanticipated and perplexing results may be obtained. The most common effect of the presence of a mixture of isoforms is reduction of the slope of the $\bar{X}$ versus $\log([X])$ plot: that is, raising the ligand saturation from 0.1 to 0.9 requires a greater than 1.91-fold increase of the logarithm of free-ligand activity. This result is easily explained as follows. If each component of the mixture binds the ligand with different affinity, but yields the same signal, then the apparent fractional saturation is given by the sum of all liganded species, divided by the sum of all species. In the simplest case of a mixture of only two isoforms, which bind the ligand with 1:1 stoichiometry:

$$P1 + X \Leftrightarrow P1X$$
$$P2 + X \Leftrightarrow P2X$$
$$\bar{X} = \text{total liganded sites/total binding sites}$$
$$= (P1X + P2X)/(P1 + P1X + P2 + P2X)$$

To derive the binding polynomial of the mixture we need to know the fraction of each isoform:

$$F_1 = (P1 + P1X)/(P1 + P1X + P2 + P2X)$$
$$F_2 = (1 - F1) = (P2 + P2X)/(P1 + P1X + P2 + P2X)$$

F1 and F2 can usually be assumed to be constant throughout the measurement (i.e., no preferential precipitation or denaturation of either P1 or P2 occurs during the experiment).
We define:

$$\bar{X}_1 = P1X/(P1 + P1X) = [X]/(K_{d,1} + [X])$$
$$\bar{X}_2 = P2X/(P2 + P2X) = [X]/(K_{d,2} + [X])$$

and we obtain:

$$\bar{X} = \bar{X}_1 F_1 + \bar{X}_2 F_2$$

that is, the apparent fractional ligand saturation of the mixture is the weighted average of the fractional saturation of its components.

A simulated experiment for $F_1 = 0.5$, and $K_{d,1} = 10 K_{d,2}$ is depicted in Figure 3.6 to visually demonstrate the features of this system, compared to the binding isotherm of a pure protein with $X_{1/2}$ close to that of the mixture.

In practice, if the researcher is unaware of the presence of a mixture of isoforms, suspicion arises because the ligand binding isotherm shows a lower than expected slope, that is, it extends over a greater than expected interval of ligand concentration. In a monomeric single binding site protein, the presence of isoforms is the most common cause of a broadened binding isotherm. Isoforms can sometimes be detected by non-denaturing electrophoresis, isoelectric focusing, or analytical chromatography. By contrast isoforms usually escape detection by the more commonly employed SDS denaturing electrophoresis, because they usually have very similar molecular weights. In proteins composed by multiple subunits there may be other causes of lowered slope of the $\bar{X}$ versus $\log([X])$ plot, as listed in Table 3.1. The reader will notice that the entries of Table 3.1 reduce to two groups: chemical heterogeneity and negative cooperativity.

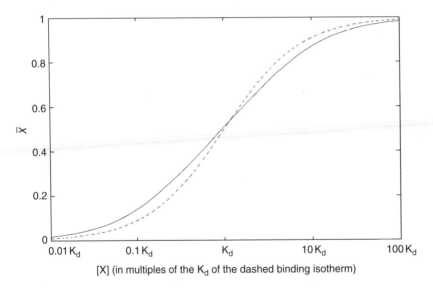

Figure 3.6 Reduced slope of the $\bar{X}$ versus $\log([X])$ plot for a mixture of isoforms (continuous line), as compared to a pure protein (dashed line). The binding curve for the two isoforms was simulated under the assumptions that they are present in equal amounts and that their K_ds are 0.3 and three times that of the simulated pure monomeric protein (dashed line).

Table 3.1 Common possible causes of broadened binding isotherms.

1) Genetic heterogeneity of the protein sample (presence of isoforms)
2) Heterogeneity of the protein sample due to the presence of different aggregation states that do not interconvert during the experiment (e.g., monomers and dimers)
3) Heterogeneity of the protein sample due to incomplete denaturation (partially denatured molecules still binding the ligand) or other accidental modifications
4) Intramolecular heterogeneity (i.e., monomeric proteins having more than a single ligand binding site; oligomeric proteins made up of different subunits)
5) Negative cooperativity

3.11   Poor or Absent Signal

It may happen that the binding reaction one is interested in yields a poor signal, or that the signal to noise ratio of the measurement is poor. This case may often merge into the problem of high affinity, since in order to increase the signal, one is forced to increase the protein concentration, and may reach the condition $[P]_{tot} > K_d$.

Remedies for this condition include: (i) selection of a different determination method, with higher signal; (ii) adoption of a ligand replacement experiment, in which the ligand giving a poor signal displaces, or is displaced by, one giving a stronger signal (see Sections 1.8 and 3.7); or (iii) if a second ligand presents heterotropic linkage with the ligand of interest, measurement of the effect of the latter on the affinity for the former (see Figure 1.8).

References

Anderson, S.R. and Antonini, E. (1968) The binding of carbon monoxide by human hemoglobin. Proof of validity of the spectrophotometric method and direct determination of the equilibrium. J Biol Chem, 243: 2918–2920.

Brunori, M. (1975) Molecular adaptation to physiological requirements: The hemoglobin system of trout. Curr Top Cell Regul, 9: 1–39.

Brunori, M., Coletta, M., Giardina, B., and Wyman, J. (1978) A macromolecular transducer as illustrated by trout hemoglobin IV. Proc Natl Acad Sci USA, 75: 4310–4312.

Brunori, M., Bonaventura, J., Bonaventura, C., Antonini, E., and Wyman, J. (1972) Carbon monoxide binding by hemoglobin and myoglobin under photodissociating conditions. Proc Natl Acad Sci USA, 69: 868–871.

Cassoly, R. and Gibson, Q. (1975) Conformation, co-operativity and ligand binding in human hemoglobin. J Mol Biol, 91: 301–313.

Edmondson, D.E., Binda, C., Wang, J., Upadhyay, A.K., and Mattevi, A. (2009) Molecular and mechanistic properties of the membrane-bound mitochondrial monoamine oxidases. Biochemistry, 48: 4220–4230.

Gill S.C. and von Hippel P.H. (1989) Calculation of protein extinction coefficients from amino acid sequence data. Anal Biochem., 182: 319–326.

Imai, K., Morimoto, H., Kotani, M., Watari, H., and Hirata, W. (1970) Studies on the function of abnormal hemoglobins. I. An improved method for automatic measurement of the oxygen equilibrium curve of hemoglobin. Biochim Biophys Acta, 200: 189–196.

Imai, K. (1982) Allosteric Effects in Hemoglobin. Cambridge University Press, Cambridge, UK.

Kariv, I., Cao, H., and Oldenburg, K.R. (2001) Development of a high throughput equilibrium dialysis method. J Pharm Sci, 90: 580–587.

López, C., Saravia, C., Gomez, A., Hoebeke, J., and Patarroyo, M.A. (2010) Mechanisms of genetically-based resistance to malaria. Gene, 467(1-2): 1–12.

Maguire, J.J., Kuc, R.E., and Davenport, A.P. (2012) Radioligand binding assays and their analysis. Methods Mol Biol, 897: 31–77.

Moore, E.G. and Gibson, Q.H. (1976) Cooperativity in the dissociation of nitric oxide from hemoglobin. J Biol Chem, 251: 2788–2794.

Record, M.T. Jr., Anderson, C.F., and Lohman, T.M. (1978) Thermodynamic analysis of ion effects on the binding and conformational equilibria of proteins and nucleic acids: The roles of ion association or release, screening, and ion effects on water activity. Q Rev Biophys, 11: 103–178.

Root, R.W. (1931) The respiratory function of the blood of marine fishes. Biol Bull, 61: 427–456.

Rossi Fanelli, A. and Antonini E. (1958) Studies on the oxygen and carbon monoxide equilibria of human myoglobin. Arch Biochem Biophys, 77: 478–492.

Sharma, V.S., Schmidt, M.R., and Ranney, H.M. (1976) Dissociation of CO from carboxyhemoglobin. J Biol Chem, 251: 4267–4272.

Part II

Ligand Binding to Multiple Binding Site Proteins

4

Proteins with Multiple Binding Sites

Many ligand-binding proteins possess multiple binding sites for the same ligand, having binding stoichiometries higher than 1:1. The thermodynamic consequences of ligand:protein stoichiometric ratios higher than 1:1 have great physiological relevance, the most significant being *homotropic cooperativity*. Homotropic cooperativity is defined as the influence of one ligand upon the binding affinity of another ligand of the same type. Homotropic cooperativity may be of the *positive* type (if the ligand affinity increases with increasing ligand saturation) or of the *negative* type (if the ligand affinity decreases with increasing ligand saturation). Positive homotropic cooperativity increases the steepness of the $\bar{X}$ versus $\log([X])$ plot and increases the responsiveness of the protein to small changes in ligand concentrations like those that are most frequently encountered in the physiological milieu, a property that J. Wyman called the "cybernetics of biological macromolecules" (Wyman, 1981).

In this chapter we analyze ligand binding to proteins having multiple binding sites. In principle, the chapter's content and the order of the subject matter is very similar to that of Chapter 1, but of course many subjects change significantly as a consequence of the change in the reaction stoichiometry, and the new property *homotropic linkage* is introduced.

The laws that we shall develop and the equations that we shall derive in this Chapter will always contain terms that depend on the number of binding sites present in the protein. As a general rule we shall take as a starting point the simplest possible case, that is, the homodimer, and we shall then generalize to higher-order oligomers.

4.1 Multiple Binding Sites: Determination of the Binding Stoichiometry

The case of proteins bearing multiple binding sites for the same ligand is common. It applies to hemoglobins, hemocyanins, and other oxygen carriers, to metal-binding proteins like transferrin, ceruloplasmin, metallothionein, and so on, to antibodies, to hormone receptors, and so on. It is particularly relevant to DNA-binding proteins, although these represent a special case that will not be treated here because their multiple DNA binding sites do not bind separate ligand molecules but DNA segments interconnected by intervening DNA.

Reversible Ligand Binding: Theory and Experiment, First Edition. Andrea Bellelli and Jannette Carey.
© 2018 John Wiley & Sons Ltd. Published 2018 by John Wiley & Sons Ltd.

The general reaction scheme is as follows:

$$P + nX \Leftrightarrow PX + (n-1)X \Leftrightarrow PX_2 + (n-2)X \Leftrightarrow \ldots \Leftrightarrow PX_n$$

In some cases the protein is made up of a single polypeptide chain having multiple binding sites. For example, calmodulin is a single polypeptide chain of MW 17,000 and has four calcium-binding sites (Babu *et al.*, 1988). Far more common, however, is the case of proteins made up of multiple subunits, each having one ligand-binding site. As well, multiple binding sites can be present on a monomeric or multimeric protein for more than one type of ligand, giving rise to heterotropic linkage phenomena (this case, dealt with in Sections 4.9 and 4.10, below, is the extension of the one considered in Section 1.9).

The first problem one encounters in these cases is to determine the stoichiometry of the reaction, that is, how many ligand molecules are bound to the protein at saturation. We encountered this problem already in Sections 1.2 and 3.8, where we remarked that at least two different kinds of information are required for its unequivocal solution, namely: (i) the molar ratio of the reactants, that can be estimated in an appropriately designed titration experiment at $[P]_{tot} \gg K_d$ (see Section 3.8), and (ii) the actual molecular weight of the protein in solution, that is, the value that reflects the correct number of monomeric units that compose the protein, to be determined by structural analysis (e.g., by analytical ultracentrifugation, gel filtration, or other structural investigation). Under some circumstances further information may be required, notably the equilibrium constant of the association-dissociation reaction of the protein subunits, because it is not uncommon that ligation determines or biases the aggregation state of the protein. For example, the hemoglobin from lamprey is dimeric or tetrameric when unliganded (deoxygenated) and dissociates to monomers in the liganded (oxygenated) state. Several hormone receptors are monomeric when unliganded and dimerize upon ligation. The case of ligand-dependent aggregation causes the reaction stoichiometry, and possibly the ligand affinity to depend on protein concentration (see Figure 3.1B). This complex phenomenon deserves a dedicated treatment; it is discussed in Chapter 5.

In many cases stoichiometry is not deliberately sought at the beginning of one's studies, but is questioned because of the finding of a ligand binding isotherm incompatible with the hypothesis of 1:1 stoichiometry. It is important to remark that the shape of the ligand binding isotherm (and especially the slope of the $\bar{X}$ versus $\log([X])$ plot) is functional information whose relationship with binding stoichiometry is not strict. The ligand-binding isotherm of an oligomeric protein made up of identical subunits with non-interacting ligand binding events is the same as that for a monomeric protein forming a 1:1 complex. Conversely a broadened binding isotherm may indicate negative cooperativity in a multimeric protein, but is also compatible with a mixture of monomeric, 1:1 isoforms (see Section 3.10). In the following sections, we shall analyze the functional behaviour of proteins capable of binding ligands with stoichiometry higher than 1:1, and shall demonstrate that the slopes of their ligand binding isotherms may be increased, decreased or unchanged with respect to that of monomeric, single-binding site proteins.

4.2   The Binding Polynomial of a Homooligomeric Protein Made Up of Identical Subunits

The case of proteins made up of identical subunits, each capable of binding one ligand, is common in biology. The functional behavior of these systems depends on whether or not the interactions among subunits are influenced by the presence of the ligand. In the present section we shall consider the simplest, but probably not the most common, case, in which the first molecule of ligand has no effect on the binding of subsequent ones. Proteins of this kind may be described as devoid of homotropic interactions, or non-cooperative. In these proteins each subunit plays its functional role without transmitting information to the neighbouring subunits. If the subunits are identical the system is described by only one equilibrium constant K_d.

In this section we shall make two points: (i) *the distribution of the ligands in a partially saturated non-cooperative homo-oligomer is statistical*, and (ii) *this distribution may be expressed by applying appropriate statistical factors to the equilibrium constants of each binding step*. In the case of non-cooperative homooligomers this treatment is trivial, but it has great pedagogical value for the study of more complex cases (e.g., cooperative homooligomers).

If one neglects the distribution of liganded and unliganded subunits in the oligomers, the binding polynomial of a multimeric protein devoid of homotropic interactions is identical to that of a monomeric, single-site protein (eqns. 1.5a, 1.5b, 1.5c, 1.6, 4.1, 4.2):

$$[M]_{tot} = [M]\left(1 + [X]/K_d\right)$$ (eqn. 4.1)

$$\bar{X} = [MX]/[M]_{tot} = [X]/\left([X] + K_d\right)$$ (eqn. 4.2)

where [M] stands for the concentration of the subunits, or monomers, irrespective of their distribution among oligomers. This binding polynomial yields the same dependence of the fractional ligand saturation on the free ligand concentration as for the case of monomeric proteins. Therefore, it also describes the same hyperbolic binding isotherm. In short, in the case of non-interacting binding events on homooligomeric proteins, the binding isotherm offers no functional evidence that the protein in solution is an oligomer, nor any information on the binding stoichiometry.

The same, however, is not necessarily true for the signal, which may be influenced not only by the extent of ligation but also by the distribution of liganded and unliganded subunits among the oligomers. For example, the ligand may quench the fluorescence of aromatic residues of the protein and the signal associated with its binding may not be the same for the partially and fully liganded protein derivatives, due to electronic influence among the fluorophores. In these cases one has access to information about how the liganded and unliganded subunits are distributed among the oligomers.

In the absence of homotropic intersubunit interactions, the liganded and unliganded subunits are statistically distributed among the oligomers (see also Appendix 4.1). The distribution of liganded subunits for such a system is binomial because each site has only two states: liganded, with probability $\bar{X}$, and unliganded, with probability $(1-\bar{X})$.

Thus, one can calculate the population of each ligation intermediate for an oligomer with n sites at any value of ligand saturation $\bar{X}$ using the general formula of the binomial distribution:

$$n!\,\bar{X}^i\left(1-\bar{X}\right)^{n-i} /\ i!\left(n-i\right)!$$

where the index i represents the number of liganded sites in the oligomer. For example, if the homooligomer has four identical subunits, does not present homotropic interactions, and has $\bar{X} = 0.4$, the above formula predicts the following relative abundance of ligation intermediates:

fraction of unliganded state, P_4: $4!\ 0.4^0\ 0.6^4 /\ 0!\ 4! = 0.1296$
fraction of singly liganded state, P_4X: $4!\ 0.4^1\ 0.6^3 /\ 1!\ 3! = 0.3456$
fraction of doubly liganded state, P_4X_2: $4!\ 0.4^2 0.6^2 /\ 2!\ 2! = 0.3456$
fraction of triply liganded state, P_4X_3: $4!\ 0.4^3 0.6^1 /\ 3!\ 1! = 0.1536$
fraction of fully liganded state, P_4X_4: $4!\ 0.4^4 0.6^0 /\ 4!\ 4! = 0.0256$
sum of all fractions $= 1.000$

$$\bar{X} = \left(0.3456 + 0.3456\mathrm{x}2 + 0.1436\mathrm{x}3 + 0.0256\mathrm{x}4\right)/4 = 0.4$$

A fully equivalent and easier to remember formula to calculate the relative abundances of all ligation intermediates makes use of the n-th power of the binomial:

$$\left[\left(1-\bar{X}\right)+\bar{X}\right]^n = \left\{\left(K_d+[X]\right)/\left(K_d+[X]\right)\right\}^n \qquad \text{(eqn. 4.3)}$$

Because the sum of $\bar{X}$ and $(1-\bar{X})$ is unity, its power is also unity, and each term of the n-th power of eqn. 4.3 represents the relative fraction of one of the ligation intermediates. For the simplest case of a homodimeric protein (n=2) the following relationships can be derived from eqn. 4.3:

$$[P] = [M_2] = \left(1-\bar{X}\right)^2 = [P]_{tot}\, K_d^2/\left([X]+K_d\right)^2 \qquad \text{(eqn. 4.4)}$$

$$[PX] = [M_2X] = 2\,\bar{X}\left(1-\bar{X}\right) = [P]_{tot}\, 2\,K_d[X]/\left([X]+K_d\right)^2 \qquad \text{(eqn. 4.5)}$$

$$[PX_2] = [M_2X_2] = \bar{X}^2 = [P]_{tot}\,[X]^2/\left([X]+K_d\right)^2 \qquad \text{(eqn. 4.6)}$$

where P represents the homodimer and M the monomer. Notice that in this system the possible dissociation of the dimer into free monomers is neglected, its consideration being deferred to Chapter 5.

Eqns. 4.4 through 4.6 contain all the information required to construct the binding polynomial of the reaction for the homodimer with equivalent, non-interacting binding events. We only need to choose the reference species (see the definition of the binding polynomial in Section 1.4). If we adopt as a reference species P, the unliganded homodimer, we obtain:

$$[PX]/[P] = 2[X]/K_d$$
$$[PX_2]/[P] = [X]^2/K_d^2 \qquad \text{(eqn. 4.7)}$$
$$[P]_{tot} = [P]\left(1+2[X]/K_d+[X]^2/K_d^2\right) = [P]\left(1+[X]/K_d\right)^2$$

There is an alternative method to construct the binding polynomial (eqn. 4.7) that has great value for the more complex cases to be analyzed in the following sections, and deserves being described in detail. One should consider as many independent binding events as there are subunits in the oligomer; thus, two *microscopic equilibrium constants* are required for a dimer, defined as $K_{d,1}$ and $K_{d,2}$. In the non-cooperative homodimer $K_{d,2}$ = $K_{d,1}$ and the two may be indicated as K_d, but if ligation-dependent intersubunit interactions occur, $K_{d,1} \neq K_{d,2}$. Each binding event is governed by the pertinent microscopic equilibrium constant times the appropriate *statistical factor*. Statistical factors introduce the statistical corrections of species distribution in the equilibrium constants, rather than in concentration of ligation intermediates, and greatly simplify the description of complex systems. For example, in the case of a homodimer one must describe two ligand dissociation reactions each governed by its own microscopic equilibrium constant and statistical factor:

PX_2 <==> PX + X governed by equilibrium constant $K_{d,2}$ and statistical factor 2
PX <==> P + X governed by equilibrium constant $K_{d,1}$ and statistical factor ½.

The concept of the microscopic equilibrium constant is somewhat abstract: it is the equilibrium constant one would measure if one could detect that reaction alone, for the single binding site involved; in the non-cooperative oligomer it corresponds to the equilibrium constant for the subunits (eqns. 4.1 and 4.2).

The statistical factors are explained by the consideration that the doubly liganded species PX_2 has two identical liganded sites that can release the ligand. Once one ligand has dissociated, the protein has only one site that can rebind it in the inverse reaction. The relative probability factors of the direct and inverse reactions are thus 2/1. The opposite applies to the mono-liganded species PX, which has one bound site from which the ligand can dissociate and two empty and indistinguishable sites to which the ligand can rebind in the inverse reaction, leading to the probability factor 1/2. An alternative, but fully equivalent, way to explain the statistical factors is that there is only a single possible type of unliganded homodimer, only a single type of doubly liganded homodimer, but two identical and indistinguishable types of singly liganded homodimers, bearing the ligand on either subunit. Thus the unique type of doubly liganded dimer may dissociate one ligand to yield two types of singly liganded dimer, whereas the two types of singly liganded dimer dissociate their ligand to form one and the same type of unliganded dimer.

By taking into account the above considerations, the following relationships can be derived:

$[PX] = [P][X]/\frac{1}{2}K_{d,1} = [P] \, 2[X]/K_{d,1}$
$[PX_2] = [PX] \, [X]/2K_{d,2} = [P][X]^2/K_{d,1}K_{d,2}$

The binding polynomial we derive from the above relationships is as follows:

$$[P]_{tot} = [P]\left(1 + 2[X]/K_{d,1} + [X]^2/K_{d,1}K_{d,2}\right) \qquad \text{(eqn. 4.8)}$$

Eqn. 4.8 has general validity and applies equally to positively cooperative, non-cooperative and negatively cooperative homodimers. In the case of the non-cooperative homodimer, the condition $K_{d,1} = K_{d,2} = K_d$ applies, leading to a simplified formula identical to eqn. 4.7. This demonstrates that the statistical factors applied to the equilibrium constants reproduce the binomial coefficients applied to the ligation intermediates.

In order to derive the fractional saturation $\bar{X}$ from the population of ligation intermediates, one sums the liganded species, each multiplied by the number of bound sites,
and divides by the binding polynomial multiplied by the total number of sites (2 for
the dimer):

$$\bar{X} = [P]\left(2[X]/K_{d,1} + 2[X]^2/K_{d,1}K_{d,2}\right)/2[P]\left(1 + 2[X]/K_{d,1} + [X]^2/K_{d,1}K_{d,2}\right)$$
$$= [X]/K_{d,1}\left(1 + [X]/K_{d,2}\right)/\left(1 + 2[X]/K_{d,1} + [X]^2/K_{d,1}K_{d,2}\right) \qquad \text{(eqn. 4.9)}$$

In the case of the non-cooperative homodimer, the above formulation reduces to
eqn. 4.2 because $K_{d,1} = K_{d,2}$, but we present it explicitly in view of the more complex cases
to be discussed below, in which $K_{d,1} \neq K_{d,2}$. Moreover, equivalence of eqns. 4.2 and 4.9 is
proof of the correctness of the calculation of the population of ligation intermediates.
We encourage the reader to demonstrate in the other cases to be presented in this
chapter that when the interaction factors or heterogeneities are neglected, eqn. 4.2 is
obtained.

If we set eqn. 4.9 equal to 0.5, we can calculate the $X_{1/2}$ of the reaction (see Appendix 4.2),
which results:

$$\sqrt{\left(K_{d,1}K_{d,2}\right)} = X_{1/2} = Xm \qquad \text{(eqn. 4.10)}$$

We further remark that:

i) the plot $\bar{X}$ versus log([X]) of a dimer is symmetric, hence $X_{1/2} = Xm$ (Appendix 4.2)
ii) in the absence of cooperativity, $K_{d,1} = K_{d,2} = K_d$, hence $\sqrt{(K_{d,1}K_{d,2})} = K_d$.

Point (i) applies to any dimer, irrespective of the presence of homotropic interactions,
and of the subunits being identical or not (see Appendix 4.2). In the case of higher order
oligomers, however, the $\bar{X}$ versus log([X]) plot may or may not be symmetric, in which
case $X_{1/2} \neq Xm$.

By using either the binomial distribution of liganded subunits (eqns. 4.4 through 4.6)
or the equilibrium constants corrected for the statistical factors (eqn. 4.7) one can
estimate the population of ligation intermediates, as shown in Figure 4.1. Note the
symmetry of all curves.

The case of homodimers, described above, is common in biology, but other possibilities
exist. Many proteins assemble into trimers, tetramers, hexamers, or higher-order oligomers. The reasoning made for the homodimer applies to any homooligomer, and its
conclusions can be generalized to a homooligomer of n identical non-interacting
subunits as follows.

i) The statistical factors to be applied to the equilibrium dissociation constants are:

$$n/1; (n-1)/2; (n-2)/3; \ldots; 1/n$$

For example, in the case of the homotetramer:

$$K_{d,4}' = 4K_d; \quad K_{d,3}' = 3/2K_d; \quad K_{d,2}' = \tfrac{2}{3}K_d; \quad K_{d,1}' = \tfrac{1}{4}K_d$$

(A)　　　　　　　　　　　　　　　　(B)

(C)　　　　　　　　　　　　　　　　(D)

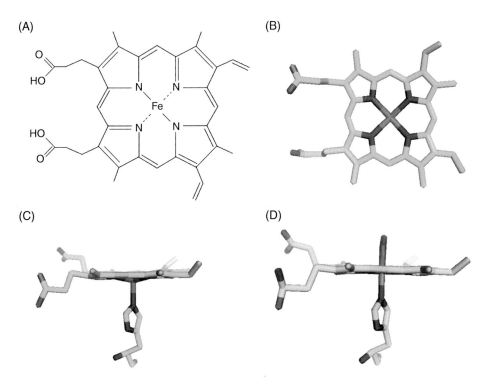

Figure 7.1 The heme.
　　Panel A: the structural formula of heme (iron protoporphyrin IX). Panel B: top view of the three dimensional structure of unliganded heme, taken from the crystal structure of the α subunits in deoxyHb (PDB code 2dn2, see Park *et al.*, 2006), in atomic colors with green carbon atoms. Panel C: side view of unliganded heme and the proximal His from the α subunits of deoxyHb (PDB code 2dn2) showing the doming of the porphyrin and the out of plane position of the five-coordinated iron. Panel D: side view of the heme taken from the crystal structure of the α subunits of oxyHb (PDB code 2dn1), showing the proximal His and the bound oxygen, in atomic colors with cyan carbon atoms. The projection view is approximately the same as that of panel C to highlight the flattening of the heme and the in plane position of the six-coordinated iron.

Reversible Ligand Binding: Theory and Experiment, First Edition. Andrea Bellelli and Jannette Carey.
© 2018 John Wiley & Sons Ltd. Published 2018 by John Wiley & Sons Ltd.

(A)   (B)

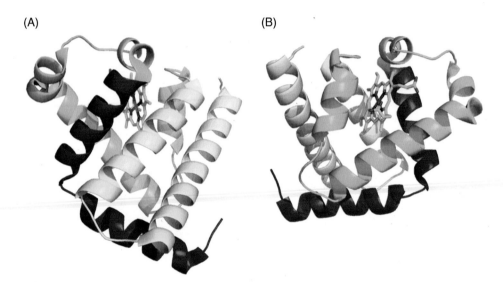

Figure 7.3 Ribbon representation of the backbone of sperm whale myoglobin.
α-helices A and B in red; C,D,E, and F in green; G and H in yellow; the heme is represented as sticks in atomic colors with green carbon atoms.

Panel A shows the macromolecule from the side farthest from the heme, and panel B from the opposite side, where the heme propionates interact with solvent. The bottom of the heme pocket is formed mostly by helices B and G, whereas its sides are formed by helices C, E, and F.

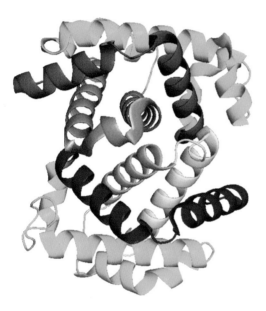

Figure 7.4 The αβ dimer of human deoxyhemoglobin.

The secondary structures of the α subunit are shown in red (helices A and B), green (helices C through F) and yellow (helices G and H); the corresponding secondary structures of the β subunit are shown in magenta, green, and orange. The figure shows the interface contacts αB-βH (red-orange), αG-βG (yellow-orange), and αH-βB (yellow-magenta). The heme-binding regions (green in both subunits) are relatively far from the αβ intersubunit interface.

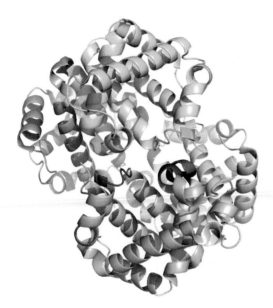

Figure 7.5 The human deoxy-hemoglobin tetramer and the a1b2 intersubunit interface. α subunits are shown in green, β subunits in pink. The FG corner and C helices of each subunit, which constitute the α1β2 interface are highlighted using stronger colors.

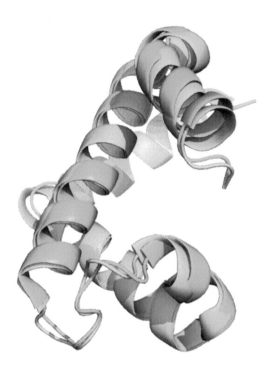

Figure 7.6 Superposition of the α subunit of deoxy-hemoglobin (green) and carbon monoxy-hemoglobin (cyan). The figure shows only residues 1-93 (helices A through E) in order to better visualize the C-helix (bottom) and F-helix and FG corner (top). Superposition was optimized for the region where ligand-linked structural changes are minimal, corresponding to the α1β1 contact region (helices B, G and H, residues 2-35 and 94-138), and the orientation emphasizes the region where the ligand-linked structural changes are largest, corresponding to the α1β2 contact region.

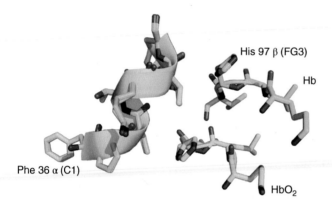

Figure 7.7 Ligand-dependent quaternary structural change at the αC-βFG contact region of the $\alpha1\beta2$ interface of human hemoglobin (detail). Upon oxygenation the FG corner of the β subunit slides by one helical turn over the C helix of the α subunit. The α C helices of the structures of HbO$_2$ (light blue, PDB file 2dn1) and Hb (green, PDB file 2dn2) were superimposed to make evident the relative movement of the β FG corner.

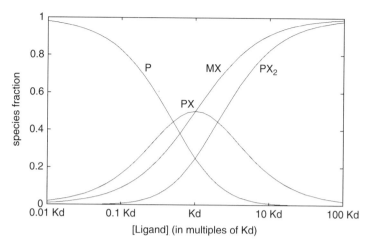

Figure 4.1 Fractional saturation of a non-cooperative, non-dissociating homodimeric protein. MX indicates the fraction of bound monomers over total monomers, irrespective of their distribution among dimers, as calculated from eqns. 4.2 or 4.9. P indicates the fraction of fully unliganded dimeric protein (eqn. 4.4); PX the fraction of singly liganded dimeric protein (eqn. 4.5); and PX_2 the fraction of doubly liganded dimeric protein (eqn. 4.6).

and the product of all statistical factors is unity. In these formulas we introduce the following convention, to be used throughout: $K_{d,i}$ indicates the microscopic equilibrium dissociation constant for site i; $K_{d,i}'$ indicates the apparent dissociation equilibrium constant for the same site. The relationship between the two is given by the statistical factor, that is, $K_{d,i}' = i/(n-i+1)K_{d,i}$. If the oligomer is non-cooperative, all the intrinsic constants are identical and we use only the term K_d, but the $K_{d,i}'$s differ because of their statistical factors (eqns. 4.10, 4.11, 4.12, 4.13).

ii) The binding polynomial for a non-cooperative oligomer made of identical subunits is:

$$[P]_{tot} = [P](1+[X]/K_d)^n \qquad \text{(eqn. 4.11)}$$

iii) The fractional saturation is:

$$\bar{X} = [X]/K_d (1+[X]/K_d)^{n-1}/(1+[X]/K_d)^n \qquad \text{(eqn. 4.12)}$$

iv) $X_{1/2}$ is :

$$X_{1/2} = \sqrt[n]{\left(K_{d,1}' \, K_{d,2}' \ldots K_{d,n}'\right)} = K_d = Xm \qquad \text{(eqn. 4.13)}$$

that is, the geometric mean of the apparent constants equals the intrinsic constant (this statement is equivalent to saying that the product of all the statistical factors is unity); moreover, in this system the $\bar{X}$ versus log([X]) plot is symmetric and the K_d equals the $X_{1/2}$ and the Xm. Heterooligomers (Section 4.3) and negatively or positively cooperative homooligomers (Section 4.4) with more than two subunits or binding sites may present an asymmetric $\bar{X}$ versus log [X] plot. In these cases we would have:

$$\sqrt[n]{\left(K_{d,1}' \times K_{d,2}' \times \ldots \times K_{d,n}'\right)} = Xm$$

but a unique K_d would not be defined, and $Xm \neq X_{1/2}$ [Wyman 1963].

We finally remark that if the researcher prefers the use of association rather than dissociation constants the statistical factors are the reciprocal of those given above. For example, the association constant for the first site of a homotetramer would be $K_{a,1} = 4 K_a$ and for the fourth site $K_{a,4} = \frac{1}{4} K_a$.

4.3 Intramolecular Heterogeneity

A protein possessing multiple ligand–binding sites may be made up of non-identical subunits, or the binding sites may be non-equivalent if they reside on the same polypeptide chain. In such cases the protein will behave like a mixture of isoforms, with a fixed number of different affinity sites. Contrary to mixtures of isoforms, that can usually be resolved biochemically, the protein will appear as a single, pure band in non-denaturing conditions (e.g., analytical chromatography, electrophoresis).

As in the case of the mixture of isoforms, the ligand-binding isotherm will be broader than that of a single binding site protein, consistent with the presence of two or more classes of binding sites with different affinities (Figure 4.2).

If there is no cooperativity, the fractional saturation of each class of binding sites can be calculated independently of the other(s), and the distribution of liganded and unliganded subunits within the oligomer will be statistical. In the simplest possible case of a non-cooperative heterodimer made up of different subunits M_a and M_b we have:

$$\bar{X}_a = [M_a X]/[M_a]_{tot} = [X]/([X] + K_{d,a})$$
$$\bar{X}_b = [M_b X]/[M_b]_{tot} = [X]/([X] + K_{d,b})$$
$$\bar{X} = (\bar{X}_a + \bar{X}_b)/2$$

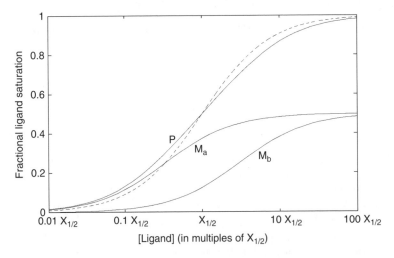

Figure 4.2 Ligand-binding isotherm of a non-cooperative heterodimer with non-equivalent binding sites. The continuous line P represents the fractional saturation of the heterodimer; M_a and M_b represent the corresponding curves for each of the constituent monomers (multiplied by 0.5 to take into account that the concentration of each monomer is half that of the heterodimer). The dashed line represents the fractional saturation of a non-cooperative homodimer (eqn. 4.2) with the same $X_{1/2}$ for comparison; notice that it is steeper than that for the heterodimer P. In this simulation $K_{d,a} = \frac{1}{3} X_{1/2}$ and $K_{d,b} = 3 X_{1/2}$.

The binding curve for the heterodimer, $\bar{X}$ versus log [X], is symmetric with

$$X_{1/2} = \sqrt{\left(K_{d,a}\ K_{d,b}\right)}$$

The distribution of liganded and unliganded monomers amongst heterodimers results from the product of the populations of the liganded and unliganded states of the two subunits:

$$\left[\bar{X}_a + \left(1 - \bar{X}_a\right)\right]\left[\bar{X}_b + \left(1 - \bar{X}_b\right)\right]$$
$$= \bar{X}_a\bar{X}_b + \bar{X}_a\left(1 - \bar{X}_b\right) + \left(1 - \bar{X}_a\right)\bar{X}_b + \left(1 - \bar{X}_a\right)\left(1 - \bar{X}_b\right)$$

(eqn. 4.14)

In this equation the term $\bar{X}_a\,\bar{X}_b$ represents the fraction of heterodimers bearing the ligand on both subunits; $\bar{X}_a(1 - \bar{X}_b)$ the fraction bearing the ligand on monomer a but not on monomer b; and so on (Figure 4.3). Notice that in this formula $\bar{X}_a$ and $\bar{X}_b$ range from 0 to 1.

It may be interesting, if only for historical reasons, to elaborate the fractional saturation of the heterodimer (curve P in Figure 4.2) according to the equations devised by Hill and by Scatchard. The resulting plots are reported in Figure 4.4 to visually demonstrate the (limited) usefulness of these representations, which both demonstrate deviations from the behaviour expected for a single-site system. Although we do not recommend usage of these representations for quantitative analysis, they may be useful to visualize a condition that deserves further scrutiny.

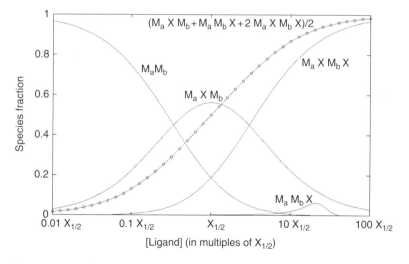

Figure 4.3 Population of ligation intermediates in a heterodimer, as calculated from eqn. 4.14, using the same parameters as in Figure 4.2. Notice that one of the two liganded intermediates, M_aM_bX, is poorly populated, because it requires two unlikely conditions to be met, namely the higher affinity subunit must be in the unliganded state and the lower affinity subunit in the liganded state. The circles represent the sum of liganded species divided by the total number of subunits; they are superimposed on the curve of overall ligand saturation (curve P in Figure 4.2).

(A)

(B)

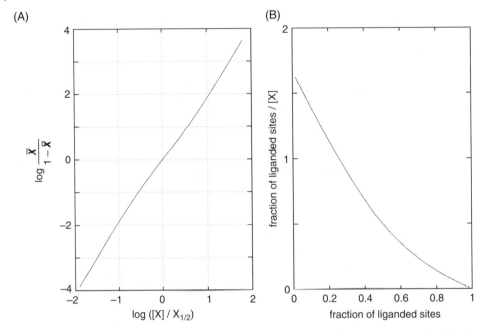

Figure 4.4 The Hill (panel A) and Scatchard plot (panel B) for data set P from Figure 4.2. In the Hill plot the heterodimer presents a slope <1, and its Scatchard plot is bent.

4.4 Oligomeric Proteins with Interacting Binding Events: Homotropic Linkage

A very interesting and quite common case is that of the homooligomeric protein whose subunits exchange information about the presence or absence of a bound ligand, that is, whose intersubunit interactions are affected by ligation. When this type of interaction occurs between identical ligand binding sites it is called *homotropic linkage*. In these systems the ligand affinity of each subunit is affected, directly or indirectly, by the ligation state of the other subunits. While the thermodynamic relationships between the binding sites can be described by generally applicable models, the structural bases of homotropic linkage are highly system-specific, and impossible to generalize. Nevertheless, they have been investigated in detail for several proteins, and we shall present the important example of hemoglobin in Chapter 7.

Homotropic linkage results in positively or negatively cooperative ligand-binding equilibria, whose quantitative description requires not only the statistical factors introduced for the non-cooperative homooligomer (Section 4.2), but also different microscopic constants. The negatively cooperative homooligomer presents a broadened ligand-binding isotherm, apparently identical to that of the heterooligomer (Section 4.3). The positively cooperative homooligomer presents a ligand-binding isotherm steeper than that of a single binding site protein. Positive cooperativity may occur also in heterooligomers; in this case the shape of the $\bar{X}$ versus $\log([X])$ plot is the result of the opposing tendencies of intramolecular heterogeneity (which tends to broaden the curve) and cooperativity (which makes the curve steeper). As in the preceding sections, we begin our analysis by considering the simplest case of a cooperative homodimer.

The reaction scheme for the cooperative homodimer is similar to the one considered in Section 4.2, except that it requires two microscopic equilibrium constants, one for ligand binding to the fully unliganded dimer, and one for ligand binding to the singly liganded dimer (or one for dissociation of the only liganded site of the mono-liganded dimer, the other for either liganded site in the fully liganded dimer):

$$PX_2 \Leftrightarrow PX + X \Leftrightarrow P + 2X$$

Let the equilibrium constants for the two reactions be defined as follows:

$$PX_2 \Leftrightarrow PX + X \quad \text{has} \quad 2\,K_{d,2}$$
$$PX \Leftrightarrow P + X \quad \text{has} \quad \tfrac{1}{2}\,K_{d,1}$$

$K_{d,2}$ and $K_{d,1}$ are two independent parameters and one should apply a least-squares minimization routine to obtain their best estimates from the experimental data, whereas in the case presented in Section 4.2 we had only one parameter since we assumed $K_{d,2} = K_{d,1}$. Apart from this point, this case is similar to that of the non-cooperative homodimer discussed in Section 4.2. In particular, Eqns. 4.8, 4.9 and 4.10 are valid and applicable, but eqn. 4.8 cannot be reduced to eqn. 4.7. Moreover, the $\bar{X}$ versus $\log([X])$ plot is symmetric (Appendix 4.2).

An important difference with the case of the non-cooperative homodimer is that the distribution of liganded and unliganded subunits among the cooperative homodimers is not statistical, because the ligation-dependent interactions among subunits cause some intermediates to be more populated than others. Thus we cannot rely on eqns. 4.3 through 4.6 to calculate the population of each ligation intermediate, and we must resort to eqn. 4.8 instead.

For the reader interested in the history of biochemistry we may add that an equation analogous to eqn. 4.8, but applied to the hemoglobin tetramer, was first developed by G. Adair in 1925. Thus, we may follow Adair and extend the above treatment to cooperative homooligomers of higher order by analyzing the original (tetramer) case. We only need to assume as many equilibrium constants as there are ligand-binding sites in the protein. The reaction scheme for a homotetramer is:

$$PX_4 \Leftrightarrow PX_3 + X \Leftrightarrow PX_2 + 2\,X \Leftrightarrow PX + 3\,X \Leftrightarrow P + 4\,X$$

and we define four microscopic equilibrium constants as:

$$K_{d,i} = [PX_{i-1}][X]/[PX_i]$$

with i=1 to 4.

The $K_{d,i}$ defined above is actually an average value because even if the subunits are identical, their contacts with other subunits in any oligomer greater than a dimer cannot be identical, and thus two oligomers bearing the same number of ligands may not be identical (see Section 4.5, below). Nonetheless, the $K_{d,i}$s can be used to construct the binding polynomial, and require multiplication by the appropriate statistical factors (eqn. 4.15).

$$[P]_{tot} = [P]\left(1 + 4[X]/K_{d,1} + 6[X]^2/K_{d,1}K_{d,2} + 4[X]^3/K_{d,1}K_{d,2}K_{d,3} + [X]^4/K_{d,1}K_{d,2}K_{d,3}K_{d,4}\right)$$

$$(\text{eqn. } 4.15)$$

The fractional saturation function is:

$$\bar{X} = \left([X]/K_{d,1} + 3[X]^2/K_{d,1}K_{d,2} + 3[X]^3/K_{d,1}K_{d,2}K_{d,3} + [X]^4/K_{d,1}K_{d,2}K_{d,3}K_{d,4}\right)$$
$$/\left(1 + 4[X]/K_{d,1} + 6[X]^2/K_{d,1}K_{d,2} + 4[X]^3/K_{d,1}K_{d,2}K_{d,3} + [X]^4/K_{d,1}K_{d,2}K_{d,3}K_{d,4}\right)$$

$$(\text{eqn. } 4.16)$$

which would reduce to eqns. 4.11 and 4.12 respectively if $K_{d,1} = K_{d,2} = K_{d,3} = K_{d,4}$. The use of statistical factors in eqns. 4.15 and 4.16 may not be obvious, and a more detailed explanation is provided in Appendix 4.1. With respect to the treatment originally proposed by Gilbert Adair for hemoglobin (Adair, 1925), we adopted a minor change because Adair used association constants, whereas, for the sake of consistency, we used dissociation constants.

In the presence of ligation-dependent intersubunit interactions, the intrinsic constants will not be equal to each other, and there is no way of predicting their progression, unless some structural hypothesis is made. Moreover, an oligomer of order higher than 2 may or may not present a symmetric $\bar{X}$ versus log([X]) plot and hence may or may not satisfy the condition $X_{1/2} = Xm$ (Appendix 4.2). Indeed, Adair's equation is free of assumptions on the structure of the macromolecule, and applies to every macromolecule provided that the binding stoichiometry is known. L. Pauling was the first to propose in 1935 a structural interpretation of Adair's equation (Pauling, 1935). He referred his model to the case of hemoglobin, at the time the best-known example of homotropic cooperativity; but he intended his model as generally applicable. He suggested that the liganded state of the protein's subunits forms "facilitating" interactions of (presumed) equal strength, with the contacting, liganded subunits. Pauling assigned a privileged status to $K_{d,1}$, which he considered the equilibrium constant of the reaction in the absence of facilitating interactions, and rewrote the K_ds of the successive binding steps as $K_{d,1}$ times the appropriate power of an interaction factor. C.D. Coryell, and later J. Wyman, remarked that in the case of hemoglobin the dissociated subunits (obtained in the presence of urea) have higher affinity than the tetramer, and thus modified Pauling's equation by taking $K_{d,4}$ as the equilibrium constant of the subunits in the absence of ligand dependent interactions. In their model the ligand dependent inter-subunit interactions constrain the affinity of the subunit and ligation releases the constraint. Irrespective of the ligand-dependent intersubunit interactions being of the facilitatory or inhibitory type, Pauling's model has special interest as a conceptual step in the understanding of cooperativity, because it acknowledged the fact that Adair's scheme required a structural interpretation, and attempted to provide one. Not surprisingly it had several successive elaborations, the most complete being that by Koshland and co-workers (Koshland *et al.*, 1966).

We develop below Pauling's versions of eqns. 4.8 and 4.9 for the cooperative homodimer, before attacking the more complex case of tetramers and higher order oligomers. To do so, we rewrite eqns. 4.8 and 4.9 using one equilibrium constant and one interaction factor; this does not change the total number of parameters required to describe the system unless one has some theoretical reason or hypothesis to constrain the interaction factor(s). Let us take $K_{d,2}$ as the equilibrium dissociation constant for the subunit in the absence of ligand-dependent interactions:

$$K_{d,2} = K_d$$
$$K_{d,1} = \varepsilon K_d$$

where ε is an interaction term whose meaning is that the free energy of dissociation of a liganded subunit is K_d if the partner subunit is liganded, and is decreased by the factor RT ln (ε) if the partner subunit is unliganded. The model implies that the free energy RT ln (ε) is accounted for by specific intersubunit (weak) bonds that form or break upon ligation. Pauling's hypothesis describes the macromolecule as a mechanical device, and cooperativity as the deterministic effect of movements and structure changes occurring inside it. We call this hypothesis *sequential* (or *intramolecular) cooperativity*, because every single macromolecule in the sample should sequentially explore every energy state, by breaking or forming the appropriate number of ligand-dependent intersubunit interactions. *The commonly adopted definition of cooperativity saying that the last ligand is bound with higher affinity than the first has its literal meaning only in the case of sequential models of cooperativity.*

The binding polynomial results:

$$[P]_{tot} = [P]\left(1 + 2[X]/\varepsilon K_d + [X]^2/\varepsilon K_d^2\right) \qquad \text{(eqn. 4.17)}$$

and the fractional ligand saturation:

$$\bar{X} = \left([X]/\varepsilon K_d + [X]^2/\varepsilon K_d^2\right)/\left(1 + 2[X]/\varepsilon K_d + [X]^2/\varepsilon K_d^2\right) \qquad \text{(eqn. 4.18)}$$

Eqns. 4.17 and 4.18 are fully equivalent to eqns. 4.8 and 4.9, respectively, but eqn. 4.17 cannot be reduced to eqn. 4.11 except in the case where ligand dependent interactions are absent ($\varepsilon = 1$), consistent with the precise structure-function relationships implied by the model.

Eqn. 4.18 allows us to determine $X_{1/2}$ for the reaction; indeed by applying the definition of $X_{1/2}$ we may equate $\bar{X} = 0.5$ and $[X] = X_{1/2}$ to obtain:

$$2X_{1/2}/\varepsilon K_d + 2X_{1/2}^2/\varepsilon K_d^2 = 1 + 2X_{1/2}/\varepsilon K_d + X_{1/2}^2/\varepsilon K_d^2$$

which yields:

$$X_{1/2} = K_d \sqrt{\varepsilon} = \sqrt{\left(K_{d,1}K_{d,2}\right)}$$

In its essence the treatment of the cooperative homodimer given above is complete. However, a brief analysis of the two opposite cases $K_{d,2} > K_{d,1}$ and $K_{d,2} < K_{d,1}$, though implicit in eqns. 4.8 and 4.9, seems appropriate, and will provide interesting information.

If $\varepsilon < 1$ we have the following relationship between the two intrinsic constants: $K_{d,1} < K_{d,2}$, meaning that the fully liganded dimer releases one molecule of ligand more readily than the singly liganded dimer releases its only molecule of ligand. The resulting equilibrium isotherm will be indistinguishable from that of the heterodimer (Section 4.3) and will be broader than that observed for a single-site protein. On the contrary, if $\varepsilon > 1$, removing one molecule of ligand from the fully liganded dimer will be more difficult than removing the only molecule of ligand bound to a singly liganded dimer, and the ligand binding isotherm will be compressed on the $\log([X])$ axis, as characteristic of positive homotropic cooperativity (Figure 4.5).

Pauling's model applied to the homodimer is hardly rewarding: it provides no obvious advantage over Adair's equation. When we turn our attention to a higher-order oligomer, the hypothesis becomes much more compelling, because one may limit the number of interaction factors taking into account the presumed or known structure of the macromolecule. We may consider just one of the cases considered by Pauling, that of a tetramer presenting what he called the tetrahedral functional geometry (each liganded subunit may form facilitating interactions with all the other three). Pauling's reaction scheme is reported in Figure 4.6. Binding of the ligand proceeds *via* four successive steps, that obey the following rule: whenever two adjacent subunits are both liganded they form a facilitating interaction whose free energy contribution is added to that of binding.

The four reaction steps may be described as follows. The first ligand molecule binds to an unliganded tetramer to form the singly liganded intermediate in a reaction

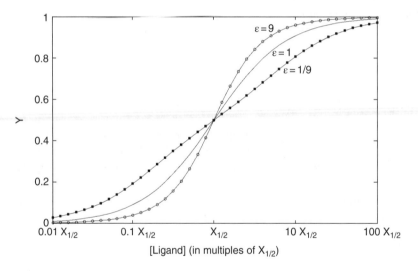

Figure 4.5 Ligand-binding isotherms for a non-interacting homodimer ($\varepsilon = 1$), and two interacting ones (with $\varepsilon = 1/9$ and $\varepsilon = 9$). It is evident that a ligand-dependent intersubunit interaction with $\varepsilon > 1$ increases the steepness of ligand-binding isotherm, whereas one with $\varepsilon < 1$ broadens the ligand-binding isotherm. The three curves were simulated using the following parameter sets: $K_d = 1$, $\varepsilon = 1$; $K_d = 3$, $\varepsilon = 1/9$; $K_d = \frac{1}{3}$ $\varepsilon = 9$ (values of Kd in arbitrary units); all parameter sets have the same $X_{1/2} = 1$ (in arbitrary units).

Figure 4.6 Pauling's reaction scheme for a cooperative homotetramer presenting the tetrahedral functional geometry.
Open circles represent unliganded subunits; closed circles represent liganded subunits.

governed by $K_1 = K_d$ (Pauling used association equilibrium constants, whereas in this book we use dissociation constants). K_d has a privileged status in this model, as it represents the affinity of the subunit in the absence of facilitating interactions. The second ligand molecule binds to a singly liganded tetramer, yielding the doubly liganded derivative. In a tetramer that presents the tetrahedral geometry, the two liganded subunits are necessarily in reciprocal contact and form a weak interaction, whose free energy adds to that of binding; the resulting equilibrium constant is $K_2 = \varepsilon K_d$. If cooperativity is positive $\varepsilon < 1$, and $K_2 < K_1$ (since we use dissociation constants, this implies that the affinity for the second ligand molecule is higher than that for the first one). The third subunit that becomes liganded forms two facilitating interactions with the already liganded ones, thus $K_3 = \varepsilon^2 K_d$. Finally, the fourth subunit that becomes liganded forms facilitating interactions with the other three, thus $K_4 = \varepsilon^3 K_d$.

The four Adair constants are thus explained using only two parameters (K_d and ε) plus an assumption on the functional geometry of the macromolecule, that is, the distribution of the facilitating interactions (in the original paper Pauling compared the tetrahedral tetramer described above to the square tetramer in which each subunit forms facilitating interactions with only two other subunits). A consequence of Pauling's hypothesis is that the progression of the Adair constants is monotonic, a characteristic that is not implied by Adair's equation, nor strictly necessary for cooperativity, which only requires that $K_{d,4}$ is smaller than the other three.

We consider Pauling's the first modern model of cooperativity, because it acknowledges the solidly based structural data and thermodynamic formulation of Adair, and offers a fundamental conceptual contribution to the interpretation of structure-function relationships in macromolecules, that is, that changes in the internal weak bonding network of the macromolecule may change its affinity for external ligands. Preceding hypotheses, like that formulated by A.V. Hill 20 years before Pauling's, were qualitatively different because they lacked the crucial information of ligand binding stoichiometry.

Unfortunately, Pauling's model has an important weakness: as pointed out by J. Wyman (1948), the asymmetry of protein monomers causes the oligomer to require as many different interaction terms as functionally relevant types of intersubunit interfaces. For a more complete and systematic reappraisal of Pauling's hypothesis and its possible variants, we refer the interested reader to the work of Koshland, Nemethy and Filmer (1966); for a history of cooperativity theories, the reader may refer to Edsall (1972, 1980) or to Bellelli (2010).

The study of the structure of biological macromolecules received an enormous impulse from 1950s; this led to some important reformulations of the problem of protein structure-function relationships, and most notably of cooperativity, which we discuss in the next two sections.

4.5   Cooperativity: Biochemistry and Physiology

The common, but at first sight puzzling, behavior of proteins endowed with *positive homotropic cooperativity* has profound implications for biochemistry and physiology that deserve consideration. The essence of positive cooperativity is an increase of ligand affinity as the ligand saturation increases. This may only occur in proteins having more than a single ligand-binding site, and its quantitative aspect is the progression of the microscopic equilibrium constants $K_{d,1}$ through $K_{d,n}$, be these real or apparent parameters. Positive cooperativity causes the ligand-binding isotherm to become steeper than in single-site proteins, or in oligomeric proteins in which all microscopic equilibrium constants are equal, an effect which is better appreciated in the $\bar{X}$ versus $\log([X])$ plot (see Figure 4.5). Positive cooperativity is somewhat paradoxical, as it implies that higher and lower affinity sites are present, and that the lower affinity sites are the first to bind the ligand, at the lowest ligand concentration. This is possible only if the high-affinity sites do not exist before the initial binding of the ligand. This in turn requires that the binding of the first ligand(s) causes some change in the macromolecule that increases the affinity for the successive ligand(s). There is no need for the successive equilibrium constants to follow a constant trend. Some intermediate binding step might even invert the general trend (i.e., $K_{d,i} > K_{d,i-1}$), as long as the last ligand bound is the one with the highest affinity.

Positive cooperativity has very important consequences in several physiological processes. In the case of carrier proteins, for example oxygen carriers, it increases the ligand affinity where the ligand is in excess, thus facilitating its uploading; and it lowers the ligand affinity where the ligand is in demand, thus facilitating its release. In the case of cooperative enzymes a similar mechanism operates, leading to improved catalytic efficiency, and faster substrate consumption, when the substrate is in excess. In the case of hormone receptors, positive cooperativity, if present, causes the saturation, hence the response of the receptor, to be expressed over a smaller change in the concentration of the hormone, making the system more responsive.

The opposite condition is called *negative homotropic cooperativity* or *anticooperativity*. This condition causes the last ligand molecule to be bound with lower affinity than the first, and results in a broadened ligand-binding isotherm. Negative cooperativity is not easily distinguished from intramolecular heterogeneity (Section 4.3), as both conditions broaden the ligand-binding isotherm. However, as a general indication, we suggest that negative cooperativity is the most likely explanation for a symmetric homooligomer presenting a broadened binding isotherm, whereas intramolecular heterogeneity is a more likely explanation for the case of heterooligomers.

A crucial consequence of the progression of the equilibrium constants (whether even or uneven) in a positively cooperative oligomeric protein is that the ligation intermediates are poorly populated, as A.V. Hill had already recognized in 1910-1913. By contrast, in a negatively cooperative oligomer the population of ligation intermediates is increased with respect to a non-cooperative oligomer. Figure 4.7 compares the population of ligation intermediates for the three binding isotherms reported in Figure 4.5 to demonstrate this point.

While it is obvious that cooperativity may be greater or smaller, there is no single agreed upon parameter to estimate the amount of cooperativity expressed by a given protein. The reason for this is that the model-free, thermodynamically sound explanation of cooperativity is Adair's (eqns. 4.15 and 4.16), which requires as many equilibrium constants as binding sites are present in the macromolecule. Consequently, cooperativity does not lend itself to be unequivocally measured by a single index or parameter. Moreover, the precise determination of the Adair K_ds for the n binding sites of a homooligomer is usually difficult and plagued by large uncertainties (Marden *et al.*, 1989).

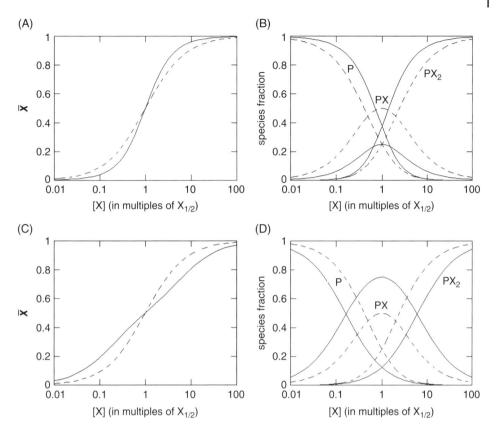

Figure 4.7 Population of unliganded, singly liganded and doubly liganded species in homodimeric proteins.

Panel A: Simulated ligand-binding isotherms for a non-cooperative homodimer ($\varepsilon = 1$; dashed line) and for a positively cooperative homodimer ($\varepsilon = 9$; continuous line). Panel B: population of the unliganded (P), singly liganded (PX) and doubly liganded (PX$_2$) intermediates for the binding isotherms from panel A (dashed lines: the three intermediates in the non-cooperative homodimer; continuous lines: the three intermediates in the positively cooperative homodimer). Notice that the singly liganded intermediate (bell-shaped curve) is less populated in the cooperative than in the non-cooperative case.

Panel C: Simulated ligand-binding isotherms for a non-cooperative homodimer ($\varepsilon = 1$; dashed line) and for a negatively cooperative homodimer ($\varepsilon = 1/9$; continuous line). Panel D: population of the unliganded, singly liganded and doubly liganded intermediates for the binding isotherms from panel C (dashed lines: the three intermediates in the non-cooperative homodimer; continuous lines: the three intermediates in the negatively cooperative homodimer). Notice that the singly liganded intermediate (bell-shaped curve) is more populated in the negatively cooperative than in the non-cooperative case.

The parameters used in this simulation are identical to those used in Figure 4.5.

Historically, the first parameter aimed at measuring the extent of cooperativity (of hemoglobin) has been the slope of the Hill plot. This parameter has no direct physical meaning, or, to be more precise, its original physical meaning, that is, that the Hill coefficient n represents the average number of subunits that constitute the macromolecule, has been proven wrong. However the slope of the Hill plot has an interesting property: in the case of positive cooperativity, it is limited between one (absence of cooperativity) and the number of interacting binding sites of the macromolecule. Thus one can easily compare the actual value of the experimentally measured Hill coefficient with its theoretical maximum, if the number of binding sites is known.

A number of researchers (reviewed by Forsén and Linse, 1995) advocated the use of the logarithm of the ratio of the equilibrium dissociation constant of the first and last binding site of the macromolecule. This parameter (multiplied by RT) represents the *cooperativity free energy*, that is, the free energy difference for the binding of the first and last ligand, and has a precise physical meaning, but in proteins having more than two binding sites it ignores the intermediate ligation steps, thus it is insufficient to predict the actual steepness of the ligand-binding isotherm except in the case of dimeric proteins.

S.J. Gill (Wyman and Gill, 1990) proposed the derivative of the $\overline{X}$ versus log([X]) plot, which he named the ligand-binding capacity of the macromolecule, as a convenient index of cooperativity. This parameter never really gained widespread usage, and its very name may suggest a different meaning, namely the total amount of ligand that a given macromolecule or biological sample is able to bind, thus causing potential confusion (Section 1.4). In this book we make use of Gill's concept, but we refer to it as the slope of the $\overline{X}$ versus log([X]) plot, in order to not confuse the reader. Interestingly Gill's parameter has a formal analogy to the Hill coefficient, in that both parameters measure the steepness of the binding isotherm, and it is easy to convert one into the other.

We may conclude this section with the following consideration. A complete quantitative description of cooperativity requires as many parameters as there are binding sites in the macromolecule. The very idea of measuring the extent of cooperativity reflects some specific interest or viewpoint of the researcher. In order to represent the different possible viewpoints of the researchers, several different indices of cooperativity have been devised. All of them simplify the essence of the phenomenon, and none fully captures it. From the viewpoint of physiology the most relevant aspect of cooperativity is the steepness of the ligand-binding isotherm, and this explains the success of the Hill coefficient. From the viewpoint of the physico- chemical properties of the macromolecule, the cooperativity free energy is possibly more meaningful than the steepness of the binding isotherm. Which index of cooperativity is most meaningful ultimately depends on the question the researcher aims at answering, rather than on the greater precision of the index *per se*.

4.6 Allostery and Symmetry: The Allosteric Model of Cooperativity

The detailed structural determinants of cooperativity differ in the various proteins that have been studied in sufficient depth. However, there are two broad groups or types of hypotheses on cooperativity, which invoke different structural principles. The prototype of one of these is due to Pauling and has been already discussed in Section 4.4 above. The prototype of the other group was put forward by J. Monod, J. Wyman, and J.P. Changeux (MWC) in 1965, and in its general outline may apply to many positively cooperative proteins (Monod *et al.*, 1965). Both types of hypotheses assume that the typical cooperative protein is an oligomer of similar or identical subunits, thus application to monomeric, multi-site cooperative proteins requires some extension. Pauling's model and its successive evolutions hypothesize that upon ligand binding to a subunit, some weak bonds between that subunit and neighboring ones form or break. Thus, this model allows each subunit within the oligomer to assume the structure and to form the bonds characteristic of its ligation state. Models of this type have been called *sequential* to imply that the changes in the structure and ligand affinity occur stepwise, within each single macromolecule. The MWC model, on the contrary, assumes *symmetry* of the oligomer as its leading structural principle, and implies that the changes in the structure and ligand affinity occur in an all-or none (*concerted*) fashion for all subunits in the macromolecule, even if some of them are liganded and the others unliganded. We devote the present section to a summary of Monod's hypothesis, and the next section to a comparison of the two.

Monod and co-workers proposed that cooperative proteins are symmetric oligomers, which can sample at least two different *states*, envisaged as different because of both their quaternary structure and ligand affinity. We shall analyze the case of a symmetric homodimer first and will then move to higher order oligomers. Monod's concepts and terms were remarkably prescient given the limited state of knowledge about protein structures in the early 1960s; for example, this model embodies the assumption that proteins are dynamic, a concept that came into widespread acceptance much later.

A protein monomer is intrinsically asymmetric, even though it may include pseudo-symmetric structural domains, but two identical monomers may form perfectly symmetric homodimers. The intersubunit interface of a symmetric homodimer is by necessity of the type that Monod called isologous, that is, it is made up by identical substructures in each monomer (e.g., specific α helices or β strands). In an isologous interface each intersubunit contact occurs twice (Figure 4.8). A crucial postulate of the MWC model is that cooperative proteins are oligomers stable in (at least) two different conformational states, both symmetric, and both populated under equilibrium conditions independently of the presence of the ligand. Monod coined the term *allostery*, a neologism from Greek allos=other and stereos=solid shape, to indicate this property. If the two (or more) structures have different affinities for the ligand, ligation will bias their equilibrium, and shift the relative populations, causing positive cooperativity.

Because of its postulates, the MWC model cannot explain negative cooperativity, nor can it be applied without extensive conceptual modifications to cooperative monomeric proteins like calmodulin. Moreover, the model is formulated under the assumption of quaternary constraint (i.e., that the oligomer has lower ligand affinity than its isolated subunits; see also Section 5.1).

The original MWC model is an abstract formulation in which structural principles (allostery and symmetry) are directly coupled to thermodynamic ones (preferential stabilization of one structure over the others). The model does not make any assumption on the structural details that cause one state to differ from the other, and whenever possible it refers to states, rather than structures. However, one can imagine that ligand binding to one subunit of the oligomer causes some structure change or strain that affects the contacts at the isologous interface, thus biasing the allosteric equilibrium. The symmetry requirement of isologous interfaces causes the subunits of a partially liganded oligomer to adopt the same structure irrespective of whether they are or are not

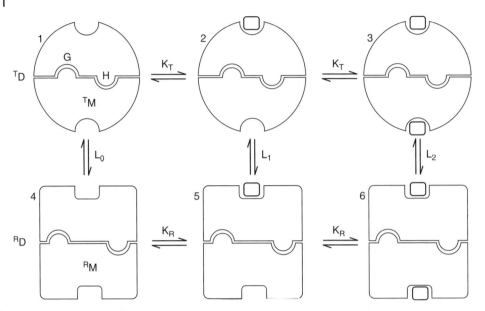

Figure 4.8 Structure- and ligand-binding reactions of an allosteric symmetric homodimer. The allosteric homodimer (D) is postulated to exist in two quaternary conformations or structures, both symmetric, called ^TD and RD respectively. Each monomer either assumes the tertiary T or R structure (^TM and ^RM respectively), depending on the quaternary structure of the whole oligomer. The inter-subunit interface is isologous, hence symmetric and is here represented with two structural features: a groove (indicated as G) and a protruding α helix (H), such that the helix of monomer 1 (H1) perfectly fits into the groove of monomer 2 (G2) and vice versa. As is characteristic of a isologous interface, the same contacts occur twice: G1-H2 and G2-H1. The structural features at the isologous interface are not superimposable in the two quaternary states (here represented as a greater G to H distance in the R-structure than in the T structure). As a consequence the mixed, non-symmetric state in which one subunit is tertiary T state and the other is tertiary R-state would never be populated. The T and R states of the dimer freely interconvert in any ligation state, and their equilibrium is described by the allosteric constant L, whose value depends on the number of bound ligands (see text). Each state binds ligands with a characteristic intrinsic affinity constant (K_R or K_T).

liganded. This view can be extended in order to accommodate the cases of allosteric proteins in which the main difference between the liganded and unliganded states is dynamic (i.e., related to internal motions of the polypeptide chain and residues) rather than static (i.e., related to the structural arrangement). This difference, however, is not substantial, because the static and dynamic properties of the macromolecule cannot be imagined as uncorrelated to each other.

A qualitative description of the molecular machinery implied by the allosteric model is as follows. The oligomer is stable in two structural conformations, both symmetric, having different ligand affinity. Thus two states, differing because of quaternary structure and functional properties are identified. These are defined as relaxed (R) and taut or tense (T). The T state is hypothesized to have stronger intersubunit contacts, that impose a structural deformation of the subunits (the "tension") and reduce their ligand affinity, whence the quaternary constraint. The two states are in equilibrium independently of the presence of the ligand (Figure 4.8). In the absence of ligand the low-affinity T state is more populated than the R-state, because its stronger contacts make it more

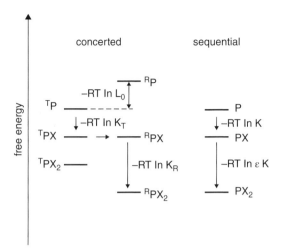

free energy

concerted   sequential

^{R}P

$-RT \ln L_0$

^{T}P   P

$\downarrow -RT \ln K_T$ $\downarrow -RT \ln K$

^{T}PX ^{R}PX   PX

^{T}PX$_2$

$-RT \ln K_R$ $-RT \ln \varepsilon K$

^{R}PX$_2$ PX$_2$

Figure 4.9 Energy diagram for the two-state concerted model, as compared to a sequential model.

stable (Figure 4.9). However, the ligand binds more strongly to the high-affinity R state, thus as ligation proceeds the equilibrium between the R and T conformation is progressively biased in favor of the R state, causing an apparent increase of ligand affinity. An energy diagram is possibly the best way to represent the effect of ligand binding to an allosteric protein (Figure 4.9).

An important point is that the effective ligand affinity of an allosteric protein represents the weighted average of the affinities of the T and R state. Thus, the MWC model ascribes cooperativity to the changes in the relative population of the two states, a phenomenon we may call *statistical cooperativity*. Contrary to sequential models, in the allosteric model there is no actual stepwise increase in the ligand affinity; rather there is a stepwise change of the relative population of the two states. *The definition of cooperativity saying that the last ligand is bound with higher affinity than the first applies to the allosteric model in the statistical sense that the first ligand has higher probability of binding to a low-affinity T-state protein and the last ligand has higher probability of binding to a high-affinity R-state protein (see eqn. 4.24).*

Although positive cooperativity may be observed also in non-symmetric oligomers, and in monomers containing multiple binding sites for the same ligand such as calmodulin, structural symmetry is a plausible explanation of cooperativity for symmetric homooligomers. Indeed, as schematically depicted in Figure 4.7, the isologous interface of a symmetric homodimer may cause the unliganded subunit of the singly liganded macromolecule to adopt the same (high-affinity) structure of the liganded partner subunit.

The algebraic formulation of the MWC model follows from the qualitative explanation given above. We derive it at first for the symmetric homodimer (Figures 4.8 and 4.9).

i) The homodimeric protein (D) can adopt two structural conformations having different affinities for the ligand:

$$K_R = \left[{}^R DX_{i-1} \right] [X] / \left[{}^R DX_i \right] \qquad \text{(eqn. 4.19)}$$

$$K_T = \left[{}^T DX_{i-1} \right] [X] / \left[{}^T DX_i \right] \qquad \text{(eqn. 4.20)}$$

where i stands for the number of bound ligands and ranges between 1 and n (i.e., the permitted values for a dimer are i=1 or i=2).

ii) The R state has higher ligand affinity, that is, smaller ligand dissociation constant, than T (eqns. 4.19, 4.20, 4.21, 4.22):

$$K_R < K_T$$

iii) K_R and K_T do not depend on the number of bound ligands, that is, within the R and T state no homotropic linkage is present; we need not define a $K_{R,1}$ and a $K_{R,2}$ and the same applies to K_T.

iv) Both states are populated and in equilibrium, irrespective of the presence of the ligand; their ratio in the absence of the ligand is governed by the allosteric constant L_0:

$$L_0 = \left[{}^T D \right] / \left[{}^R D \right] \qquad \text{(eqn. 4.21)}$$

v) In the absence of the ligand the T (lower-affinity) conformation is favored; thus:

$$L_0 \gg 1$$

vi) In the presence of ligand, L_i depends on the number of ligands bound, in a way that is fully defined by the other parameters, as one can easily derive from the reaction scheme in Figure 4.8 or the energy diagram in Figure 4.9:

$$L_i = L_0 \, K_R^i / K_T^i = L_0 \, c^i \qquad \text{(eqn. 4.22)}$$

with $c = K_R / K_T$. The above relationship is general, that is, within the framework of the original formulation of the allosteric model it applies to an oligomer of any number of subunits.

In order to define the binding polynomial for an allosteric homodimer, we need to choose a reference species, which in the original formulation of the model was the unliganded R state protein, here ${}^R D$. It might be argued that, given the above premises, a more rational choice would have been the fully liganded state ${}^R DX_2$; however, for the sake of consistency, we maintain the original formulation. We obtain:

$$[{}^R DX] = [{}^R D] 2[X]/K_R$$
$$[{}^R DX_2] = [{}^R D][X]^2/K_R^2$$
$$[{}^T D] = [{}^R D] L_0$$
$$[{}^T DX] = [{}^R D] L_0 \, 2[X]/K_T$$
$$[{}^T DX_2] = [{}^R D] L_0 \, [X]^2/K_T^2$$
$$[D]_{tot} = \left(1 + [X]/K_R\right)^2 + L_0 \left(1 + [X]/K_T\right)^2$$

and, generalizing for a n-subunit oligomer:

$$[P]_{tot} = \left(1 + [X]/K_R\right)^n + L_0 \left(1 + [X]/K_T\right)^n \qquad \text{(eqn. 4.23)}$$

We remark that this is simply the sum of the two binding polynomials for the two non-interacting n-subunit oligomers in the R and T states, the latter multiplied by L_0 as a scaling factor (see eqn. 4.7 and 4.11).

The fractional ligand saturation of the homodimer is the ratio between the sum of bound and free sites in the two allosteric conformations:

$$\overline{X} = \frac{[X]/K_R\left(1+[X]/K_R\right)+L_0[X]/K_T\left(1+[X]/K_T\right)}{\left(1+[X]/K_R\right)^2+L_0\left(1+[X]/K_T\right)^2}$$

and, generalizing for a n-subunit oligomer:

$$\overline{X} = \frac{[X]/K_R\left(1+[X]/K_R\right)^{n-1}+L_0[X]/K_T\left(1+[X]/K_T\right)^{n-1}}{\left(1+[X]/K_R\right)^n+L_0\left(1+[X]/K_T\right)^n}$$

If one equates [X] to $X_{1/2}$ and $\overline{X}$ to 0.5 and solves the above eqn. for $X_{1/2}$, one obtains:

$$X_{1/2} = K_R{}^n \sqrt{\left[\left(1+L_0\right)/\left(1+L_n\right)\right]}$$

For the allosteric model to yield a positively cooperative ligand binding isotherm one further condition should be satisfied, that is, $L_n < 1$. Together with conditions (i) to (vi), this ensures that ligation causes the allosteric structural change, and that in the fully liganded state the population of the R conformation exceeds that of the T conformation. Because $L_n = Lc^n$ (eqn. 4.22), fulfillment of this condition depends on all three parameters of the model. We can easily calculate how the ligand concentration affects the population of the R (or T) state:

$$\left[{}^R D\right]_{tot} = \left[{}^R D\right]\left(1+[X]/K_R\right)^2$$
$$\left[{}^T D\right]_{tot} = \left[{}^R D\right]L_0\left(1+[X]/K_T\right)^2$$
$$\left[{}^R D\right]_{tot}/\left(\left[{}^R D\right]_{tot}+\left[{}^T D\right]_{tot}\right)=\left(1+[X]/K_R\right)^2/\left\{\left(1+[X]/K_R\right)^2+L_0\left(1+[X]/K_T\right)^2\right\}$$

A (simulated) comparison of the ligand binding isotherm and the relative population of R state are reported in Figure 4.10.

The aim of every model of cooperativity is to explain the progression of the intrinsic equilibrium constants (i.e., to define something like the ε factor we introduced in Section 4.4). The explanation provided by the allosteric model is rigorous but quite counterintuitive, since both the T and R states bind ligands non-cooperatively, thus the model has no place for an effective change of any intrinsic ligand binding constant: it has only K_R and K_T. However, one can derive from the MWC model the apparent stepwise Adair constants to be used in eqns. 4.8 or 4.15, which correspond to the weighted averages of K_R and K_T for every ligation intermediate:

$$\begin{aligned}K_i &= \left(\left[{}^R PX_{i-1}\right]+\left[{}^T PX_{i-1}\right]\right)[X]/\left(\left[{}^R PX_{i-1}\right]+\left[{}^T PX_{i-1}\right]\right)\\ &= K_R\left(1+L_{i-1}\right)/\left(1+L_i\right)\end{aligned}$$

(eqn. 4.24)

for i=1 to n.

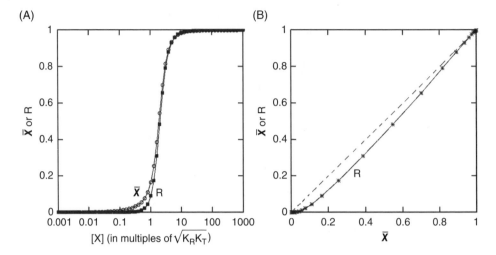

Figure 4.10 Relationships between the fractional ligand saturation and the fraction of protein in the R state. Panel A: Fraction of bound sites (open symbols) or of R state (closed symbols), as a function of ligand concentration. Panel B: Fraction of bound sites (dashed line) or fraction of R state (continuous line marked with asteriscs) as a function of the fraction of bound sites.

4.7 Two Alternative Concepts of Cooperativity

The structural bases of the MWC model, with its descendants and derivatives, are essentially alternative and incompatible with those of Pauling's sequential model or its variants. It is perfectly conceivable that some cooperative proteins are Monod-like and others Pauling-like, but intermediate approaches are scarcely viable. In this section we analyze the contact points between the two types of models and their key differences, starting from the important concept of ligand-induced fit.

Proteins are complex and dynamic structures. Overall, they appear as compact, packed to the extent that in their interior the amino acid residues contact each other and largely exclude molecules as small as water. Even though proteins exhibit significant fluctuations that allow ligands to enter and exit, their average structure remains packed. Stable, non-fluctuating cavities may be present, as was demonstrated by experiments in which hemoproteins were crystallized under high pressures of inert gases (usually xenon) (Schoenborn *et al.*, 1965), but are neither large nor common. Even the ligand-binding sites, in the absence of the ligand, may tend to some extent to collapse or to admit water, in order to maintain close packing and to satisfy hydrogen bond donors and acceptors.

The binding of a ligand may cause changes in the protein structure and its dynamics because finding room for the ligand may require small displacements of amino acid side chains or slight distortions of the tertiary structure. These changes may extend to variable distance from the binding site, and in oligomeric proteins they may even extend to unliganded subunits across the intersubunit interface. The case of hemoglobin has been studied in great detail, thus we refer the reader to Chapter 7 for a structural analysis of that example.

The fact that ligand binding induces structural changes in the protein was called *ligand-induced fit* by Koshland *et al.* (1966). However, the concept, if not the name, is implicitly present in the model of cooperativity by Linus Pauling (1935). The concept of induced fit has often been interpreted as alternative to that of conformational equilibrium postulated by the MWC model. Undoubtedly ligand-induced fit and conformational equilibrium may appear as quite different concepts. Induced fit implies that the protein has two different structures, one for the liganded and one for the unliganded state; thus the relationship between structure and ligation state is deterministic. The concept of conformational equilibrium implies that the two structures are in equilibrium, even in the absence of the ligand, and the ligand prefers one over the other biasing their equilibrium, thus the relationship between structure and ligation state is statistical.

What is often overlooked is that in sequential models induced fit operates at the level of the subunit, whereas in concerted models conformational equilibrium occurs at the level of the quaternary assembly. Thus, the MWC model does not require the absence of induced fit at the tertiary structural level, within each quaternary state. Indeed, the structural reason why ligation of a subunit biases the conformational equilibrium of the oligomer (eqn. 4.22) is that ligation causes a structural change in the subunit that conflicts with the structural requirements of the T quaternary assembly. This phenomenon has been observed by x-ray crystallography in liganded derivatives of T-state hemoglobin (Section 7.9).

Sequential models of cooperativity are based on the concept of induced fit, and usually postulate that ligation of one subunit may affect its nearest neighbors, but not necessarily all the subunits in the oligomer. A necessary consequence is that a partially liganded oligomer may be asymmetric, as it may contain subunits having different structures. Cooperativity occurs because each ligation step changes the structure of one subunit by an induced fit mechanism and favors the formation or breakage of intersubunit bonds. In the preceding sections, we defined as *intramolecular* this type of cooperativity. A sequential model of cooperativity requires as many microscopic binding constants as there are binding sites in the macromolecule, or, more often, one binding constant for the first site plus one or more interaction factors ε (Section 4.4) for the successive sites. The number of parameters may be reduced if structure based assumptions are made to lower the number of interaction factors. For example, L. Pauling (1935) assumed only one interaction factor for the second, third, and fourth binding site of hemoglobin, raised to an appropriate power factor (Section 4.4).

Models of cooperativity based on allostery postulate that the structural changes responsible for the increase of ligand affinity occur in a concerted (i.e., all-or-none) fashion. The liganded subunits, rather than affecting the contacting unliganded subunits, change the relative stability of the two allosteric conformations. The crucial, and often overlooked, implication is that in allosteric models, cooperativity is a *statistical* phenomenon, which one could not properly attribute to the single macromolecule, if not in terms of probability. Statistical cooperativity results from a ligand-induced shift in the population of different states of the macromolecule.

The difference between intramolecular and statistical cooperativity is clearly a matter of concepts, but is reflected in a number of experimentally accessible details that we can summarize as follows. Sequential models admit, while the MWC model forbids, asymmetric ligation intermediates, in which subunits having different structures coexist in the same oligomer. By contrast, the MWC model requires, and sequential models forbid, that the high-affinity structure of the macromolecule is populated also in the absence of the ligand (to a little extent). The allosteric model predicts non-cooperative binding under experimental conditions in which the quaternary structural change are prevented (e.g., ligand binding to protein crystals), whereas no clear prediction can be derived from sequential models under the same conditions.

In our view the real alternative is between sequential, intramolecular cooperativity, and allosteric, statistical cooperativity. The more limited concept of induced fit *per se* is not alternative, and may be complementary, to allostery. Indeed, although it is possible that the functional behaviour of a multiple ligand-binding site protein may entirely depend on an induced fit mechanism in the absence of allostery, it is inconceivable that allosteric phenomena may occur in the absence of ligand-induced localized structural changes, even though these may be inhibited or minimized by the symmetry requirements of an isologous interface (Figure 4.8). Thus, some form of ligand-induced fit is a plausible component of allosteric phenomena, which may be invoked to explain the destabilization of the T structure caused by ligand binding (eqn. 4.22).

The role of ligand-induced fit in the allosteric model is apparently denied by the symmetry requirement of the allosteric model. Symmetry dictates that the oligomer is either T state or R state, with all its constitutive subunits sharing the same tertiary structure, irrespective of the presence or absence of the ligand. However, minor tertiary structural changes induced by ligand binding (and structurally documented in those cases which lent themselves to analysis by x-ray crystallography; see Chapter 7) are responsible for structural tension or strain at the intersubunit interfaces, and make the liganded subunits fit poorly in the T state or the unliganded ones fit poorly in the R state. The thermodynamic counterparts of the strained structures of ligation intermediates are: (i) their reduced population, which is at the basis of statistical cooperativity; and (ii) the destabilization of the T state induced by ligation (or of the R state by ligand dissociation; see eqns. 4.22 and 4.24). In summary, in the allosteric model induced fit phenomena constitute the structural bases of the effect of ligation in inducing the allosteric structural change. The structural model of cooperativity in hemoglobin developed by M.F. Perutz (1970), while allosteric in its essence, makes ample reference to such induced fit concepts as localized tertiary structural changes and interface strain (more on this in Chapter 7).

A parallel case may be made for the possibility of structural symmetry in induced-fit models of cooperativity. These models do not forbid structural symmetry, even though they do not require it either. That is, sequential models may admit that the ligation intermediates have a hybrid, non-symmetric structure (schematized for a homodimer in Figure 4.11), forbidden by the premises of the allosteric model. Alternatively, an induced fit cooperative scheme for a fully symmetric homodimer would be similar to the one depicted in Figure 4.7, except that species 4 and 6 would not exist, and tension at the interface would occur only in the singly liganded intermediate species. Both the symmetric and the non-symmetric, cooperative reaction schemes for a homodimer would be fully described by eqns. 4.17 and 4.18, with the only difference that in the symmetric scheme the equilibrium constants would represent the weighted average of the populations of the two alternative conformations.

The flexibility of induced-fit models comes at a price. Because these models, by using more than one interaction factor, are compatible with virtually any possible combination of structural and functional features, they have poor predictive capabilities. For example, as stated above, they are compatible with symmetric and non-symmetric

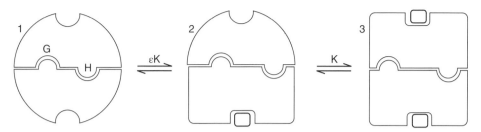

Figure 4.11 An induced-fit structural model of a cooperative homodimer may admit the asymmetric ligation intermediate (here represented with an intermediate G to H distance; this structural feature would not be admissible within the two-state framework depicted in Figure 4.9).

Table 4.1 Possible causes of cooperativity.

Family	Mechanism of cooperativity
Intramolecular cooperativity (deterministic cooperativity)	Sequential models
	Cooperativity is caused by ligand-dependent intersubunit interactions facilitating or inhibiting ligand binding (Section 4.4).
Statistical cooperativity	Allosteric (MWC) model and its variants (Section 4.6)
	Cooperativity caused by differential binding of heterotropic ligands (Section 4.9)
	Cooperativity due to ligand-linked dissociation (Chapter 5)

structures; or with any type of even or uneven progression of the ligand-binding constants; or with negative homotropic cooperativity (whereas non-symmetric structures, uneven progression of the binding constants, and negative cooperativity are all forbidden by the MWC model).

In conclusion, it is our opinion that there is a fundamental dichotomy in the explanation of cooperativity, and that this dichotomy is more marked at the abstract level of the two hypotheses of sequential and intramolecular *versus* allosteric and statistical (see Table 4.1). At the level of the structure of the macromolecule, the allosteric model is not incompatible with induced-fit type local structural changes of the protein, nor is the sequential model incompatible with molecular symmetry or with a sort of allosteric equilibrium for the ligation intermediates. These mechanistic details do not make the allosteric model a special case of a sequential model, or, vice versa, the sequential model a complicated variant of the allosteric model. The crucial feature of the allosteric model, which lacks a counterpart in sequential models, is the allosteric equilibrium in the fully liganded or fully unliganded states: that is, its statistical nature. Convincing proof of the existence of the fully liganded protein in the T conformation or of the fully unliganded in the R conformation would rule out the sequential hypothesis. Moreover, the allosteric model is incompatible with the presence of non-symmetric protein states (e.g., with state 2 in Fig. 4.11), and convincing demonstration of a ligation intermediate in a mixed structural conformation would rule out the allosteric model in favor of the sequential one. Obviously, examples are known of proteins whose functioning is satisfactorily described as allosteric, and proteins whose functioning is satisfactorily described as sequential.

Several authors tried to combine the features of sequential and allosteric models, often invoking the concept of "nesting," developed by Wyman and co-workers (Robert *et al.*, 1987) to explain the case of giant oxygen carriers, in which a cooperative substructure of a handful of subunits is embedded in a larger cooperative superstructure. Examples of such hybrid models are the so-called cooperon model (Brunori *et al.*, 1986) in which the substructure (e.g., the $\alpha\beta$ heterodimer within the hemoglobin tetramer) exhibits intramolecular cooperativity, whereas the superstructure has allosteric properties; and the structure-based model of hemoglobin proposed by Di Cera *et al.* (1987a), in which intramolecular cooperative interactions occur between the α subunits within the T allosteric state. These models do not represent a synthesis of the two types identified above: rather they sum up intramolecular and statistical cooperativity, and require a greater number of parameters than those required by the parent models. Thus they may be justified in some cases but are not a breakthrough, as Pauling's and Monod's models were at the time of their formulation.

Allostery is not the only molecular phenomenon that may cause cooperativity because of population selection: heterotropic ligands (Section 4.9) and ligand-linked dissociation (Chapter 5) may achieve the same result, also in non-allosteric oligomers (Table 4.1).

4.8   Ligand Replacement in Oligomeric Proteins

Identical linkage phenomena occur when two different ligands compete for the same binding site, and have been described in Section 1.8 for monomeric proteins. The description may be extended to oligomeric proteins, with one important warning: ligand replacement experiments probe the properties of the last binding site only. This occurs because in the presence of two ligands, both at significant concentrations, the fraction of unliganded or partially liganded states is negligible.

The reaction scheme for a n-meric protein in which ligand X is being replaced by ligand Y is therefore as follows:

$$PZ_{n-1}X + Y \Leftrightarrow PZ_{n-1}Y + X$$

where Z_{n-1} represents any possible combination of n−1 molecules of X and Y.

As a consequence of this property, the replacement experiment is governed by the partition constant:

$$K_{p,n} = K_{d,X,n} / K_{d,Y,n}$$

and essentially no information is gathered on the partition constant of the first n-1 sites, unless very special experiments are devised in which a significant fraction of the binding sites is empty. For example, S.J. Gill measured the O_2/CO partition for the first binding site of hemoglobin by simultaneously diluting both ligands (at a constant ratio) in the gas phase (Di Cera *et al.*, 1987b).

Since the properties of only one binding site (the last one) are explored, *ligand replacement in oligomeric proteins is usually non-cooperative*, even though both ligands may exhibit strong positive (or negative) cooperativity. Exceptions to this rule may occur if the two ligands differ greatly in their ability to promote the tertiary and quaternary structural changes responsible for cooperativity, which usually implies that their cooperativity in separate ligand binding experiments greatly differs. This case is rare in the pertinent literature; for example, it has been observed for oxygen and carbon monoxide binding to octopus hemocyanin (Connelly *et al.*, 1989).

4.9 Heterotropic Linkage in Multimeric Proteins

Heterotropic effects can occur when a protein binds two ligands at different binding sites; they were treated for the case of monomeric proteins in Section 1.9 and are here extended to the case of oligomeric proteins. In oligomeric proteins this case, discovered by C. Bohr in 1904 but first analyzed in modern terms much later (Wyman 1948, 1964), is quite complex, and may introduce positive statistical cooperativity for a ligand that would bind non-cooperatively in the absence of a heterotropic effector.

As usual we consider first the simplest possible case, that is, the non-cooperative homodimer, already analyzed in Section 4.1, to which we add the heterotropic ligand Y that binds with a stoichiometry of 1:1 to the dimer (Figure 4.12).

In the absence of the heterotropic ligand Y the homodimer behaves as described by eqns. 4.1 through 4.9 and its affinity for X is dictated by $K_{X,1}$ and $K_{X,2}$ (plus their statistical factors). In the absence of X, the complex of the protein with the allosteric effector, PY, presents the dissociation equilibrium constant K_Y. If the concentration of Y is so high that the concentration of species of P lacking bound Y is negligible, then the equilibrium between PY and X will be governed by the dissociation constants $^Y K_{X,1}$ and $^Y K_{X,2}$. If $^Y K_{X,i} > K_{X,i}$ the heterotropic linkage between Y and X is of the negative type (each ligand decreases the affinity of the protein for the other); if $^Y K_{X,i} < K_{X,i}$ the linkage between Y and X is of the positive type (each ligand increases the affinity of the protein for the other). In this example we make no assumption on P and PY to present homotropic cooperativity for ligand X; that is, the binding isotherms of P in the absence of Y and of PY in the presence of saturating concentrations of Y may be hyperbolic ($K_{X,1}=K_{X,2}$ and $^Y K_{X,1}=^Y K_{X,2}$), cooperative or anti-cooperative (negatively cooperative).

The system requires seven equilibrium constants, five of which are independent of the others, and two correspond to combinations of the other five. We are free to choose which constants we define as independent, and we select: K_Y, $K_{X,1}$, $K_{X,2}$, $^Y K_{X,1}$, and $^Y K_{X,2}$, all defined as dissociation constants.

The constants for dissociation of Y from the complexes PYX and PYX$_2$ can be derived from the other five:

$$^X K_Y = K_Y{}^Y K_{X,1}/K_{X,1}$$

$$^{X2} K_Y = {}^X K_Y \left({}^Y K_{X,2} / K_{X,2} \right) = K_Y{}^Y K_{X,1}{}^Y K_{X,2}/K_{X,1}K_{X,2}$$

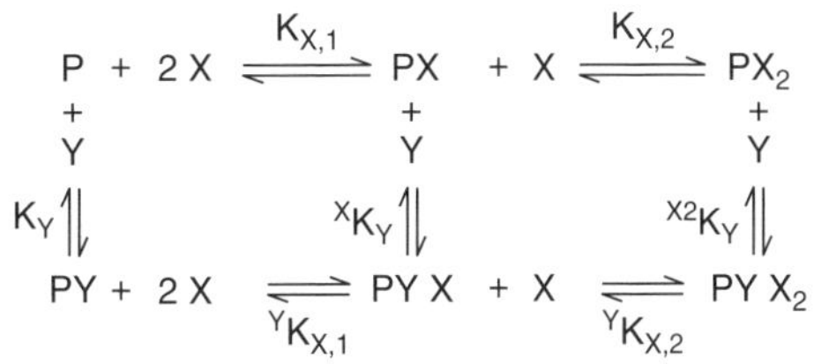

Figure 4.12 Binding of ligands Y and X to different binding sites of a homodimeric protein. The scheme assumes binding stoichiometries 2:1 for ligand X and 1:1 for ligand Y.

K_Y and $^{X2}K_Y$ are easily accessible to experimental determination, because they define the affinity for ligand Y in the absence of ligand X and in the presence of saturating concentrations of ligand X respectively. By contrast $^X K_Y$ defines the affinity for ligand Y of a ligation intermediate of ligand X, which can never be obtained in a pure state, uncontaminated by P and PX_2.

We can write the concentrations of the six chemical species in the reaction scheme, taking the concentration of species P as a reference:

$$[PX] = [P]2[X]/K_{X,1}$$
$$[PX_2] = [P][X]^2/K_{X,1}K_{X,2}$$
$$[PY] = [P][Y]/K_Y$$
$$[PXY] = [PY]2[X]/^Y K_{X,1} = [P]2[X][Y]/K_Y{}^Y K_{X,1}$$
$$[PX_2Y] = [PY][X]^2/^Y K_{X,1}{}^Y K_{X,2} = [P][X]^2[Y]/K_Y{}^Y K_{X,1}{}^Y K_{X,2}$$

From the above we derive the following binding polynomial, and the fractional saturation of each ligand:

$$[P]_{tot} = [P](1+2[X]/K_{X,1} +[X]^2/K_{X,1}K_{X,2} +[Y]/K_Y(1+2[X]/^Y K_{X,1} +[X]^2/^Y K_{X,1}{}^Y K_{X,2}))$$

(eqn. 4.25)

$$\bar{X} = \frac{[X]/K_{X,1} +[X]^2/K_{X,1}K_{X,2} +[Y][X]/K_Y{}^Y K_{X,1} +[Y][X]^2/K_Y{}^Y K_{X,1}{}^Y K_{X,2}}{1+2[X]/K_X +[X]^2/K_{X,1}K_{X,2} +[Y]/K_Y +2[Y][X]/K_Y{}^Y K_{X,1} +[Y][X]^2/K_Y{}^Y K_{X,1}{}^Y K_{X,2}}$$

(eqn. 4.26)

$$\bar{Y} = \frac{[Y]/K_Y(1+2[X]/^Y K_{X,1} +[X]^2/^Y K_{X,1}{}^Y K_{X,2})}{1+2[X]/K_X +[X]^2/K_{X,1}K_{X,2} +[Y]/K_Y +2[Y][X]/K_Y{}^Y K_{X,1} +[Y][X]^2/K_Y{}^Y K_{X,1}{}^Y K_{X,2}}$$

(eqn. 4.27)

The half saturating concentrations result:

$$X_{1/2} = Xm = \sqrt{(K_{X,1}K_{X,2})}\sqrt{\{(1+[Y]/K_Y)/(1+[Y]/^{X2} K_Y)\}}$$

(eqn. 4.28)

$$Y_{1/2} = Ym = K_Y(1+2[X]/K_{X,1} +[X]^2/K_{X,1}K_{X,2})/(1+2[X]/^Y K_{X,1} +[X]^2/^Y K_{X,1}{}^Y K_{X,2})$$

(eqn. 4.29)

The above equations are of general validity and allow the researcher to determine the affinity for Y from the dependence of $X_{1/2}$ on the concentration of Y or, vice versa, the affinity for X from the dependence of $Y_{1/2}$ on the concentration of X (eqns. 4.26, 4.27, 4.28, 4.29). Often eqns. 4.28 and 4.29 are used in their logarithmic form, for example:

$$\log(Xm) = \log(Xm^0) + \tfrac{1}{2}\log\{(1+[Y]/K_Y)/(1+[Y]/^{X2} K_Y)\}$$

The plot of $\log(X_{1/2})$ as a function of $\log([Y])$ is sigmoidal and in its central region may be approximated to a straight line whose slope is limited by the ratio of the stoichiometric coefficients of X and Y, with positive sign in the case of positive heterotropic interaction, and negative sign otherwise. For example, the logarithmic form is used in the classical Bohr plot of the dependence of O_2 affinity of hemoglobin on pH, and its maximum slope is -0.5 implying that linkage is of the negative type, and the apparent stoichiometric ratio between O_2 and hydrogen ions is 2:1 (actually 4:2, because on average two protons are released as the deoxygenated tetramer binds four oxygen molecules).

Interesting properties of eqns. 4.28 and 4.29 (or their logarithmic forms) are: (i) they make use of $X_{1/2}$ (or Xm), a robust parameter that can be easily determined also in the absence of refined structural interpretations of one's system; and (ii) they estimate the minimum stoichiometric ratio of the two ligands, in the absence of structural information. When the protein is greater than a dimer, its $\bar{X}$ versus $\log([X])$ plot can be non-symmetric. In this case $X_{1/2} \neq$ Xm and in the above equations one should use Xm rather than $X_{1/2}$ (see below).

Heterotropic effectors can modify the cooperativity of ligand binding, if present, and can also cause cooperativity to appear in a protein that would otherwise be devoid of it (Table 4.1). To further illustrate this point, we consider the case of a non-cooperative homodimer.

If the homodimer is non-cooperative, the condition $K_{X,1}=K_{X,2}$ applies and we may define a single constant for the ligand affinity in the absence of Y, which we call K_X. The same occurs for $^YK_{X,1}$ and $^YK_{X,2}$.

The binding polynomial simplifies to:

$$[P]_{tot} = [P]\left\{\left(1+[X]/K_X\right)^2 + [Y]/K_Y\left(1+[X]/^YK_X\right)^2\right\} \qquad \text{(eqn. 4.30)}$$

The similarity of eqn. 4.30 to the binding polynomial of the allosteric model (eqn. 4.23) is immediately apparent and implies that under the appropriate experimental conditions this system will exhibit positive cooperativity for X, even though the protein is devoid of cooperativity for X when fully unliganded to Y and when fully liganded to Y. Moreover, ligand Y will introduce positive cooperativity regardless of whether its linkage to X is of the negative or positive type. Indeed, to present positive homotropic cooperativity for X this system requires only that the fractional saturation of P with Y changes significantly as a consequence of the addition of X. Similarly to the MWC model, a heterotropic interaction that obeys the above equation causes the system to exhibit cooperativity of the type that was called statistical in Section 4.7. Statistical cooperativity is due to the ligand-induced shift in the relative population of two states of the macromolecule, one bound to the heterotropic effector Y, and the other free of it. At variance with intramolecular cooperativity, statistical cooperativity can never be negative.

The fractional saturation for each ligand can be simulated using eqns. 4.25 through 4.28 irrespective of whether the conditions $K_{X,1}=K_{X,2}$ and $^YK_{X,1}=^YK_{X,2}$ apply or not. However, if the above conditions apply, the specially interesting case occurs where cooperativity for ligand X entirely depends on the heterotropic effector Y. The binding isotherms for X simulated under these conditions are reported in Figure 4.13, and the corresponding dependencies of $X_{1/2}$ on [Y] and of $Y_{1/2}$ on [X] are reported in Figure 4.14.

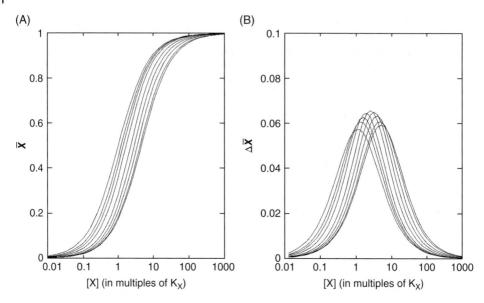

Figure 4.13 Positive homotropic cooperativity induced by the differential binding of the heterotropic effector. The protein is a non-cooperative dimer ($K_{X,1}=K_{X,2}$ and $^YK_{X,1}=^YK_{X,2}$) Parameters: $^YK_X = 5\,K_X$ (this entails $^{X2}K_Y = 25\,K_Y$, see text); concentrations of the effector Y (from left to right, in multiples of K_Y) 0, 0.5, 1, 2, 5, 10, 20, 50, 100. Panel A: $\overline{X}$ versus log([X]) curves. Panel B derivatives of the curves shown in panel A; notice that the maximum slope (indicative of cooperativity) peaks at intermediate concentrations of the effector, and tends toward that of a non-cooperative isotherm at very high and very low concentrations of the effector.

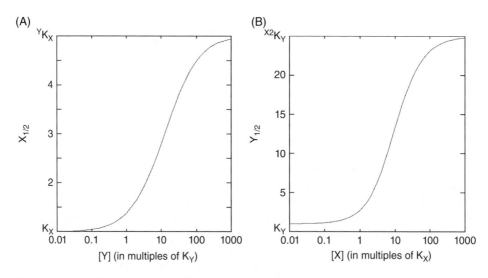

Figure 4.14 Negative heterotropic linkage in a non-cooperative homodimer.
Panel A: $X_{1/2}$ as a function of [Y], as calculated from eqn. 4.27.
Panel B: $Y_{1/2}$ as a function of [X], as calculated from eqn. 4.28.

Figure 4.13A shows the effect of increasing concentrations of the heterotropic ligand Y on the binding isotherm for X in a homodimeric non-cooperative protein, and Figure 4.13B reports the $\Delta \bar{X}$ versus log([X]) plot for those isotherms. In the latter plot cooperativity appears as an increase in the maximum of the bell-shaped curve, and is seen to peak at intermediate concentrations of Y and to decrease afterward.

Cooperativity in this system occurs because the heterotropic effector Y is bound or released as a response to the binding of ligand X. In other words, a necessary requirement for this type of cooperativity is that the concentration of Y is high enough to saturate either P or PX_2, depending on their relative affinities for Y, but not both, such that ligand X changes the ratio of the Y-liganded and the Y-unliganded populations.

Generalizing the above equations for an oligomer of n sites, cooperative or non-cooperative, that binds only one molecule of the heterotropic ligand Y we obtain:

$$Xm = ^n\sqrt{\left(K_{X,1}K_{X,2}\ldots\right)} \, ^n\sqrt{\left\{\left(1+[Y]/K_Y\right)/\left(1+[Y]/^{Xn}K_Y\right)\right\}}$$

or, in logarithmic form,

$$\log Xm = \log Xm^0 + 1/n\log\left\{\left(1+[Y]/K_Y\right)/\left(1+[Y]/^{Xn}K_Y\right)\right\}$$

where $\log Xm^0$ is the log Xm measured in the absence of effector Y.

We remark that in the n-sites oligomer we can no longer equate $X_{1/2} = {}^n\sqrt{(K_{X,1}K_{X,2}\ldots)}$ because this would require that the $\bar{X}$ versus log([X]) plot is symmetric, a condition that necessarily occurs in the monomer and the dimer, and may or may not occur in higher order oligomers. However, the thermodynamic parameter Xm is the geometric mean of the binding constants irrespective of the binding isotherm being symmetric or asymmetric, and it is this parameter that we use in the above equation. We also remark that in many macromolecular assemblies higher than the dimer it is empirically observed that the binding curve approaches symmetry and thus that $X_{1/2}$ approaches Xm.

Cooperativity, if present, may be increased or decreased by heterotropic effectors; or the heterotropic effector may cancel a negative homotropic cooperativity that is present in its absence. For example, T. Yonetani pointed out that human hemoglobin presents only a moderate degree of cooperativity under conditions where no heterotropic effector is present, but cooperativity is strongly increased under physiological conditions where heterotropic effectors are present (Yonetani *et al.*, 2002). The most obvious example is provided by the Bohr effect: at pH=9.2, where both the oxygenated and deoxygenated states of hemoglobin are deprotonated, the Hill coefficient and free energy of cooperativity of human hemoglobin are 2.35 and 1.82 kcal/mol O_2 respectively; at pH=7.4 and in the presence of chloride the same parameters are raised to 3.02 and 3 respectively (Imai, 1982). A similar effect is induced by 2,3 diphosphoglycerate (DPG), which under physiological conditions binds to tetrameric hemoglobin with a stoichiometry of 1:1, and thus has a stoichiometric ratio O_2:DPG = 4:1. The opposite case, that is, that the heterotropic effector *decreases* cooperativity, though less common, is also documented in hemoglobin and possibly in other proteins. It is explained as a consequence of extreme quaternary constraint, inhibiting the ligand-induced fit (in sequential models) or the quaternary structural change (in the allosteric model).

4.10 Heterotropic Linkage and the Allosteric Model

In the original framework of the MWC model, the effect of heterotropic ligands was accounted for under the hypothesis that they would have different affinity for the T and R state and would thus bias the allosteric equilibrium, but would not change either $K_{T,X}$ or $K_{R,X}$. The elegance of this hypothesis will be better appreciated if one compares the complexity of the reaction scheme (Figure 4.15) with the simplicity of the equations necessary for its quantitative description.

Indeed, as shown in Figure 4.15 the minimal MWC scheme for a cooperative homodimer capable of binding ligands X and Y has no fewer than 12 reaction intermediates! Yet all eight binding steps of ligand X are governed by either K_R or K_T as occurs in the absence of the heterotropic effector Y. The equilibrium between the T and R states in the absence of Y is governed by the allosteric constant L_0, and only two additional constants are required, that describe the binding of the heterotropic effector to the T and R states:

$$^T K_Y = \left[^T P\right][Y] / \left[^T PY\right]$$
$$^R K_Y = \left[^R P\right][Y] / \left[^R PY\right]$$

We remark that the allosteric constant for the protein bound to the effector Y depends on L_0 and the two constants defined above:

$$^Y L_0 = L_0 {}^R K_Y / {}^T K_Y$$

The allosteric constants of all ligation intermediates L_1, L_2, $^Y L_1$ and $^Y L_2$ may be defined using equations analogous to eqn. 4.22.

The binding polynomial of this system is defined by eqn. 4.23, except that we need to define an apparent allosteric constant, which takes into account L_0, [Y], $^Y K_R$ and $^Y K_T$:

$$L_{0,app} = \frac{\left[^T P\right] + \left[^T PY\right]}{\left[^R P\right] + \left[^R PY\right]} = L_0 \frac{1 + [Y]/^Y K_T}{1 + [Y]/^Y K_R}$$

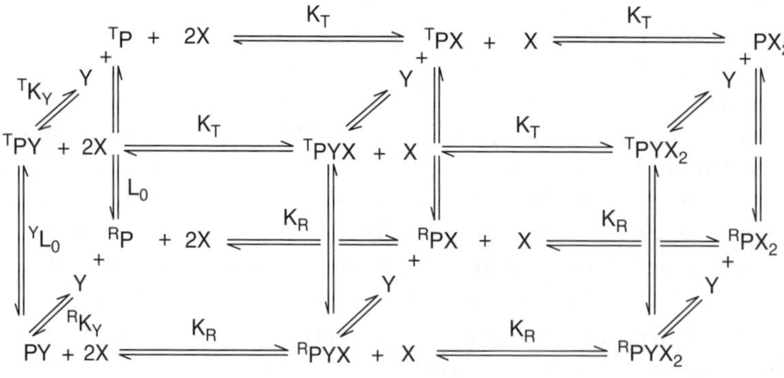

Figure 4.15 Reaction scheme of a two-state allosteric homodimer that binds two ligands at different binding sites. The dimer binds two molecules of ligand X and one of ligand Y, at different sites.

We then need to substitute $L_{0,app}$ for L_0 in eqn. 4.23 to obtain the binding polynomial and fractional saturation with ligand X of the protein.

Unfortunately, this very ingenious formulation is often insufficient to account for the experimental data. For example, in hemoglobin most heterotropic effectors not only change $L_{0,app}$: they also affect at least K_T, and in some cases K_R as well (Imai, 1982, 1983). As expected, because of differential binding to the two states, heterotropic effectors may significantly affect the cooperativity of hemoglobin (Yonetani *et al.*, 2002; Imai, 1982).

References

Adair, G.S. (with the collaboration of A.V. Bock and H. Field Jr.) (1925) The hemoglobin system. VI The oxygen dissociation curve of hemoglobin. J Biol Chem, 63: 529–545.

Babu, Y.S., Bugg, C.E. and Cook, W.J. (1988) Structure of calmodulin refined at 2.2: A resolution. J Mol Biol, 204: 191–204.

Bellelli A. (2010) Hemoglobin and cooperativity: experiments and theories. Curr Protein Pept Sci, 11: 2–36.

Brunori, M., Coletta, M. and Di Cera, E. (1986) A cooperative model for ligand binding to biological macromolecules as applied to oxygen carriers. Biophys Chem, 23: 215–222.

Connelly, P.R., Gill, S.J., Miller, K.I., Zhou, G. and van Holde, K.E. (1989) Identical linkage and cooperativity of oxygen and carbon monoxide binding to Octopus dofleini hemocyanin. Biochemistry, 28: 1835–1843.

Di Cera, E., Robert, C.H. and Gill, S.J. (1987a) Allosteric interpretation of the oxygen binding reaction of human hemoglobin tetramers. Biochemistry, 26: 4003–4008

Di Cera, E., Doyle, M.L., Connelly, P.R. and Gill, S.J. (1987b) Carbon monoxide binding to human hemoglobin A0. Biochemistry, 26: 6494–6502.

Edsall, J.T. (1972) Blood and hemoglobin: Tthe evolution of knowledge of functional adaptation in a biochemical system, part I: The adaptation of chemical structure to function in hemoglobin. J Hist Biol, 5: 205–257.

Edsall, J.T. (1980) Hemoglobin and the origins of the concept of allosterism. Fed Proc, 39: 226–235.

Imai, K. (1982) Allosteric Effects in Haemoglobin. Cambridge University Press: Cambridge, UK.

Imai, K. (1983) The Monod-Wyman-Changeux allosteric model describes haemoglobin oxygenation with only one adjustable parameter. J Mol Biol, 167: 741–749.

Koshland, D.E., Jr, Némethy, G. and Filmer, D. (1966) Comparison of experimental binding data and theoretical models in proteins containing subunits. Biochemistry, 5: 365–385.

Forsén, S. and Linse, S. (1995) Cooperativity: Over the Hill. Trends Biochem Sci, 20: 495–497.

Marden, M.C., Kister, J., Poyart, C. and Edelstein, S.J. (1989) Analysis of hemoglobin oxygen equilibrium curves. Are unique solutions possible? J Mol Biol, 208: 341–345.

Monod, J., Wyman, J. and Changeux, J.P. (1965) On the nature of allosteric transitions: A plausible model. J Mol Biol, 12: 88–118.

Pauling, L. (1935) The oxygen equilibrium of hemoglobin and its structural interpretation. Proc Natl Acad Sci USA, 21: 186–191.

Perutz, M.F. (1970) Stereochemistry of cooperative effects in haemoglobin. Nature, 228: 726–739.

Robert, C.H., Decker, H., Richey, B., Gill, S.J. and Wyman J. (1987) Nesting: Hierarchies of allosteric interactions. Proc Natl Acad Sci USA, 84: 1891–1895.

Schoenborn, B.P. (1965) Binding of xenon to horse haemoglobin. Nature, 208: 760–762.

Weber, G. (1992) Protein Interactions. Chapman and Hall.

Wyman, J. (1948) Heme proteins. Adv Protein Chem, 4: 407–531.

Wyman, J. (1963) Allosteric effects in hemoglobin. Cold Harbor Spring Symposia on Quantitative Biology, XXVIII: 483–489.

Wyman, J. (1964) Linked functions and reciprocal effects in hemoglobin: a second look. Adv Protein Chem, 19: 223–286.

Wyman, J. (1981) The cybernetics of biological macromolecules. Biophys Chem, 14: 135–146.

Wyman, J. and Gill, S.J. (1990) Binding and Linkage. Mill Valley, CA: University Science Books.

Yonetani, T., Park, S.I., Tsuneshige, A., Imai, K. and Kanaori, K. (2002) Global allostery model of hemoglobin. Modulation of O(2) affinity, cooperativity, and Bohr effect by heterotropic allosteric effectors. J Biol Chem, 277: 34508–34520.

Appendix 4.1 Statistical Distribution of the Ligand Among the Binding Sites: Statistical Factors

As described in Section 4.2, if a non-cooperative protein is made up by identical subunits, each binding one molecule of the ligand with the same affinity, the fraction of bound over total sites will be indistinguishable from that of the isolated subunits (provided that the subunits can be isolated). The system will be composed by a statistical mixture of ligation intermediates, that is, the liganded and unliganded subunits will be randomly distributed to form fully unliganded proteins, singly liganded proteins, and so on, and the composition of the mixture is governed by a binomial distribution of liganded and unliganded subunits among the protein oligomers.

We may clarify the relationships between the binomial coefficients used to calculate the fraction of ligation intermediates and the statistical factors to be applied to the equilibrium constants. The homodimer is too simple a system for a meaningful example, thus we shall consider a symmetric homotetramer, whose binomial coefficients are:

for the unliganded species (n=4, i=0): $n!/(i!(n-i)!) = 1$
for the singly liganded species (n=4, i=1): 4
for the doubly liganded species (n=4, i=2): 6
for the triply liganded species (n=4, i=3): 4
for the fully liganded species (n=4, i=4): 1

The statistical coefficients for the equilibrium constants are defined considering the ratio of the possibilities for the forward and backward reactions:

$K_{d,1}$ refers to the reaction $P_4X \Longleftrightarrow P_4 + X$: statistical factor = ¼
$K_{d,2}$ refers to the reaction $P_4X_2 \Longleftrightarrow P_4X + X$: statistical factor = ⅔
$K_{d,3}$ refers to the reaction $P_4X_3 \Longleftrightarrow P_4X_2 + X$: statistical factor = 3/2
$K_{d,4}$ refers to the reaction $P_4X_4 \Longleftrightarrow P_4X_3 + X$: statistical factor = 4

When we calculate the concentrations of the ligation intermediates using the concentration of the fully unliganded species ($[P_4]$) as the reference, we find that:

$[P_4]$ is the reference species, does not depend on any equilibrium constant, and is assigned a statistical coefficient of 1

$[P_4X]$ is a function of $[X]/K_{d,1}$ and its binomial coefficient is the inverse of the statistical factor of $K_{d,1}$ (i.e., 4);

$[P_4X_2]$ is a function of $[X]^2/K_{d,1}K_{d,2}$ and its binomial coefficient is the inverse of the product of the statistical factors of $K_{d,1}$ and $K_{d,2}$ (i.e., 6);

$[P_4X_3]$ is a function of $[X]^3/K_{d,1}K_{d,2}K_{d,3}$ and its binomial coefficient is the inverse of the product of the statistical factors of $K_{d,1}$, $K_{d,2}$ and $K_{d,3}$ (i.e., 4);

$[P_4X_4]$ is a function of $[X]^4/K_{d,1}K_{d,2}K_{d,3}K_{d,4}$ and its binomial coefficient is the inverse of the product of the statistical factors of $K_{d,1}$, $K_{d,2}$, $K_{d,3}$ and $K_{d,4}$ (i.e., 1);

Thus, the binomial coefficients of ligation intermediates correspond to the inverse of the products of the statistical factors of the pertinent equilibrium dissociation constants (or to the products of the statistical factors of the association constants).

Appendix 4.2 Symmetry of the $\bar{X}$ Versus Log([X]) Plot: The Concept of Xm

We observed in Section 1.4 that the $\bar{X}$ versus log($[X]$) plot of a protein that binds a ligand with 1:1 stoichiometry is symmetric with respect to the point (log($X_{1/2}$), 0.5). In this appendix we demonstrate that this property also applies to all proteins possessing two binding sites for the same ligand, whether they be homodimers, heterodimers, or monomeric, two-site proteins. Moreover, this property is independent of the presence or absence of cooperativity. This property is not shared by higher order oligomers, that is, the $\bar{X}$ versus log($[X]$) plot of a trimeric or tetrameric protein may or may not be symmetric.

The relevance of the symmetry of the $\bar{X}$ versus log($[X]$) plot lies in the fact that it is a condition for the equivalence of $X_{1/2}$ with Xm, the ligand concentration required to express half the free energy of the reaction. Xm is an important thermodynamic parameter, which is not easily derived from the plot, thus it is interesting to know if it can be equated to $X_{1/2}$ (or approximated to it for proteins with stoichiometries higher than 2:1). A schematic depiction of the symmetry condition and its relevance in determining Xm is reported in Figure 4.16.

As one can easily derive from the figure 4.16, the symmetry condition is that the value of $\bar{X}$ calculated for each ith submultiple of $X_{1/2}$ equals the value of $(1 - \bar{X})$ calculated for each ith multiple of $X_{1/2}$, that is:

$$\bar{X}_{[X]=X_{1/2}/i} = 1 - \bar{X}_{[X]=iX_{1/2}}$$

To demonstrate that the above condition applies to any protein that binds ligands with 2:1 stoichiometry, we need first to derive $X_{1/2}$ from the fractional ligand saturation as given by eqn. 4.9, that is:

$$\bar{X} = \left([X]/K_{d,1} + [X]^2/K_{d,1}K_{d,2}\right) / \left(1 + 2[X]/K_{d,1} + [X]^2/K_{d,1}K_{d,2}\right)$$

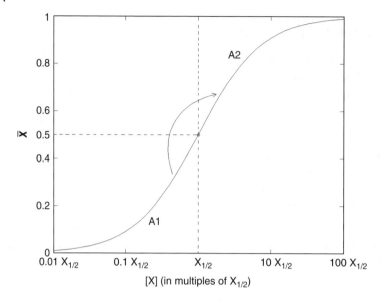

Figure 4.16 A ligand-binding isotherm is symmetric if the lower half of the curve can be perfectly superimposed to the upper half by 180° rotation around the point log($X_{1/2}$),0.5. If this occurs, the areas A1 and A2 are equal.

We equate [X] to $X_{1/2}$ and $\bar{X}$ to 0.5, and solve for $X_{1/2}$:

$$0.5 = \left(X_{1/2}/K_{d,1} + X_{1/2^2}/K_{d,1}K_{d,2}\right)/\left(1 + 2X_{1/2}/K_{d,1} + X_{1/2^2}/K_{d,1}K_{d,2}\right)$$
$$2X_{1/2}/K_{d,1} + 2X_{1/2^2}/K_{d,1}K_{d,2} = 1 + 2X_{1/2}/K_{d,1} + X_{1/2^2}/K_{d,1}K_{d,2}$$

which yields:

$$X_{1/2} = \sqrt{\left(K_{d,1}K_{d,2}\right)}$$

The next step is to demonstrate that the symmetry condition applies to eqn. 4.9. Thus, we need to express [X] in eqn. 4.9 as a function of $X_{1/2}$:

$$[X] = i\sqrt{(K_{d,1}K_{d,2})}$$

$$\bar{X}_{[X]=i\sqrt{(Kd,1Kd,2)}} = \frac{i\sqrt{K_{d,2}}/\sqrt{K_{d,1}} + i^2}{1 + 2i\sqrt{K_{d,2}}/\sqrt{K_{d,1}} + i^2}$$

$$[X] = \sqrt{(K_{d,1}K_{d,2})}/i$$

$$1 - \bar{X}_{[X]=\sqrt{(Kd,1Kd,2)}/i} = 1 - \frac{\sqrt{K_{d,2}}/i\sqrt{K_{d,1}} + 1/i^2}{1 + 2\sqrt{K_{d,2}}/i\sqrt{K_{d,1}} + 1/i^2} = \frac{1 + \sqrt{K_{d,2}}/i\sqrt{K_{d,1}}}{1 + 2\sqrt{K_{d,2}}/i\sqrt{K_{d,1}} + 1/i^2}$$

To demonstrate that the symmetry condition applies we need only to multiply the last equation by i^2/i^2.

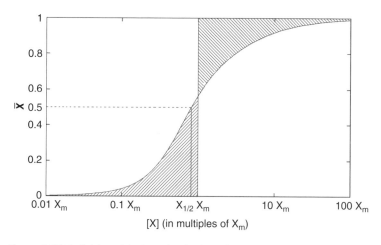

Figure 4.17 Definition of the Xm: the shadowed areas are equal. From Wyman 1963, modified.

Wyman (1963) pointed out that symmetry of the $\bar{X}$ versus $\log([X])$ plot has the important consequence that the free energy change (per site) due to ligation results:

$$\Delta F = -RT \ln X_{1/2}$$

In oligomers greater than a dimer, the $\bar{X}$ versus $\log([X])$ plot may be symmetric or asymmetric, depending on the binding constants of the different sites. If the $\bar{X}$ versus $\log([X])$ plot is asymmetric, the free energy of binding has no simple relationship with $X_{1/2}$. However, a point can be defined for which the above relationship is valid; this point is called Xm and has the property:

$$\Delta F = -RT \ln Xm$$

Unfortunately, in those cases in which Xm $\neq X_{1/2}$, finding Xm is not straightforward, as it requires the (numerical) integration of the binding curve, as shown in Figure 4.17.

5

Ligand-Linked Association and Dissociation

The majority of proteins capable of binding more than one molecule of ligand are oligomers of single-site subunits. Quite often the oligomer is fully assembled at physiological concentration, and does not exhibit a significant tendency to dissociate into its constituent subunits. However, in some cases, the oligomer may undergo reversible association-dissociation equilibria, which may or may not be physiologically relevant. The ligand-binding equilibria of oligomers capable of reversible dissociation into the constituent subunits may be very complex, especially if the liganded and unliganded states have different tendency to dissociate, a condition that occurs when the isolated subunits and the oligomer have different ligand affinity.

Examples of ligand-dependent association-dissociation phenomena are provided by the receptors of several hormones (Ward *et al.*, 2007; Germain and Bourguet, 2013; Kovacs *et al.*, 2015), which tend to be monomeric when unliganded and dimeric when liganded; or by some hemoglobins (e.g., from lamprey), which are monomeric when liganded and dimeric or tetrameric when unliganded (Briehl, 1963; Andersen, 1971), and so on. A less extreme case is that of vertebrate hemoglobins, in which a tetramer-dimer equilibrium is present and is biased in the direction of dissociation by heme ligands (see Chapter 7). The dependence of the aggregation state on ligation (and vice versa) was called *polysteric linkage* by Colosimo *et al.* (1976).

Recording the aggregation state of the protein as a function of ligation, or the ligand affinity as a function of the oligomerization state, is not always possible, but, when possible, provides relevant information on the reaction mechanism. Moreover, the researcher must be aware of the possible relationship between association and ligation since *ligand-dependent changes in the association state of the oligomer may cause unexpected results, typically a dependence of ligand affinity on protein concentration.*

Given the complexity of this type of reaction, we shall focus this chapter on the ligand-linked association-dissociation equilibria in homodimers; in Chapter 7, however, we shall discuss the case of the hemoglobin tetramers. We believe that it has greater pedagogical value to concentrate on one and the same system (the homodimer), and explore the effect of different types of linkage, than to select the best-characterized example for each, at the expense of the possibility of fruitfully comparing the different types of linkage.

We start with a definition of the important concepts of quaternary constraint and quaternary enhancement, whose meaning can be fully appreciated only in the case of reversibly dissociating oligomers (Section 5.1). Next, we shall analyze the case of a

Reversible Ligand Binding: Theory and Experiment, First Edition. Andrea Bellelli and Jannette Carey.
© 2018 John Wiley & Sons Ltd. Published 2018 by John Wiley & Sons Ltd.

reversibly dissociating homodimer whose dissociation state is not affected by ligation (Section 5.2). Strictly speaking this is not a case of ligand-linked association-dissociation, but it provides the background required to understand the behavior of a model system under the simplest possible assumptions. The following sections of this chapter deal with the case of the non-cooperative dimer whose reversible dissociation into subunits is linked to ligand binding, and two special variants of this case (Sections 5.3 through 5.5). Finally, we shall consider the more complex case of the cooperative homodimer whose dissociation equilibrium is linked to ligand binding (Section 5.6). For simplicity in this complex matter, we shall assume that the experiments can be carried out under optimal conditions, and in particular that $[P]_{tot} << K_d$.

5.1   Quaternary Constraint and Quaternary Enhancement

In those cases in which the ligand-binding isotherm can be recorded for the isolated subunits and for the oligomer, one may directly obtain the information necessary to compare the ligand affinity of the two states. The term "subunit" in this context requires an extended interpretation because it must also include those cases in which a higher-order oligomer dissociates into lower-order oligomers (e.g., hemoglobin tetramers that dissociate into dimers). Several possibilities must be considered.

  i) The ligand affinity of the isolated subunits is higher than the average affinity of the oligomer, that is, aggregation decreases the affinity for the ligand. This condition is called *quaternary constraint*, and a typical example is provided by vertebrate hemoglobins (Figure 5.1A).
 ii) The ligand affinity of the isolated subunits is identical to that of the oligomer; if the oligomer is non-cooperative its ligand binding isotherm is superimposable to that of the isolated subunits.
iii) The ligand affinity of the isolated subunits is lower than the average affinity of the oligomer, that is, aggregation increases the affinity for the ligand. This condition is called *quaternary enhancement*. An example of quaternary enhancement is provided (with some caution) by arthropod hemocyanins (Figure 5.1B). A special, and not uncommon, example of quaternary enhancement is provided by those proteins whose ligand-binding sites are formed at the subunit interfaces such that an isolated subunit presents only an incomplete binding site. Unfortunately, this type of protein rarely dissociates into functional subunits under physiological conditions.

The most straightforward cases of quaternary enhancement and quaternary constraint are observed when the ligand-binding isotherm of the isolated subunits superimposes to either the lower or the higher asymptote of the oligomer binding curve. We call these cases *pure quaternary enhancement* and *pure quaternary constraint*, respectively. However intermediate cases are known in which the ligand affinity of the isolated subunits coincides neither with the first nor with the last binding site of the oligomer and lies in between (e.g., some hemocyanins; Savel-Niemann *et al.*, 1988).

From a historical point of view, the concept of quaternary enhancement (if not the name) was implicit in the model of cooperativity devised by L. Pauling (1935); the possibility that the same model could be also formulated in terms of quaternary constraint was considered by Coryell [1939], and, more thoroughly, by Wyman (1948). The allosteric

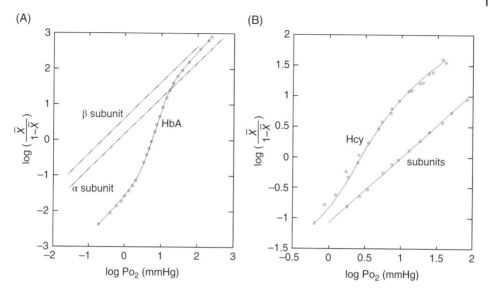

Figure 5.1 Quaternary enhancement and quaternary constraint. Panel A: Oxygen-binding isotherm of human hemoglobin at pH = 7.4 and T = 25 °C, and its isolated α and β subunits (redrawn after Imai, 1982). This is a case of quaternary constraint, because the oligomer has lower affinity than its isolated monomers. Panel B: Oxygen-binding isotherm of the hexameric hemocyanin from the lobster *Panulirus interruptus* at pH = 9.6 and its isolated subunits. In the presence of 10 mM $CaCl_2$ the hexamer (Hcy) is populated; in its absence the oligomer dissociates into its constituent subunits (redrawn after Kuiper *et al.*, 1977). This is a case of quaternary enhancement, because the oligomer has higher affinity than its isolated monomers.

The original figures for both panels reported Hill plots, and we maintained this representation. Even though in general we discourage the use of this type of representation (see Chapter 1), in the present case it has the advantage of highlighting the lower and upper asymptotes of the binding isotherm, which can be compared with the binding isotherms of the isolated subunits.

model (Monod *et al.*, 1965) was formulated on the explicit assumption of quaternary constraint. Thorough analyses of the implications of thermodynamic linkage between ligand binding and subunit association equilibria have been presented by several authors (e.g., Ackers *et al.*, 1975; Edelstein and Edsall, 1986; Weber, 1992).

5.2 The Reversibly Dissociating Homodimer Devoid of Ligand-Linked Association Equilibria

The simplest imaginable system presenting both ligand-binding and association-dissociation equilibria is the non-cooperative homodimer, whose reaction scheme is as follows (Figure 5.2).

Figure 5.2 Reaction scheme for a dissociating homodimer.

$$M_2X_2 \;\underset{}{\overset{J_d}{\rightleftharpoons}}\; M_2X + X \;\underset{}{\overset{J_d}{\rightleftharpoons}}\; M_2 + 2X$$
$$K_2 \big\updownarrow \qquad\qquad K_1 \big\updownarrow \qquad\qquad K_0 \big\updownarrow$$
$$2MX \;\underset{}{\overset{K_d}{\rightleftharpoons}}\; M + MX \;\underset{}{\overset{K_d}{\rightleftharpoons}}\; 2M + 2X$$

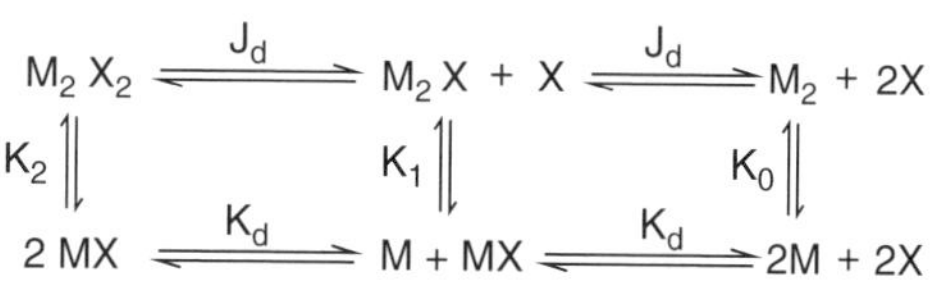

The microscopic equilibrium constants of this system are defined as:

$$J_d = [M_2X][X]/[M_2X_2] = [M_2][X]/[M_2X]$$
$$K_d = [M][X]/[MX]$$
$$K_0 = [M]^2/[M_2]$$
$$K_1 = [M][MX]/[M_2X]$$
$$K_2 = [MX]^2/[M_2X_2]$$

In this section we consider the case of absence of linkage between ligand binding and association-dissociation. This case requires $K_d = J_d$. Because ligand affinity is the same for the isolated subunits and the oligomer, ligation does not bias the association-dissociation equilibria, and the concentration of the oligomer does not bias the ligand affinity ($X_{1/2}$) of the system. Under these assumptions, we also have that $K_0 = K_1 = K_2$, and the system is fully described by only two equilibrium constants, K_d (for the dissociation of the protein-ligand complex) and K (for the dissociation of the homodimer into monomers). Both constants, being defined in direction of dissociation, have units of M (when analyzing such complex systems it is important to systematically verify that the units of the terms we define are as expected).

Since the homodimer has no cooperativity, and there is no linkage between ligand binding and oligomerization, the binding polynomial and fraction of ligand saturation for the subunits, irrespective of their occurrence in the monomeric or homodimeric state, is expected to be:

$$[M]_{tot} = [M](1+[X]/K_d); \quad \bar{X} = [X]/(K_d+[X]) \tag{eqn. 5.1}$$

and the dissociation equilibrium of the homodimer obeys Ostwald's law, irrespective of the ligand concentration:

$$K = 4\alpha^2([M]_{tot}/2)/(1-\alpha) \tag{eqn. 5.2}$$

where α indicates the degree of dissociation (i.e., the ratio between the moles of dissociated oligomers and the moles of total oligomers). Ostwald's law is written for the concentration of the oligomer, hence the term $[M]_{tot}/2$, which represents the total protein concentration expressed on a dimer basis.

The system is more complex than it may appear from the above equations, because its binding polynomial should inform us on the relative concentrations of all the protein species, expressed as functions of two variables: [X] and $[M]_{tot}$. Thus, we adopt [M] as the reference concentration and we write:

$$[MX] = [M][X]/K_d$$
$$[M_2] = [M]^2/K$$
$$[M_2X] = [M][MX]/K = [M]^2[X]/KK_d$$
$$[M_2X_2] = [MX]^2/K = [M]^2[X]^2/KK_d^2$$
$$[M]_{tot} = [M](1+[X]/K_d)+2[M]^2/K(1+[X]/K_d)^2 \tag{eqn. 5.3}$$

The characteristic of oligomers that dissociate into n subunits is that the concentration of the species used as reference appears in a n-th degree expression (with n=2 for the dimer, as in eqn. 5.3); thus, it will not simplify out when we use the binding polynomial to calculate the fractional ligand saturation $\bar{X}$. Given that we can no longer simplify for the concentration of the reference species, we solve eqn. 5.3 to express [M] as a function of $[M]_{tot}$ and [X]:

$$2[M]^2 / K\left(1+[X]/K_d\right)^2 + [M]\left(1+[X]/K_d\right) - [M]_{tot} = 0 \qquad \text{(eqn. 5.4)}$$

which yields:

$$[M] = K K_d \frac{-1+\sqrt{\left(1+8[M]_{tot}/K\right)}}{4\left(K_d+[X]\right)} \qquad \text{(eqn. 5.5)}$$

Some controls of the consistency of eqns. 5.1 through 5.5 are in order. The fractional ligand saturation one would obtain from eqn. 5.3 is identical to eqn. 5.1:

$$\bar{X} = \frac{[MX]+[M_2X]+2[M_2X_2]}{[M]+[MX]+2\left([M_2]+[M_2X]+[M_2X_2]\right)}$$

$$= \frac{[X]/K_d\left\{1+2[M]\left(1+[X]/K_d\right)/K\right\}}{\left(1+[X]/K_d\right)\left\{1+2[M]\left(1+[X]/K_d\right)/K\right\}} = \frac{[X]}{K_d+[X]}$$

Moreover, eqn. 5.5 can be demonstrated to be fully consistent with eqn. 5.2, as follows. The degree of dissociation α is defined as the ratio between the concentrations of the dissociated oligomers and the total oligomers.

$$\alpha = \frac{\left([M]+[MX]\right)/2}{\left([M]+[MX]\right)/2 + [M_2]+[M_2X]+[M_2X_2]}$$

$$= \frac{[M]\left(1+[X]/K_d\right)}{[M]\left(1+[X]/K_d\right)+2[M]^2\left(1+[X]/K_d\right)^2/K} = \frac{1}{1+2[M]\left(1+[X]/K_d\right)/K}$$

We substitute [M] with eqn. 5.5 and obtain:

$$\alpha = \frac{2}{1+\sqrt{\left(1+8[M]_{tot}/K\right)}} \qquad \text{(eqn. 5.6)}$$

α is limited between zero at high $[M]_{tot}$ and 1 at low $[M]_{tot}$. We name $\gamma = \sqrt{(1 + 8[M]_{tot}/K)}$, and replace eqn. 5.6 into 5.2:

$$\frac{4\alpha^2[M]_{tot}/2}{1-\alpha} = \frac{8[M]_{tot}/(1+\gamma)^2}{1-2/(1+\gamma)} = \frac{8[M]_{tot}}{(\gamma+1)(\gamma-1)} = \frac{8[M]_{tot}}{1+8[M]_{tot}/K-1} = K$$

This demonstrates that eqn. 5.6 is fully consistent with eqn. 5.2.

The above equations give formal expression to the premise that *in the absence of ligand-linked association-dissociation, the degree of ligand saturation is independent of protein concentration, and the degree of dissociation of the oligomer does not depend on ligand concentration.*

There is a precise pedagogical reason to present this long and tedious analysis for a trivial case, which was fully described already from eqns. 5.1 and 5.2: because the case is trivial, we have an easy test of all our equations. In particular we know that eqns. 5.3 and 5.5 simplify to yield 5.1 and 5.2. In the next examples, we shall not have this safety check and our test will be that the equations we develop should reduce to eqns. 5.3 and 5.5 if the appropriate conditions are met. We strongly encourage the reader to routinely test the models he/she develops in this way, because the quantitative description of systems presenting ligand-linked association-dissociation equilibria is error-prone.

5.3 Ligand-Linked Association-Dissociation in the Non-Cooperative Homodimer

If an oligomeric protein undergoes a reversible association-dissociation equilibrium under the experimental conditions adopted by the researcher (i.e., at the protein concentration required by the sensitivity of the chosen technique), and the ligand affinity of the oligomer and the monomer differ, then ligation will bias the protein association-dissociation equilibrium. In particular, if the ligand has higher affinity for the isolated subunits than for the oligomer, ligation promotes dissociation of the oligomer into subunits and one has a case of quaternary constraint. On the contrary, if the ligand has higher affinity for the oligomer, ligation promotes the association of the subunits, and one has a case of quaternary enhancement. In both cases, cooperativity will be introduced in the system, even though both the oligomer and the monomer themselves are non-cooperative. We shall analyze here the simplest possible case, that of a reversibly dissociating non-cooperative homodimer; more complex cases will be considered in the following sections.

The reaction scheme is identical to the one described in Section 5.2, but requires two different ligand dissociation constants, one for the isolated monomer (K_d), the other for the monomer in the oligomer (J_d). Thus, the (microscopic) equilibrium constants of this system are:

$$J_d = [M_2X][X]/[M_2X_2] = [M_2][X]/[M_2X]$$
$$K_d = [M][X]/[MX]$$
$$K_0 = [M]^2/[M_2]$$

We remark that the dissociation constants for the liganded states of the dimer will not be equal to K_0, but will follow from the preceding ones (and the appropriate statistical coefficients):

$$K_1 = [M][MX]/[M_2X] = \left([M]^2[X]/K_d\right)/\left(2[M_2][X]/J_d\right) = K_0 J_d / 2K_d$$
$$K_2 = [MX]^2/[M_2X_2] = \left([M]^2[X]^2/Kd^2\right)/\left([M_2][X]^2/Jd^2\right) = K_0 J_d^2 / K_d^2$$

We call the reader's attention to the term $J_d/2K_d$ that represents the ratio of the apparent dissociation constants, $J_d/2$ being the product of J_d by the statistical factor for the first binding site. If higher-order oligomers are considered, one should use the appropriate statistical factors. In the term J_d^2 / K_d^2, the product of the statistical factors for the first and second site of the homodimer does not appear because it is unity.

The binding polynomial of this system using the unliganded monomer as the reference species is:

$$[M]_{tot} = ([M]+[MX]) + 2([M_2]+[M_2X]+[M_2X_2])$$
$$= [M](1+[X]/K_d) + 2[M]^2 / K_0 (1+[X]/J_d)^2 \qquad \text{(eqn. 5.7)}$$

where the first term refers to the monomer and the second to the dimer.

We calculate the relationship between the concentration of the unliganded monomer and the total protein concentration (on a monomer basis), as we did in the case of eqn. 5.5 (above), and we obtain:

$$[M] = \frac{-(1+[X]/K_d) + \sqrt{\left\{(1+[X]/K_d)^2 + 8[M]_{tot}(1+[X]/J_d)^2 / K_0\right\}}}{4(1+[X]/J_d)^2 / K_0} \qquad \text{(eqn. 5.8)}$$

As expected, eqns. 5.7 and 5.8 would reduce to 5.3 and 5.5 respectively if $J_d = K_d$, that is, if the ligand affinities are the same for the monomer and dimer.

The concentration of the unliganded monomer (the reference species) is a function of both the concentrations of the free ligand and of the protein, and the exact expression of the binding polynomial for this system would require substitution of eqn. 5.8 into 5.7. However, we prefer to keep them separate for readability, and just inform the reader that the term [M] in eqn. 5.7 should be calculated using eqn. 5.8 for each value of [X] and [M]$_{tot}$.

The degree of dissociation α and the degree of ligand saturation $\bar{X}$ are both functions of [X] and [M]$_{tot}$ (and the three independent equilibrium constants).

The fractional ligand saturation of this system is:

$$\bar{X} = \frac{[MX]+[M_2X]+2[M_2X_2]}{[M]+[MX]+2[M_2]+2[M_2X]+2[M_2X_2]}$$
$$= \frac{[M][X]/K_d + 2[M]^2 / K_0([X]/J_d + [X]^2/J_d^2)}{[M](1+[X]/K_d) + 2[M]^2 / K_0(1+[X]/J_d)^2} \qquad \text{(eqn. 5.9)}$$
$$= \frac{[X]/K_d + 2[M]/K_0([X]/J_d + [X]^2/J_d^2)}{(1+[X]/K_d) + 2[M]/K_0(1+[X]/J_d)^2}$$

As noticed for the binding polynomial, to effectively expresses $\bar{X}$ as a function of [X] and [M]$_{tot}$, the independent variables of this system, we should replace every instance of [M] in eqn. 5.9 with eqn. 5.8. Contrary to the analogous equation developed in the preceding Section, the fractional ligand saturation of this system depends on the total protein concentration.

Figure 5.3 reports a family of simulated ligand-binding isotherms and the corresponding values of $X_{1/2}$ and α calculated using eqn. 5.8 and 5.9 for the case of $J_d>K_d$. Figure 5.4 reports the same information for the opposite case, $J_d<K_d$, to allow easy comparison.

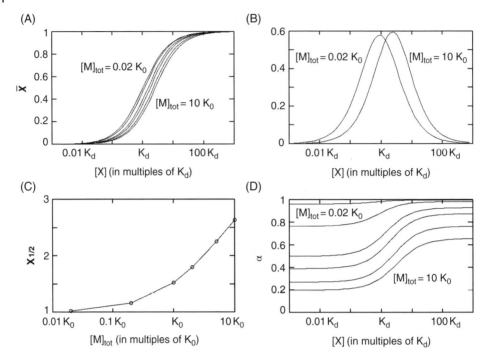

Figure 5.3 Simulation of the oxygen-binding isotherms of a non-cooperative homodimer in equilibrium with its constituent monomers, in the presence of *quaternary constraint* ($J_d>K_d$). Panel A: $\overline{X}$ versus log ([X]) plot; the different curves for each family represent different protein concentrations (here expressed as multiples of the homodimer dissociation K_0: $[M]_{tot} = 0.02\,K_0, 0.2\,K_0, K_0, 2\,K_0, 5\,K_0,$ $10\,K_0$). In this set of curves, calculated for $J_d = 5\,K_d$, dilution of the protein causes a left shift of the ligand binding isotherm (i.e., an increase of the apparent ligand affinity), given that the monomers have higher affinity than the homodimers. Panel B: derivative of the two isotherms from panel A, for the two extreme values of protein concentration ($[M]_{tot} = 0.02\,K_0$ and $10\,K_0$). Both curves present a small degree of positive cooperativity, more evident at higher protein concentration. Cooperativity depends on the change of the oligomerization state as a function of ligation, because both the dimer and the monomer in themselves are non-cooperative. Panel C: plot of the half saturating ligand concentration $X_{1/2}$ as a function of protein concentration (from the data in panel A). Panel D: changes of the degree of dissociation α of the homodimers at constant protein concentration, as a function of ligand concentration (from the data in panel A).

The apparent ligand affinity (Figures 5.3C and 5.4C), expressed as $X_{1/2}$, equals the dissociation equilibrium constant of the monomer (K_d) at low protein concentration ($[M]_{tot} \ll K_0$) and tends toward the dissociation equilibrium constant of the dimer (J_d) at high protein concentration ($[M]_{tot} \gg K_0$). Finally, if the system presents quaternary constraint, the fraction of dimers that dissociate into monomers at constant protein concentration increases with increasing the concentration of the ligand (Figure 5.3D) because $K_0<K_1<K_2$. In other words, quaternary constraint dictates that the higher the ligand saturation $\overline{X}$, the higher the degree of dissociation α. The opposite occurs in systems presenting quaternary enhancement (Figure 5.4D).

The ligand-dependent association-dissociation equilibrium introduces positive cooperativity in these systems (Figures 5.3B and 5.4B) because of statistical reasons. Other examples of statistical cooperativity are provided by the MWC model (Section 4.6) and

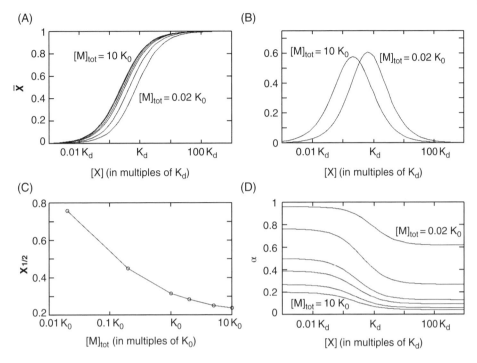

Figure 5.4 Simulation of the oxygen-binding isotherms of a non-cooperative homodimer in equilibrium with its constituent monomers, in the presence of *quaternary enhancement* ($J_d < K_d$). Panel A: $\bar{X}$ versus log ([X]) plot; the different curves for each family represent different protein concentrations (here expressed as multiples of the homodimer dissociation K_0: $[M]_{tot} = 0.02\,K_0, 0.2\,K_0$, $K_0, 2\,K_0, 5\,K_0, 10\,K_0$). In this set of curves, calculated for $J_d = 0.2\,K_d$, dilution of the protein causes a right shift of the ligand binding isotherm (i.e., a decrease of the apparent ligand affinity), given that the monomers have lower affinity than the homodimers. Panel B: derivative of the two isotherms from panel A, for the two extreme values of protein concentration ($[M]_{tot} = 0.02\,K$ and $10\,K$). Both curves present a small degree of positive cooperativity, more evident at lower protein concentration. Cooperativity depends on the change of the oligomerization state as a function of ligation, because both the dimer and the monomer in themselves are non-cooperative. Panel C: plot of the half saturating ligand concentration $\mathbf{X_{1/2}}$ as a function of protein concentration (from the data in panel A). Panel D: changes of the degree of dissociation α of the homodimers at constant protein concentration, as a function of ligand concentration (from the data in panel A).

by the effect of heterotropic ligands on (non-cooperative) oligomers (Section 4.9), as listed in Table 4.1. In all these cases cooperativity occurs because preferential binding of the ligand to the species for which it has higher affinity causes the relative population of this species to increase at the expense of the other species, causing an increase of the overall binding affinity as ligation proceeds. If the homodimer has lower affinity for the ligand than the monomer ($J_d > K_d$; Figure 5.3), an increase in ligand concentration will promote dissociation of the dimer and will increase the population of high-affinity subunits, yielding positive cooperativity (i.e., a steeper ligand-binding isotherm). In this system cooperativity, defined as the steepness of the $\bar{X}$ versus log([X]) plot, presents a bell-shaped dependence on protein concentration, with a maximum at the protein concentration where the largest ligand-dependent change in the degree of dissociation occurs (Figure 5.3D). We may better figure out the positive cooperativity of this system if we imagine that when the first ligand binds to a low-affinity homodimer, dissociation to monomers is promoted because $K_1 > K_0$ and the second ligand binds to a high-affinity monomer.

If, on the contrary, the homodimer has higher affinity for the ligand than the monomer ($J_d < K_d$), an increase of the ligand concentration will promote formation of the homodimer at the expense of the isolated subunits, hence an increase of the ligand affinity. In this case ligation of a low-affinity monomer promotes its association with an unliganded monomer because $K_1 < K_0$. The unliganded monomer in the singly liganded dimer binds the second ligand with higher affinity than it would have done as an isolated monomer because $J_d < K_d$.

It is interesting to compare the effect of the ligand on the dissociation coefficient α, at constant protein concentration, in the two cases (Figures 5.3D and 5.4D) and the effect of protein concentration on $X_{1/2}$ (Figures 5.3C and 5.4C). Indeed, the effects are opposite to each other, but both are compatible with positive cooperativity. This is a thermodynamic consequence of the fact that the ligand always prefers the higher affinity state of the protein, irrespective of its aggregation state. We may generalize this point and state that *statistical cooperativity is always positive, irrespective of being based on quaternary enhancement or quaternary constraint.* Negative cooperativity implies some ligand-induced intramolecular heterogeneity that is incompatible with statistical cooperativity, and is not present in the simple reaction schemes considered in this section. The assumption of a single ligand dissociation constant for the homodimer (i.e., $J_{d,1} = J_{d,2}$) presumes equivalence of the two binding sites and denies structural or ligand-induced intramolecular heterogeneity.

An important and often neglected point is that, due to the linkage between ligand binding and aggregation state of the oligomer, one can derive the dissociation constant of the oligomer (K_0) from the effect of protein concentration on ligand affinity, and vice versa, one can derive the ligand affinity (K_d, J_d) from a measurement of the dependence of the aggregation state of the protein on ligand concentration. *In a sense polysteric linkage is similar to heterotropic linkage, except that the heterotropic ligand in this case is the protein monomer itself.* As a practical example of the study of cooperativity carried out by exploiting association-dissociation equilibria, we shall present the case of hemoglobin in Section 7.8.

All the above analysis was carried out without any assumption on K_d, J_d, K_0, and $[M]_{tot}$; in the following sections we analyze two special cases, both having biological relevance. In Section 5.4, we consider the case where ligand binding to the dimer is negligible; in Section 5.5, association of unliganded monomers into homodimers is negligible. Both cases are variants of the scheme presented in this Section.

5.4 Oligomers That Dissociate Into Monomers Upon Ligand Binding

In this section we consider a special case of the non-cooperative homodimer considered above, that of a quaternary constraint so extreme that the ligand affinity of the homodimer is negligible and the ligand binds only to the dissociated monomers (eqn 5.10). An example of a similar condition is provided by the hemoglobin from the lamprey (Briehl, 1963; Andersen, 1971), with the caveat that the isolated subunits of this protein aggregate to dimers, but also to tetramers, a possibility that our example does not contemplate.

The reaction scheme is as follows:

$$M_2 + 2X \Leftrightarrow 2M + 2X \Leftrightarrow 2MX \qquad \text{(eqn. 5.10)}$$

This system is fully described by only two equilibrium constants:

$$K = [M]^2 / [M_2]$$
$$K_d = [M][X]/[MX]$$

Let the total concentration of monomers be

$$[M]_{tot} = [M] + [MX] + 2[M_2] = [M](1 + [X]/K_d) + 2[M]^2 / K \qquad \text{(eqn. 5.11)}$$

Solving for [M] we obtain

$$[M] = \left\{ -(1 + [X]/K_d) + \sqrt{\left((1 + [X]/K_d)^2 + 8[M]_{tot} / K \right)} \right\} / (4/K) \qquad \text{(eqn. 5.12)}$$

and we observe that eqn. 5.12 can also be obtained from eqn. 5.8 by posing $[X]/J_d = 0$.

The fractional saturation of this system is:

$$\bar{X} = \left([M][X]/K_d \right) / \left([M] + [M][X]/K_d + 2[M]^2 / K \right) \qquad \text{(eqn. 5.13)}$$

From eqns. 5.11 and 5.13 we derive:

$$X_{1/2} = K_d \left(1 + \sqrt{(1 + 4[M]_{tot} / K)} \right) / 2 \qquad \text{(eqn. 5.14)}$$

which demonstrates that $X_{1/2}$ depends on the square root of the total concentration of the protein.

We can write the inverse dependence, that is, the dependence of the degree of dissociation of the homodimer on ligand concentration.

We define

$$\alpha = \left([M] + [MX] \right) / \left(2[M_2] + [M] + [MX] \right) = \left(1 + [X]/K_d \right) / \left(1 + [X]/K_d + 2[M]/K \right)$$
$$= \frac{2(1 + [X]/K_d)}{(1 + [X]/K_d) + \sqrt{\left\{ (1 + [X]/K_d)^2 + 8[M]_{tot} / K \right\}}}$$

$$\text{(eqn. 5.15)}$$

and we remark that

i) α depends on the concentrations of both the protein and the ligand.
ii) In the absence of ligand, $\alpha = 2 / \{1 + \sqrt{(1 + 8[M]_{tot}/K)}\}$ (which is identical to eqn. 5.6).
iii) In the presence of a large excess of ligand, $\alpha = 1$.

Figure 5.5A reports a set of ligand-binding isotherms calculated according to eqns. 5.11 through 5.13 at ligand concentrations between $K_d/1000$ and $1000 \, K_d$ and total

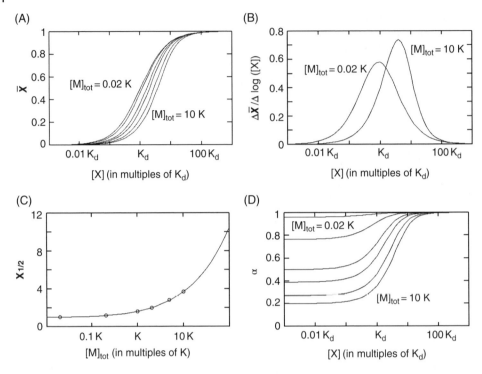

Figure 5.5 Effect of ligand concentration and total monomer concentration on a protein that forms homodimers only in the unliganded state (see text). The curves in panel A are calculated for $[M]_{tot}$ = 10 K, 5 K, 2 K, K, 0.2 K and 0.02 K, using eqns. 5.12 and 5.13; the corresponding degree of dissociation of the dimer, in the absence of the ligand, is α = 0.2, 0.27, 0.39, 0.5, 0.77, 0.96, whereas at saturating ligand concentration no homodimer is present. Panel B: derivatives of two isotherms from panel A, at $[M]_{tot}$ = 10 K and 0.02 K, highlighting the cooperativity observed at concentrations of $[M]_{tot}$ that exceed K. Panel C: dependence of $X_{1/2}$ on the protein concentration, calculated using eqn. 5.14; the points are the $X_{1/2}$ values of the isotherms reported in panel A. Panel D: dependence of the degree of dissociation α on ligand concentration, at constant $[M]_{tot}$ for the same experimental conditions as in the isotherms in panel A, calculated using eqn. 5.15.

monomer concentrations between 0.02 K and 10 K. Fig. 5.5B reports the derivative of the $\bar{X}$ versus $\log([X])$ isotherm for the two extreme protein concentrations, and Figures 5.5C and 5.5D report $X_{1/2}$ and α for the same set of binding isotherms as Figure 5.5A.

In spite of its simplicity, this system presents some interesting features. (i) An increase in total protein concentration reduces ligand affinity. This result is a consequence of the quaternary constraint, because the protein shifts from a condition in which the (ligand-binding) monomer is prevalent, both in the presence and in the absence of the ligand, to one where the (non-binding) homodimer is prevalent, at least in the absence of the ligand. (ii) The ligand promotes dissociation of the oligomer, and at $\bar{X} = 1$ the dimer has been completely dissociated into the constituent monomers. (iii) The binding process is cooperative, as demonstrated by the steep $\Delta \bar{X}$ versus $\log([X])$ plot (Figure 5.5B). The last point is perhaps the most surprising, since the chemical species capable of binding the ligand is a monomer and the homodimeric state in this oversimplified model does not bind the ligand at all.

Comparison with Figure 5.3 shows that the more marked cooperativity of this system (Figure 5.5B versus 5.3B) parallels the more marked ligand-dependence of the degree of dissociation α (Figure 5.5D versus 5.3D).

5.5 Monomers That Self-Associate to Homodimers Upon Ligation

The special case of homodimers whose constituent monomers have negligible ligand affinity is the opposite of the preceding one and occurs quite frequently in biology. For example, steroid hormone receptors are monomeric when unliganded and associate to homodimers upon ligation. For the sake of precision, we must add that the events of ligation and dimerization of steroid hormone receptors are coupled to other molecular events that will be ignored in the analysis presented here (Figure 5.6). The reaction scheme is as follows:

$$2\,M + 2\,X \Leftrightarrow 2\,MX \Leftrightarrow M_2X_2 \qquad \text{(eqn. 5.16)}$$

and again the system is described by just two equilibrium constants:

$$K = [MX]^2 / [M_2X_2]$$
$$K_d = [M][X]/[MX]$$

The binding polynomial is:

$$[M]_{tot} = [M] + [MX] + 2[M_2X_2] = [M](1+[X]/K_d) + 2[M]^2[X]^2/KK_d^2 \qquad \text{(eqn. 5.17)}$$

which we solve for [M]:

$$[M] = KK_d^2 \frac{-(1+[X]/K_d) + \sqrt{\{(1+[X]/K_d)^2 + 8[M]_{tot}[X]^2/KK_d^2\}}}{4[X]^2} \qquad \text{(eqn. 5.18)}$$

Using eqns. 5.17 and 5.18 we can calculate the fractional saturation:

$$\bar{X} = ([X]/K_d + 2[M][X]/KK_d^2)/(1+[X]/K_d + 2[M][X]^2/KK_d^2) \qquad \text{(eqn. 5.19)}$$

The fundamental properties of this system are as follows.

i) Ligand binding is strongly cooperative, unless the protein concentration is so low as to prevent self-association of the liganded monomers.
ii) Quaternary enhancement is demonstrated by the observation that ligand dissociation from the dimeric state, which would populate the singly liganded homodimer (M_2X), is necessarily followed by dissociation of the dimer.
iii) In contrast to the case considered in Section 5.4, ligand affinity increases with protein concentration (see Figures 5.5C and 5.6C).

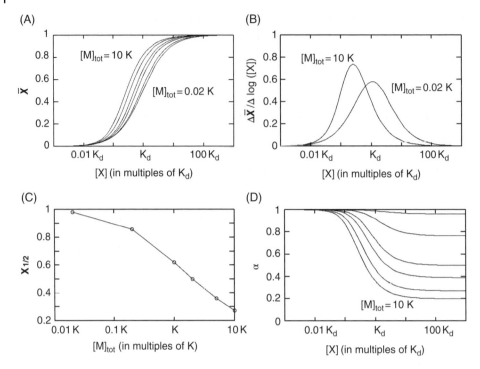

Figure 5.6 Effect of ligand concentration and total monomer concentration on a protein that forms homodimers only in the liganded state (see text). The curves in panel A are calculated for $[M]_{tot} = 10\,K$, $5\,K$, $2\,K$, K, $0.2\,K$, and $0.02\,K$, using eqns. 5.17, 5.18, and 5.19; the corresponding degree of dissociation of the dimer, in the presence of saturating concentrations of the ligand, is $\alpha = 0.2, 0.27, 0.39, 0.5, 0.77, 0.96$, whereas in the absence of the ligand no homodimer is present. Panel B: derivatives of two isotherms from panel A, at $[M]_{tot} = 10\,K$ and $0.02\,K$; cooperativity and affinity increase as the concentration of $[M]_{tot}$ increases. Panel C: dependence of $X_{1/2}$ on the protein concentration, as calculated from the isotherms reported in panel A. Panel D: dependence of the degree of dissociation α on ligand concentration, at the same $[M]_{tot}$ of the isotherms in panel A, calculated from the binding polynomial (eqn. 5.17).

5.6 Ligand-Linked Association-Dissociation in Cooperative Proteins

This is the most complex, but probably also the most informative, among the cases discussed in this chapter. The prototype example is provided by mammalian hemoglobins, which are cooperative tetramers capable of reversible dissociation into non-cooperative dimers. This definition applies to all vertebrate hemoglobins, with the caution that in some of these, for example, those from fishes, protein dissociation may be negligible at the concentrations required for determination of ligand affinity. We shall discuss in Chapter 7 some beautiful experiments carried out to explore the relationships between cooperativity and oligomerization in hemoglobin.

In this section, for the sake of consistency, we shall consider the simple case of a cooperative homodimeric protein that dissociates into monomers. In order to make our hypothetical model as general as possible, we do not make any assumption on the relationship between the ligand binding constants; thus, no simplification is possible and the equations we develop apply to every possible type of quaternary constraint or

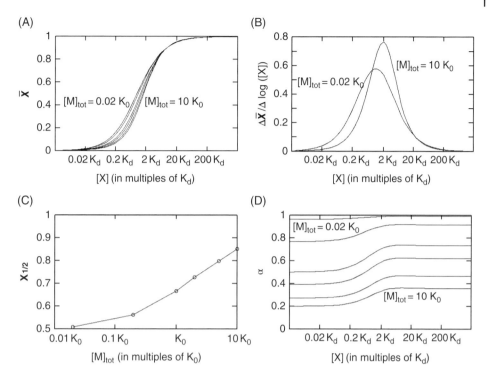

(A) (B) (C) (D)

Figure 5.7 Effect of ligand concentration and total monomer concentration on a cooperative homodimer that reversibly dissociates into monomers. The curves in panel A are calculated for $[M]_{tot} = 10\,K_0, 5\,K_0, 2\,K_0, K_0, 0.2\,K_0$ and $0.02\,K_0$, using eqns. 5.20, 5.21 and 5.22. The ligand dissociation constants are: $K_d=0.5$, $J_{d,1}=3$, $J_{d,2}=0.33$ in arbitrary units. In all the equations used in this plot the ligand dissociation constants appear in the dimensionless $[X]/K_d$ or $[X]/J_d$ ratio; thus it is unnecessary to give a definite concentration unit, provided that the same units are used for K_d, $J_{d,1}$, $J_{d,2}$; as a frame of reference the reader may assume the units to be μM. Panel B: derivatives of two isotherms from panel A, at $[M]_{tot} = 10\,K_0$ and $0.02\,K_0$; at the lowest protein concentration very little dimer is present and binding is essentially non-cooperative, whereas at the highest protein concentration significant cooperativity is present because the cooperative homodimer is the predominant species. Panel C: dependence of $\mathbf{X_{1/2}}$ on the protein concentration, as calculated from the isotherms reported in panel A. Panel D: dependence of the degree of dissociation α on ligand concentration, at the same $[M]_{tot}$ as the isotherms in panel A, calculated from the binding polynomial (eqn. 5.20).

quaternary enhancement (Figure 5.7). The equilibrium constants are defined as in Section 5.3, and we do not postulate identity of any of them, even though in the cases of pure quaternary constraint or quaternary enhancement one might be entitled to equate K_d to either $J_{d,2}$ or $J_{d,1}$, respectively:

$$J_{d,2} = [M_2X][X]/[M_2X_2]$$
$$J_{d,1} = [M_2][X]/[MX]$$
$$K_d = [M][X]/[MX]$$
$$K_0 = [M]^2/[M_2]$$
$$K_1 = [M][MX]/[M_2X] = K_0 J_{d,1}/2K_d$$
$$K_2 = [MX]^2/[M_2X_2] = K_0 J_{d,1} J_{d,2}/K_d^2$$

We solve the binding polynomial of this system as follows:

$$[M]_{tot} = [M](1+[X]/K_d)+2[M]^2(1+2[X]/J_{d,1}+[X]2/J_{d,1}J_{d,2})/K_0 \quad \text{(eqn. 5.20)}$$

$$[M] = \frac{-(1+[X]/K_d)+\sqrt{\left\{(1+[X]/K_d)^2+8[M]_{tot}\left(1+2[X]/J_{d,1}+[X]^2/J_{d,1}J_{d,2}\right)/K_0\right\}}}{4\left(1+2[X]/J_{d,1}+[X]^2/J_{d,1}J_{d,2}\right)/K_0}$$

$$\text{(eqn. 5.21)}$$

The reader may verify that eqns. 5.20 and 5.21 reduce to 5.6 and 5.7, respectively, if $J_{d,1}=J_{d,2}$ and to 5.4 and 5.5, respectively, if $J_{d,1}=J_{d,2}=K_d$.

$$\bar{X} = \frac{[M][X]/K_d+2[M]^2/K_0\left([X]/J_{d,1}+[X]^2/J_{d,1}J_{d,2}\right)}{[M](1+[X]/K_d)+2[M]^2/K_0\left(1+2[X]/J_{d,1}+[X]^2/J_{d,1}J_{d,2}\right)} \quad \text{(eqn. 5.22)}$$

An important consequence of the above equations is that one may infer the cooperativity of the oligomer from the dissociation constants of its ligation intermediates. Indeed in the above system it is straightforward to measure the dimer-monomer dissociation constants of the unliganded and fully liganded derivatives (K_0 and K_2), for example, by ultracentrifugation. The technique may actually estimate the average molecular weight in solution at different protein concentrations, from which the degree of dissociation can be calculated. The ratio K_2/K_0 yields $J_{d,1}J_{d,2}/K_d^2$, which equals unity in the absence of polysteric linkage, is greater than unity in the case of quaternary constraint, and is lower than unity in the case of quaternary enhancement.

If the system allows the researcher to measure the dimer-monomer dissociation constant of the singly liganded intermediate, the ratio $2K_1/K_0$ yields $J_{d,1}/K_d$. Therefore, the measurement of K_0, K_1 and K_2 allows one to derive the term $\varepsilon = J_{d,1}/J_{d,2}$ which we used in eqns. 4.17 and 4.18 to describe the cooperativity of the homodimer. We may recall that $\varepsilon > 1$ implies positive cooperativity while $\varepsilon < 1$ implies negative cooperativity. Unfortunately, the pure singly liganded dimer M_2X is populated only to a limited extent because it is in equilibrium with the unliganded and the doubly liganded dimers (M_2 and M_2X_2). Thus, the determination of its dissociation into monomers will be possible only in very special cases; an example of such special cases will be described in Chapter 7.

Special cases occur if the protein presents a pure quaternary enhancement ($K_d=J_{d,1}$) or pure quaternary constraint ($K_d=J_{d,2}$). In these cases either K_1 or K_2 is identical to the dissociation constant of the preceding step (i.e., either to K_0 or K_1), except for the statistical factor, in particular:

if $K_d=J_{d,1}$ (quaternary enhancement) $K_1 = \frac{1}{2}K_0$
if $K_d=J_{d,2}$ (quaternary enhancement) $K_2 = 2K_1$

The reader may verify that no similar condition applies to the case of the non-cooperative homodimer (Section 5.3), in which the ratio of each dissociation constant and the successive one equals J_d/K_d times the appropriate statistical factor.

Conversely, if the dissociation constant of the oligomer does not change as a consequence of the binding of one ligand, except for the amount corresponding to the statistical factor, this is proof that ligand binding to the oligomer occurs with the same affinity as ligand binding to the monomer.

5.7 One Ligand *Per* Dimer: Ligand-Binding Sites at Intersubunit Interfaces

A very interesting case of ligand-linked association-dissociation is provided by those proteins in which the ligand-binding site is located at the interface between two subunits. A typical example is provided by hemoglobin and 2,3 diphosphoglycerate (DPG), whose binding site is located at the interface between the two β subunits (Arnone, 1972). Other examples are provided by hormone receptors (e.g., Waters and Brooks, 2011). The reaction scheme is depicted in Figure 5.8.

This case has great biological relevance and outstanding theoretical interest. For the sake of clarity, we shall refer to the oligomeric protein as the receptor (R), and to the monomeric partner as the effector (E), because we want to explore the consequences of exchanging the roles of target and ligand between the two reaction partners. For example, in the case of hemoglobin and DPG the αβ dimer of hemoglobin would be indicated as R, and DPG as E; the hemoglobin tetramer would be indicated as R_2. Moreover, to make the case simpler and more compelling, we neglect the contribution of the dimeric, ligand-free protein to the binding polynomial, even though this species under physiological conditions is usually not negligible, and may be predominant.

From the viewpoint of ligand-binding theory this case can be treated as E binding two ligands (2 R), or alternatively as R that binds a ligand (E) and then dimerizes with a second molecule of unliganded R. Binding is cooperative, that is, $K_{d,2} < K_{d,1}$.

The binding polynomial of this system depends on the type of experiment one carries out, that is, whether E is the target (at constant concentration) titrated with R (at variable concentration) or vice versa. We shall consider and compare the two possibilities,

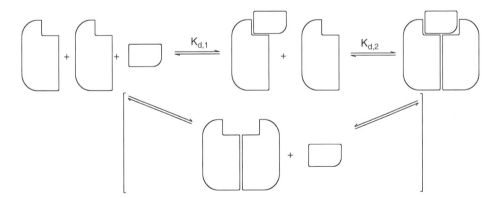

Figure 5.8 Dimerization of a protein induced by a ligand that binds at the intersubunit interface. The figure highlights the asymmetric structure of both the protein subunits and the ligand; by contrast, the homodimer is symmetric. The unbound dimeric protein (square brackets) is not considered in the treatment reported in the text.

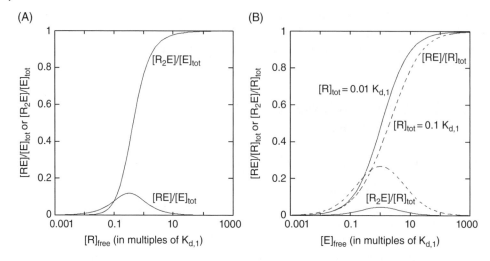

Figure 5.9 Binding of a single molecule of effector to a self-associating receptor (see text).Panel A: titration experiment of the effector (E) with excess receptor (R). Panel B: titration experiment of the receptor at two different concentrations ([R]$_{tot}$ as indicated in the panel) with the effector. The equilibrium constants are the same in the two experiments, with K$_{d,1}$=10 K$_{d,2}$. Note that the intermediate and final species are not the same in the two experiments: Panel A the intermediate species, populated at lower [R]$_{free}$ is RE and the final species, at high [R]$_{free}$ is R$_2$E, whereas the opposite occurs in panel B.

which highlight a difficulty anticipated in Section 1.1, namely that inverting the roles of the ligand and the target in a system with stoichiometry different from 1:1 does not lead to superimposable, or even similar binding, isotherms.

We schematize the reaction as follows:

$$E + 2R \Leftrightarrow RE + R \Leftrightarrow R_2E$$

and consider first an experiment in which E at fixed concentration is titrated with R. We assume that the following conditions apply: (i) cooperative binding of R to E (K$_{d,1}$>K$_{d,2}$); (ii) [E]$_{tot}$ << K$_{d,2}$; [R]$_{tot}$ >> K$_{d,2}$. The reader may appreciate that the above are the ideal conditions defined in Chapter 1, thus we are assuming a favorable and simple combination of parameters and experimental conditions.

We need not develop this system as it is fully described by eqns. 4.8 and 4.9. We only need to point out that: (i) the binding curves one obtains do not depend on the concentration of the target E; (ii) at low [R] species RE is populated; (iii) at high [R] species [R$_2$E] predominates and replaces both E and RE; (iv) the dependence of the fractional saturation of E, which we call $\bar{R}$, on [R]$_{free}$ is sigmoidal, as expected because of the cooperativity of the system; (v) in this system we shall consider, along with $\bar{R}$, the fractions [RE]/[E]$_{tot}$ and [R$_2$E]/[E]$_{tot}$, because these two species are likely to play different physiological roles (Figure 5.9A).

We now turn to the opposite experiment, in which R is the target whose concentration is kept constant, whereas [E] is varied. The equilibrium constants K$_{d,1}$ and K$_{d,2}$ are the same as in the preceding case, but the experimental conditions are: [R]$_{tot}$ << K$_{d,2}$; [E]$_{tot}$ >> K$_{d,2}$. It is likely that physiologists would prefer this experiment, and its analysis, over the preceding one, because in the organism the concentration of E usually varies to a much greater extent than that of R, and because the response of the target cell depends on the fraction of R that has been recruited, rather than on the fraction of E that has been bound. This applies not only to hormone-receptor complexes but also to the fraction of hemoglobin that is bound to DPG.

The behavior of this system in this experiment is much different from that in the preceding experiment because: (i) the binding curves one obtains depend on the concentration of the target R; (ii) at low [E] the predominant species is R_2E, provided that $[R]_{tot}$ is sufficiently high; (iii) at high [E] the predominant species is RE because it binds with higher stoichiometry, E:R being 1:2 for R_2E and 1:1 for RE; (iv) the binding curve exhibits negative cooperativity, as a consequence of the preceding point (iii).

The binding polynomial that describes this experiment, taking as a reference the concentration of the effector-free receptor [R], is as follows:

$$[R]_{tot} = [R] + [RE] + 2[R_2E] = [R](1 + [E]/K_{d,1}) + 2[R]^2[E]/K_{d,1}K_{d,2}$$

This binding polynomial is second degree with respect to the reference species, as are all the binding polynomials considered in this chapter. Thus the ligand binding isotherm depends on $[R]_{tot}$. This dependence may be made explicit as follows:

$$[R] = \frac{-(1+[E]/K_{d,1}) + \sqrt{\left\{(1+[E]/K_{d,1})^2 + 8[R]_{tot}[E]/K_{d,1}K_{d,2}\right\}}}{4[E]/K_{d,1}K_{d,2}}$$

We can now calculate the fractions $[RE]/[R]_{tot}$ and $[R_2E]/[R]_{tot}$ for any value of [E] and $[R]_{tot}$ (see Figure 5.9B).

We may comment on two apparently paradoxical results of the simulations reported in Figure 5.9, namely (i) the intermediate and final species populated in the two titrations differ; and (ii) titration of E with R yields a cooperative binding isotherm, whereas titration of R with E yields an anti-cooperative one. Both these results are consequences of the reaction stoichiometry, E:R=1:2. Indeed, point (i) is explained by the consideration that during a titration the first complex formed is that with the lowest ligand: target stoichiometric ratio, and the last complex formed is that with the highest stoichiometric ratio. Thus, when E is the target and R is the ligand, the first complex formed is RE (stoichiometric ratio 1:1) and the final one is R_2E (stoichiometric ratio 2:1). When R is the target and E the ligand, the first complex formed is R_2E (stoichiometric ratio 1:2), and the final one is RE (stoichiometric ratio 1:1). The explanation of point (ii) follows from the same reasoning. Revision of Hill's theory of cooperativity (Hill, 1913) in modern terms tells us that the Hill coefficient varies between 1 and the stoichiometric ratio. Thus, if one titrates E (=target) with R (=ligand) the Hill coefficient of the binding isotherm is $1<n<2$, that is, greater than unity. This is the hallmark of positive cooperativity. If, on the contrary, one titrates R (=target) with E (=ligand), the ligand:target ratio is 1:2 for the intermediate species R_2E and 1:1 for the end species RE. This leads to the Hill coefficient of the binding isotherm $0.5<n<1$, that is, lower than unity. This is the hallmark of negative cooperativity. In summary, if the stoichiometric ratio of a ligand:target complex is 2:1, inverting the roles of the target and the ligand also inverts the stoichiometric ratio to 1:2, and may change positive cooperativity, if present, into negative cooperativity.

5.8 Ligand-Linked Association-Dissociation in the Framework of the Allosteric Model

The MWC model was formulated under the hypothesis of perfect quaternary constraint (i.e., the isolated subunits are supposed to have the same ligand affinity as the high-affinity R state of the oligomer). Quaternary constraint is required because the unliganded subunits must form a more stable oligomer than liganded ones, in order to satisfy the condition $L_0 \gg 1$ (see Section 4.6). Even though the possible ligand-linked dissociation of the allosteric oligomer was not taken into explicit consideration in the original formulation of the model (Monod *et al.*, 1965), precise predictions can be derived, as pointed out by Edelstein and Edsall (1986). Once again we shall present the case for the homodimer, whose first ligation step is depicted below (Figure 5.10).

Ligand-linked dissociation within the MWC model is best described by taking into account separately the thermodynamic cycles it gives rise to.

i) A thermodynamic cycle is accessible to the unliganded protein: an R-state dimer might convert to T state and dissociate to monomers that reassociate to the R-state dimer. The free energy change of a cycle is zero, thus we derive the following relationship:

$$L_0 = {}^D K_R / {}^D K_{T,0}$$

that is, the allosteric constant equals the ratio of the dimer dissociation constants of the two states.

The same applies to the liganded states:

$$L_i = {}^D K_R / {}^D K_{T,i}$$

ii) A thermodynamic cycle is accessible to ligand binding within the R state, since an R-state protein molecule could bind the ligand, dissociate into monomers, the liganded monomers could release the ligand and the two unliganded monomers could reassociate into an unliganded R-state dimer. Under the assumption of pure quaternary constraint the ligand affinity of the monomer is the same as that of the R-state dimer and this, by necessity, implies that the equilibrium constant for the dissociation into monomers of the R-state dimer is independent of its ligation state,

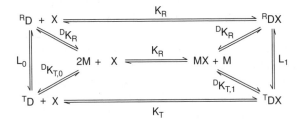

Figure 5.10 Ligand-linked dissociation within the framework of the allosteric two-state model. The protein is a homodimer, but only the first ligand-binding step is represented in the figure. The scheme assumes pure quaternary constraint, that is, the monomers have the same ligand affinity as the R-state dimer.

that is, we need not define $^D K_{R,0}$, $^D K_{R,1}$, and $^D K_{R,2}$, and we can make use of a single constant $^D K_R$ for all ligation states.

iii) A thermodynamic cycle equivalent to the one considered above is accessible to T-state dimers. In this case, however, the ligand equilibrium constant is not the same for the dimer and the monomer and we have:

$$^D K_{T,1} = {}^D K_{T,0} K_T / K_R$$

that is, the equilibrium constant for the dissociation of T-state dimers into monomers increases with ligation (because $K_T > K_R$).

We can generalize this finding and write:

$$^D K_{T,i} = {}^D K_{T,0} \left(K_T / K_R \right)^i$$

The above points suggest the following relationships:

$$L_i = {}^D K_R / {}^D K_{T,i} = {}^D K_R / {}^D K_{T,0} \left(K_T / K_R \right)^i = L_0 \left(K_R / K_T \right)^i$$

The above relationships fully describe the phenomenon of ligand-linked dissociation within the framework of the two-state model, and allow one to predict the relative population of monomers at any level of ligand saturation and protein concentration.

Unfortunately, the experimental determination of the above parameters may be quite difficult. A considerable simplification is obtained under the assumption that in the absence of the ligand the population of the R state is negligible (because $L_0 \gg 1$). This implies that the oligomer to monomer dissociation constant measured under these conditions approximates $^D K_{T,0}$. If the opposite assumption can also be made (i.e., that at saturating concentrations of the ligand the population of the T state is negligible), one can approximate the dissociation constant measured under these conditions to $^D K_R$. From these two parameters the value of L_0 can be calculated. In some cases, experimental approaches may be devised to determine the dissociation constants of ligation intermediates (see Chapter 7) that would allow one to estimate also the ratio K_T / K_R.

5.9 Practical Considerations

Ligand-linked association-dissociation should be considered when dependence of the ligand affinity on protein concentration is observed. Two recommendations follow: (i) More trivial causes of such a dependence should be ruled out. For example, if the explored range of protein concentration includes conditions in which $[P]_{tot} \simeq K_d$ and the researcher has not adopted the correction for $[X]_{free} \neq [X]_{tot}$, as described in Section 1.2, the total concentration of X required to half saturate the protein will depend on protein concentration. Notice that in this case the true $X_{1/2}$, defined as the free concentration of X required to half saturate the protein, is independent of protein concentration, but the approximation $[X]_{tot} \approx [X]$ fails. (ii) Direct evidence should be sought for dissociation of the protein at the concentration used, for example, a measurement of the molecular weight in solution is carried out by analytical ultracentrifugation, or a fluorescence signal due to the burial or exposure of an interface aromatic residue is detected.

If ligand-linked association-dissociation is present, it is important to explore a range of protein concentrations, and to formulate a plausible reaction scheme, before attempting a quantitative analysis of the experimental data. Indeed, as demonstrated above, polysteric linkage causes cooperativity, and the ligand-binding isotherms cannot be fitted by simple hyperbolas even though the isolated monomers and the oligomers may both be non-cooperative. Simulating the behavior of the system at one protein concentration using an equation that empirically accounts for cooperativity (e.g., eqns. 4.8 and 4.9) yields meaningless results and leads to misleading conclusions. This emphasizes the importance of collecting as many relevant data as possible (e.g., on the aggregation state of the oligomer in the presence and in the absence of ligand), to formulate the correct binding model, and to derive the appropriate equations.

Whenever possible it is strongly advisable to explore a protein concentration range wide enough to cover conditions in which the fraction of the oligomer is negligible and conditions in which the fraction of the monomer is negligible, to verify that the monomer obeys the equations developed in Chapter 1 and the oligomer obeys those developed in Chapter 4.

References

Ackers, G.K., Johnson, M.L., Mills, F.C., Halvorson, H.R. and Shapiro, S. (1975) The linkage between oxygenation and subunit dissociation in human hemoglobin. Consequences for the analysis of oxygenation curves. Biochemistry, 14: 5128–5134.

Andersen, M.E. (1971) Sedimentation equilibrium experiments on the self-association of hemoglobin from the lamprey Petromyzon marinus. A model for oxygen transport in the lamprey. J Biol Chem, 246: 4800–4806.

Arnone, A. (1972) X-ray diffraction study of binding of 2,3-diphosphoglycerate to human deoxyhaemoglobin. Nature, 237: 146–149.

Briehl, R.W. (1963) The relation between the oxygen equilibrium and aggregation of subunits in Lamprey hemoglobin. J Biol Chem, 238: 2361–2366.

Colosimo, A., Brunori, M. and Wyman J. (1976) Polysteric linkage. J Mol Biol, 100: 47–57.

Coryell, C.D. (1939) The existence of chemical interactions between the hemes in ferrihemoglobin (methemoglobin) and the role of interactions in the interpretation of ferro-ferrihemoglobin electrode potential measurements. J Phys Chem, 43: 841–852.

Edelstein, S.J. and Edsall, J.T. (1986) Linkage between ligand binding and the dimer-tetramer equilibrium in the Monod-Wyman-Changeux model of hemoglobin. Proc Natl Acad Sci USA, 83: 3796–3800.

Germain, P. and Bourguet, W. (2013) Dimerization of nuclear receptors. Methods Cell Biol, 117: 21–41.

Hill, A.V. (1913) XLVII. The combinations of haemoglobin with oxygen and with carbon monoxide. Biochem J, 7: 471–480.

Imai, K. (1982) Allosteric effects in hemoglobin. Cambridge, UK: Cambridge University Press.

Kovacs, E., Zorn, J.A., Huang, Y., Barros, T. and Kuriyan, J. (2015) A structural perspective on the regulation of the epidermal growth factor receptor. Annu Rev Biochem, 84: 739–764.

Kuiper, H.A., Antonini, E. and Brunori, M. (1977) Kinetic control of co-operativity in the oxygen binding of *Panulirus interruptus* hemocyanin. J Mol Biol, 116: 569–576.

Monod J., Wyman, J. and Changeux, J.P. (1965) On the nature of allosteric transitions: a plausible model. J Mol Biol, 12: 88–118.

Pauling, L. (1935) The oxygen equilibrium of hemoglobin and its structural interpretation. Proc Natl Acad Sci USA, 21: 186–191.

Savel-Niemann, A., Markl, J. and Linzen B. (1988) Hemocyanins in spiders. XXII. Range of allosteric interaction in a four-hexamer hemocyanin. Co-operativity and Bohr effect in dissociation intermediates. J Mol Biol, 204: 385–395.

Ward, C.W., Lawrence, M.C., Streltsov, V.A., Adams, T.E. and McKern, NM. (2007) The insulin and EGF receptor structures: new insights into ligand-induced receptor activation. Trends Biochem Sci, 32: 129–137.

Waters, M.J. and Brooks, A.J. (2011) Growth hormone receptor: structure function relationships. Horm Res Paediatr, 76(Suppl 1): 12–16.

Weber G. (1992) Protein Interactions. Chapman and Hall.

Wyman J. (1948) Heme proteins. Adv Protein Chem, 4: 407–531.

6
Kinetics of Ligand Binding to Proteins with Multiple Binding Sites

Kinetic experiments may be of particular relevance for proteins with multiple binding sites, because ligation intermediates that would be scarcely populated under equilibrium conditions can be significantly populated in a carefully designed kinetic experiment. A typical example is that of partially liganded species in cooperative proteins. However, kinetic experiments may prove not unequivocal and difficult to interpret, unless additional information from equilibrium experiments and structure determination is available. Thus, kinetic experiments require careful planning and extensive comparison with other information. Nevertheless, in favorable cases their information content can add substantially to the understanding of complex systems.

In this chapter we shall be concerned with the interpretation of stepwise ligand binding to and release from oligomeric proteins using the homodimer as a theoretical model. We shall next consider what type of information can be provided by the available experimental methods and instrumentation. The most versatile instrument to study the time courses of ligand binding is the stopped-flow apparatus described in Section 2.5. However, in the case of multiple site proteins special techniques may prove more effective to populate ligation intermediates: ligand pulse methods, photochemical methods and T jump (described in Section 2.6). Finally, we shall consider the kinetic aspects of cooperativity, especially, but not only, in view of the two-state allosteric model.

In many cases the overall kinetic schemes to be adopted for a quantitative description of the experimental results require numerical integration; this is not a problem with present-day personal computers, but requires the reader to be familiar with some software capable of performing this task. The simulations presented in this book have been mostly carried out using the *lsode* function of the standard *Octave* package, which, besides being open source, has the advantage of running on several operating systems.

6.1 Stepwise Ligand Binding to Homooligomeric Proteins

Ligand binding to, and release from, multiple-site proteins occurs stepwise, in a sequence of single ligand reactions. Each ligand-binding step may occur in a simple second-order process, or in a more complex sequence in which the initial second order event is followed by a first-order internal rearrangement. In the latter case the order of the reaction will depend upon which of these steps is rate-limiting. Ligand dissociation usually occurs in a single first-order step. In many cases, it is possible to devise experimental conditions in which either the association or dissociation event can be neglected, by arranging the experimental

Reversible Ligand Binding: Theory and Experiment, First Edition. Andrea Bellelli and Jannette Carey.
© 2018 John Wiley & Sons Ltd. Published 2018 by John Wiley & Sons Ltd.

conditions to overwhelmingly populate either the fully liganded or the fully unliganded protein. We shall first deal with these conditions to describe the experimental approach and data analysis. We shall deal next with experiments leading to complex mixtures, with the *proviso* that a full analysis of the latter requires more complex numerical procedures, and software capable of numerical integration of the kinetic differential equations.

Let us start with the simplest possible case, that of the ligand association with a homodimeric protein in a reaction explored under experimental conditions where ligand dissociation may be neglected. The reaction scheme is:

$$P + 2X \rightarrow PX + X \rightarrow\!\!\!> PX_2$$

In this reaction scheme we have two sequential second-order binding processes that may be approximated to pseudo-first order processes if $[X]_{tot} \gg [P]_{tot}$ (see Section 2.1). The kinetic differential equations, as written for the apparent pseudo-first order rate constants, are:

$$\delta[P]/\delta t = -[P]k_{a,1}'$$
$$\delta[PX]/\delta t = [P]k_{a,1}' - [PX]k_{a,2}'$$
$$\delta[PX_2]/\delta t = [PX]k_{a,2}'$$

We follow the same notation adopted in Chapter 2. Thus, $k_{a,i}$ is the microscopic second-order association rate constant of the ith reaction step and $k_{a,i}'$ is its apparent pseudo-first order equivalent. In the case of a symmetric homodimer made up of identical subunits, the apparent kinetic rate constants include the appropriate statistical factors, which equal the number of reacting sites, that is, $k_{a,1}' = 2k_{a,1}[X]$ and $k_{a,2}' = k_{a,2}[X]$. The analytical integration of a series of exponential decays was developed by H. Bateman (1910) to describe the decay of radioactive isotopes. Bateman's equations can be applied to a binding experiment carried out under pseudo-first order conditions, and can be simplified by introducing the assumptions $[P]_{t=0}=[P]_{tot}$, $[PX]_{t=0}=0$, $[PX_2]_{t=0}=0$, as they are reasonable for a rapid mixing experiment (Section 2.5). We obtain:

$$[P]_t = [P]_{tot} e^{-k_{a,1}'t} \qquad\qquad\qquad (eqn.\ 6.1)$$

$$[PX]_t = [P]_{tot} k_{a,1}' \left\{ e^{-k_{a,1}'t} / \left(k_{a,2}' - k_{a,1}' \right) + e^{-k_{a,2}'t} / \left(k_{a,1}' - k_{a,2}' \right) \right\} \qquad (eqn.\ 6.2)$$

$$[PX_2]_t = [P]_{tot} - [P]_t - [PX]_t \qquad\qquad\qquad (eqn.\ 6.3)$$

A simulated time course of ligand association, calculated using eqns. 6.1 through 6.3 under the assumption that no homotropic interactions are present (i.e., $k_{a,1} = k_{a,2}$), is shown in Figure 6.1. It can be easily demonstrated that in this case the time course of ligation of the binding sites is exponential if one does not consider the distribution of the liganded subunit among the dimers (see the time course marked with asterisks in Figure 6.1; see also Appendix 6.1).

Homotropic interactions between binding sites of the homodimer (i.e., positive or negative cooperativity) yield deviations of the time course of ligand saturation from a simple exponential, and may take the form of a decelerating or accelerating (autocatalytic) time course (Figure 6.2). Because the equilibrium constants equal the ratio of the

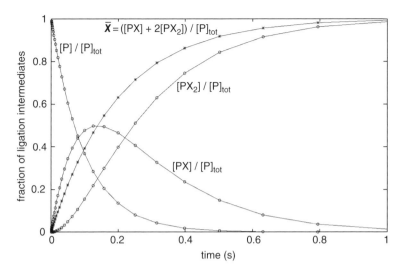

Figure 6.1 Simulated time courses of ligation intermediates and fractional ligand saturation for the non-cooperative homodimer.

Rate constants: $k_{a,1}[X] = k_{a,2}[X] = 5\,s^{-1}$ ($k_{a,1}' = 10\,s^{-1}$, $k_{a,2}' = 5\,s^{-1}$); initial conditions: $[P] = [P]_{tot}$; $[PX] = 0$; $[PX_2] = 0$.

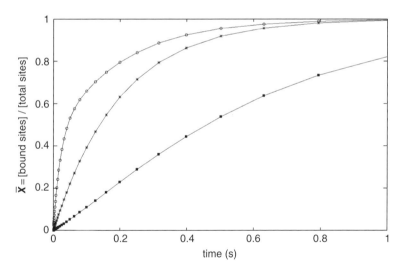

Figure 6.2 Simulations of the time courses of ligation to three homodimeric proteins.

Asterisks: the two sites of the homodimer are non cooperative, with $k_{a,1}' = 10\,s^{-1}$ and $k_{a,2}' = 5\,s^{-1}$ (i.e., the microscopic second-order rate constants of the unliganded and singly liganded species are the same and the difference between the apparent rate constants is introduced by the statistical factors); the time course is exponential.

Circles: the two sites of the homodimer are negatively cooperative, with $k_{a,1}' = 50\,s^{-1}$ (microscopic value $25\,s^{-1}$) and $k_{a,2}' = 5\,s^{-1}$. The five-fold lower value of the microscopic kinetic rate constant of the second site causes the time course to decelerate progressively as ligation proceeds. This time course is indistinguishable from a two-exponential process, but it was obtained using Bateman's equation, rather than the sum of two exponentials.

Squares: the two sites of the homodimer are positively cooperative, with $k_{a,1}' = 2\,s^{-1}$ (microscopic value $1\,s^{-1}$) and $k_{a,2}' = 5\,s^{-1}$. The five-fold higher microscopic kinetic rate constant of the second site causes the time course to accelerate progressively as ligation proceeds, which represents the kinetic counterpart of equilibrium cooperativity. This feature is called, somewhat improperly, autocatalysis.

dissociation and association kinetic constants, there is a necessary relationship between equilibrium cooperativity and kinetic autocatalysis. Thus, positive equilibrium cooperativity, that is, greater ligand affinity for the second binding site, must be associated with the second ligand-binding event occurring with a greater association rate constant, or lower dissociation rate constant, or both, than the first ligand-binding event. The same applies to negative cooperativity and a decelerating ligand association time course. Although the presence of autocatalysis in the time course of ligand association or dissociation strongly suggests equilibrium cooperativity, the opposite finding (i.e., absence of autocatalysis) should be interpreted with caution, because absence of cooperativity is not the only possible explanation of absence of autocatalysis. For example, in an allosteric cooperative protein the structure changes associated to equilibrium cooperativity may be slower than ligand binding and release, thus preventing the appearance of autocatalysis (see also Section 6.5, below).

6.2 Ligand Association to Heterooligomeric Proteins

The case of a heterodimeric protein is different from that of a homodimeric protein unless the two different subunits may be considered functionally equivalent, in which case their structural differences are neglected (see next chapter for an example). If each subunit binds independently of the other, with its own kinetic rate constant, and there are no cooperative interactions, the overall time course is the sum of two exponentials (see Figure 6.3). The population of ligation intermediates (which in this case are four in number rather than three as in the preceding case) can be derived from purely statistical

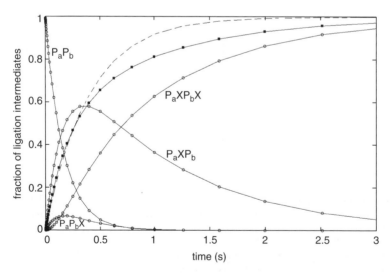

Figure 6.3 Simulated time course of ligand binding to a heterodimeric protein P_aP_b.
   Pseudo-first order rate constants used in the simulation are: $k_a'=5\,s^{-1}$ and $k_b'=1\,s^{-1}$. Circles identify the simulated time course of each indicated reaction intermediate. Squares identify the overall time course of ligation $\bar{X}_t = \left([P_aXP_b] + [P_aP_bX] + 2[P_aXP_bX] \right)/2[P_aP_b]_{tot}$. The dashed line is an exponential with the same half time as the overall time course of ligation.

considerations, provided that the two sites are effectively independent of each other. Moreover, Bateman's equations do not apply to this case, and no statistical factor is required.

The fractional saturation of each type of subunit, here referred to as P_a and P_b, can be calculated independently of its distribution among the heterodimers, as follows:

$$[P_aX]_t / [P_a]_{tot} = 1 - e^{-ka't}$$

$$[P_bX]_t / [P_b]_{tot} = 1 - e^{-kb't}$$

The fraction of liganded sites at time t is:

$$\bar{X}_t = \left([P_aX]_t / [P_a]_{tot} + [P_bX]_t / [P_b]_{tot} \right) / 2$$

The fraction of ligation intermediates is statistical and can be calculated by multiplication of the fractions of liganded or unliganded subunits (see also Figure 6.3):

$$[P_aP_b]_t / [P_aP_b]_{tot} = \left(e^{-ka't} \right)\left(e^{-kb't} \right)$$
$$[P_aXP_b]_t / [P_aP_b]_{tot} = \left(1 - e^{-ka't} \right)\left(e^{-kb't} \right)$$
$$[P_aP_bX]_t / [P_aP_b]_{tot} = \left(e^{-ka't} \right)\left(1 - e^{-kb't} \right)$$
$$[P_aXP_bX]_t / [P_aP_b]_{tot} = \left(1 - e^{-ka't} \right)\left(1 - e^{-kb't} \right)$$

6.3 Study of the Time Course of Ligand Dissociation

As already discussed in Chapter 2, several methods are available to record the time course of dissociation of a protein-ligand complex. All methods can be applied to oligomeric proteins essentially in the same manner as they are applied to monomeric ones, but the interpretation of the data requires some caution. Most of these methods were pioneered on oxy-hemoglobin or carbon monoxy-hemoglobin, and we shall often use these systems as privileged examples.

The main difficulty in the study of the kinetics of ligand dissociation from multi-site proteins is that autocatalysis, if present, is a subtler feature than equilibrium cooperativity, which may be difficult to appreciate if the experiment is not properly designed. An important requirement for the study of autocatalytic reactions is that the reaction is followed over an extended interval of ligand saturation, ideally from 100% to zero, because the slow initial process may cover only a small fraction of the signal change. While the design of an experiment aimed at following the ligand association in the stopped flow is usually straightforward, except for the case of rapidly binding, low-affinity ligands, the design of a meaningful ligand dissociation experiment may be complex. We shall consider the usefulness and limits of experimental designs based either on (i) extrapolation from association experiments and dilution; (ii) rapid chemical destruction of the free ligand; (iii) competition with another ligand (or protein); or (iv) ligand pulse methods. In the following section, we shall discuss in deeper detail the challenge inherent in the study of the kinetics of autocatalytic reactions.

Rapid *dilution of a protein-ligand mixture*, usually in a stopped-flow apparatus, is the simplest method to induce ligand dissociation. It can be used with good results in the case of non-cooperative proteins: for example, Prinz and Maelicke (1983) used dilution and competition to induce dissociation of acetylcholine analogues from acetylcholine receptor. The dilution method, however, has severe limitations. First, and foremost, the maximum dilution factor one can realize in a stopped flow apparatus is 1:5 or 1:10, thus the extent of the observable reaction is limited. Even if the reaction is so slow that a stopped flow is unnecessary, there are practical limits to the extent of dilution one may practically achieve. In optimal cases one may expect to start from a ligand saturation of approximately 80% before mixing and to reach an end point of approximately 20%. The experiment is best designed based on an accurately determined ligand-binding isotherm, and we refer the reader to Figure 4.5 for a visual estimate of the effect of a 10-fold dilution of the free ligand. Unfortunately, the ligand-saturation change one may obtain with this method is unlikely to be suitable for appreciating autocatalysis, if present. An improvement of the experimental design is obtained if one explores a widely spaced set of initial ligand saturations. The second limit of the dilution experiment is that the time course one records monitors the approach to an equilibrium condition in which a significant amount of free ligand is present. As a consequence, the apparent rate constant is the sum of the (pseudo-first order) association rate constant plus the dissociation rate constant (see eqn. 2.11 in Section 2.2). This may not be a disadvantage, but points out that the information one obtains from a dilution experiment is conceptually similar to that obtained from an association experiment carried out under similar conditions. Needless to say, if autocatalysis is present, estimating the dissociation rate constant from regression of a series of binding or dilution experiment, as shown in Figure 2.3 for single binding site proteins, is subject to large errors. Finally, dilution experiments are unsuitable for high-affinity ligands, because in these cases the ligand released by the complex raises significantly the concentration of the free ligand and severely reduces the saturation change one can effectively monitor.

The method of *rapid ligand removal* is conceptually straightforward: the researcher uses a stopped-flow apparatus to mix the protein ligand complex with a chemical reagent or enzyme capable of rapidly destroying the free ligand and follows the time course of (irreversible) ligand release from the protein under conditions in which the free-ligand concentration in solution equals zero. Unfortunately, not all ligands can be rapidly destroyed, thus this method is not widely applicable. It is highly desirable that the rate of ligand destruction is much higher than that of its release from the protein-ligand complex, otherwise destruction of the released ligand will compete with rebinding, complicating the interpretation of the experimental data. In some cases rapid removal of the ligand may be achieved by means of enzymes that use the ligand as a substrate (provided that the reaction product is sufficiently modified not to bind to the protein). Important properties of this method are: (i) the whole reaction from almost 100% saturation down to zero can be explored; and (ii) the end point of the time course is not an equilibrium mixture of the protein and the ligand, thus the ligand association reaction does not contribute, and the time course is governed only by k_d. Because of the above properties, this method is most suitable to reveal autocatalysis, if present.

A typical example is that of oxy-hemoglobin (Hb), where the reagent of choice is sodium dithionite, a reductant that reduces dioxygen to hydrogen peroxide first and then to water, and also reduces any trace of ferric Hb that may form during the reaction to ferrous Hb. Dithionite may not be an ideal reagent, because of its several oxidation products (sulfur oxides). However, it reacts rapidly with dioxygen (the most effective reducing species being its dissociation product SO_2^- (Lambeth and Palmer, 1973), so that the free gas concentration in solution during the time course of the reaction is essentially zero. Thus, if the dithionite concentration used is high enough (10-50 mM), oxygen rebinding to hemoglobin is negligible. These characteristics are important because they greatly simplify the interpretation of the time courses. Oxygen dissociation from oxy-hemoglobin is autocatalytic, as one may expect given that the equilibrium isotherm is cooperative. Indeed, in the case of oxy-hemoglobin cooperativity stems mostly from a difference in the dissociation rate constants of the first and last molecule of oxygen, the association rate constants being much less sensitive to the presence or absence of other bound ligands.

Ligand competition is a very effective method for studying ligand release. As a general rule it is much easier to find a competing ligand than a reagent capable of destroying the ligand. For example, in the case of oxy-hemoglobin the ideal reagent to use is CO, which has higher affinity than oxygen. However, the kinetics of ligand replacement are complex (as already discussed in Section 2.6) because their apparent rate constant depends on all four kinetic rate constants involved, that is, those of association and dissociation of both ligands. Thus, several time courses should be collected, in which the concentrations of both ligands are systematically varied. Moreover, the complexes of the two ligands should provide different signals (e.g., they should have different absorbance, fluorescence, or other real-time detectable features).

A further property of the ligand replacement method, when applied to oligomeric proteins with multiple ligand-binding sites, is that the protein replaces one ligand at a time, thus, one explores the dissociation rate constants of the last site only. For example, the reaction of oxygen replacement by CO in hemoglobin obeys the following kinetic scheme:

$$HbX_3O_2 + CO \Leftrightarrow HbX_3 + CO + O_2 \Leftrightarrow HbX_3CO + O_2$$

where HbX_3 represents any triply liganded species, irrespective of each molecule of X being O_2 or CO. The reaction scheme shows that during the time course no partially liganded species is populated, the tetramer always having the general formula HbX_3 or HbX_4. Thus, in a replacement experiment only the rate constants of association and dissociation of the nth site of an n-subunit oligomer are tested, and no information can be collected on the rate constants for partially liganded intermediates. For example, the replacement of oxygen with CO in hemoglobin is non-autocatalytic. One should not view this property of replacement reactions as a limit, though. Rather, ligand replacement is the method of choice to determine the dissociation rate constants of the last site of the macromolecule (e.g., in hemoglobin it can be used to determine $k_{d,4}$). A more complex behavior can be anticipated in the uncommon case that the two ligands are non-equivalent with respect to the cooperativity of the protein, as occurs in the case of those hemocyanins that bind oxygen cooperatively and CO non-cooperatively (see Section 4.7).

Competition between two proteins for the same ligand is an approach that can be applied to few protein-ligand complexes, as it requires not only two proteins capable of binding the same ligand (which is easy to find), but also that the two binding processes yield different signals upon ligation of the same ligand (which is quite uncommon).

For example, although we have a host of different hemoglobins, from different animals, we cannot detect one competing with another for oxygen or CO, because the spectroscopic properties of their ligand complexes are extremely similar. V.S. Sharma demonstrated long ago that the heme-containing polypeptide microperoxidase, obtained by proteolytic digestion of horse heart cytochrome-c, binds CO with greater affinity than hemoglobin (Sharma *et al.*, 1975, 1976). The absorbance spectrum of the CO complex of microperoxidase is sufficiently different from that of hemoglobin to allow the reaction to be followed by spectroscopy in a stopped-flow apparatus, but the method is far from widely applicable.

Ligand pulse methods are quite informative, but can be applied only to a subset of the cases in which rapid ligand removal is possible. Ligand binding pulse was pioneered by Q.H. Gibson (Gibson, 1973), who mixed in a stopped flow a solution of deoxy-hemoglobin and dithionite with a solution of oxygen. Because oxygen reacts with hemoglobin at a faster rate than it is destroyed by dithionite, oxy-hemoglobin is formed to a variable extent during the dead time of the mixing apparatus, and its dissociation may be followed. The very special advantage of this method is that one can arbitrarily vary the amount of complex formed to study the ligand dissociation of virtually any mixture of ligation intermediates, thus overcoming the limits of kinetic methods in detecting cooperativity (Figure 6.4).

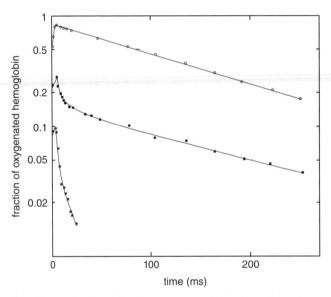

Figure 6.4 Time courses of oxygen dissociation from hemoglobin, as a function of inital saturation (oxygen pulse). Experimental conditions: observation wavelength 438 nm; T=20 °C; 0.05 M phosphate buffer pH=7; hemoglobin concentration 17 μM (after mixing); oxygen concentrations, after mixing: 150 μM (circles); 63.5 μM (square); 26.6 μM (asterisks); dithionite concentration (before mixing) 0.1%. In an oxygen pulse experiment a solution of deoxy hemoglobin containing sodium dithionite is rapidly mixed with an oxygen containing solution; the protein transiently forms an oxygen complex, which subsequently dissociates because the gas is consumed by excess dithionite. The experiment demonstrates that the higher the fraction of oxyhemoglobin formed, the slower the dissociation of the complex. This is a consequence of cooperativity: at high saturation the high affinity, slowly dissociating state of the complex is prevalent, whereas at low saturation the low affinity, rapidly dissociating state predominates. Data from Gibson (1973).

6.4   Practical Problems in the Study of Ligand-Binding Kinetics with Oligomeric Proteins

The study of the kinetic properties of oligomeric proteins is intrinsically complex, and as a general rule comparison of multiple experiments is required to draw any significant conclusion (see below). Before considering the relevance of unifying kinetic and equilibrium data, we shall dedicate a few words to the analysis of some practical difficulties and limitations of this type of study.

The complexity, in our view, resides in the kinetic consequences of homotropic interactions, namely kinetic cooperativity (autocatalysis). In general terms, it is obvious that equilibrium cooperativity must have a kinetic counterpart. This stems from the consideration that, the equilibrium constants being the ratio of the kinetic rate constants of the forward and backward reactions, the change in the equilibrium constant, from K_1 to K_n or from K_T to K_R, must be reflected in a change of either or both the association or the dissociation kinetic constant. Unfortunately, ligation-dependent changes in the apparent kinetic rate constants are less informative than changes in the equilibrium constants, and more difficult to measure. The reason for this apparent paradox is evident from Bateman's integrated equations (eqns. 6.1 through 6.3), which show that no ligation step successive to the first one depends on a single kinetic rate constant; rather, the apparent rate constants of all successive steps are functions of the preceding rate constants. Thus if in eqn. 6.1 through 6.3 we imagine that $k_{a,2}$ is infinitely larger than $k_{a,1}$, its contribution tends to cancel out, and the overall time course approaches an exponential with rate constant equal to $2k_{a,1}$ (or n times $k_{a,1}$ in the case of an oligomer of n subunits; see Figure 6.5). As a consequence, in a ligand association experiment initiated by rapid mixing no information can be collected about the difference in the rate constant between the first and last ligation steps. This is at variance with equilibrium experiments, where one may measure, in spite of some uncertainty, the difference in affinity between the first and the last binding site. Moreover, in an equilibrium experiment, the derivative of the $\bar{X}$ versus log [X] plot (or the slope of the Hill plot) gives an indication about the minimum number of interacting sites, whereas kinetic autocatalysis provides virtually no information on this point except that the protein must be at least dimeric, the effective observable increase in the kinetic rate constant due to autocatalysis being much smaller than the number of sites.

6.5   Advanced Techniques for the Study of Ligation Intermediates

The study of ligand association and dissociation kinetics by means of rapid mixing methods, with the exception of the ligand-pulse experiment, is intrinsically limited by the fact that in sequentially ordered kinetic processes, each step is rate limited by the preceding ones (Section 6.5). No similar limitation occurs in equilibrium studies where the apparent affinity of the first binding site does not limit that of the successive sites. In other words, the two kinetic rate constants required to describe an autocatalytic time course of ligand binding are limited between k_1 and nk_1, whereas the asymptotes of the Hill plot (whatever their experimental error) are determined by K_1 and K_n.

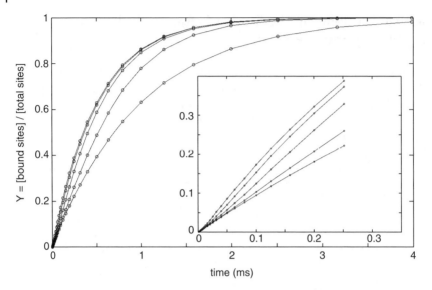

Figure 6.5 Simulated time courses of ligand binding to a cooperative homodimer: extent of autocatalysis.

Pseudo-first order apparent rate constants (including the statistical factors): $k_{a,1}' = 2\,s^{-1}$ throughout; $k_{a,2}' = 1,3,10,30$, and $100\,s^{-1}$ (from the slowest to the fastest time course, respectively). The simulation shows how kinetic cooperativity tends to plateau as the difference between $k_{a,1}'$ and $k_{a,2}'$ increases. The slowest time course is exponential (absence of kinetic cooperativity); faster time courses are autocatalytic until the fastest one again approaches an exponential with apparent overall rate constant twice that for the first binding site (see text for an explanation). Notice that in the fastest time course the microscopic rate constant of the second site is 100 times greater than that of the first site.

Methods exist that overcome this limitation of kinetic studies and provide direct experimental access to the kinetics of successive binding events. We shall consider in this section two such methods, namely photolysis and T jump. These methods have very stringent requirements and, as a general rule, they complement, but do not replace, rapid mixing experiments.

Photochemical methods (Gibson, 1959), already presented in Chapter 2, if applicable to the system of interest, may provide very significant information for multiple binding sites proteins.

The power of photochemical methods lies in the fact that by varying the intensity of the photolytic pulse one can achieve virtually any level of photodissociation, from 100% down to the practical detection limit. Thus there is no difficulty in isolating and studying (mostly) the final binding step. If we neglect, for the present time, the problem offered by geminate rebinding (discussed in Section 2.5), the distribution of photochemical intermediates is statistical (i.e., binomial). If the intensity of the photodissociating pulse and the photochemical yield of the protein-ligand complex are such that a fraction F of the total bound ligand is removed, and the fraction of liganded sites remaining is $F' = 1-F$, then the distribution of the ligand among the protein molecules will be binomial, that is:

$$\left(F + F'\right)^n$$

where n represents the number of binding sites of the macromolecule. For example, if the protein is a homodimer, the development of the above binomial distribution yields the following population of the photochemical intermediates: F^2=fraction of fully unliganded dimers; $2FF'$=fraction of singly liganded dimers; F'^2=fraction of fully liganded dimers. After the photolysis pulse is switched off, ligand rebinding to the unliganded sites occurs, and one can vary the relative population of singly and fully unliganded intermediates by varying the intensity of the pulse. In principle, the ligand recombination time course of the fully unliganded dimer should be superimposable with that recorded by rapid mixing, provided that the light-excited unliganded state fully relaxes to ground state before rebinding. In contrast, the time course of recombination of the singly liganded dimer should be determined solely by $k_{a,2}$, unencumbered by the contribution of $k_{a,1}$ observed in the sequential reaction scheme (eqn. 6.2). Figure 6.6 shows a simulation of this experiment for one of the homodimers from Figure 6.5.

The reader may wonder why we make in this book such an extensive usage of simulated, rather than real experimental data: can we not reproduce the actual time courses of hemoglobin and CO? We would have liked to do so, but unfortunately, real-world experiments are too complex to constitute valid pedagogical examples, at least unless a simulated simple case has been analyzed in detail first. However, we discuss the real case of hemoglobin and related hemoproteins in the next chapter.

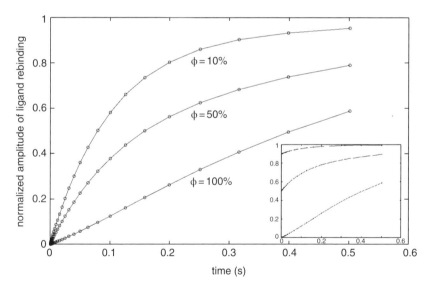

Figure 6.6 Simulated time courses for a photolysis experiment.

The simulated protein is a homodimer with $k_{a,1}'=2\,s^{-1}$ and $k_{a,2}'=10\,s^{-1}$; that is, the protein presents a significant positive kinetic cooperativity, with a 10-fold difference between the intrinsic kinetic rate constants for the two sites, before correction for statistical coefficients. If the light pulse has sufficient intensity to allow a photochemical yield of 100%, no singly liganded dimer is present at the beginning of rebinding and the progress curve is superimposable to that obtained by rapid mixing (third trace from left in Figure 6.5). Lower pulse intensities result in partial photochemical dissociation of the complex (as indicated by φ 10% and 50%) and a residual population of the singly liganded intermediate, which rebinds at a faster rate than the unliganded dimer because of cooperativity. Inset: effective amplitude of the simulated signal for the traces in the main Figure.

T jump is a completely different kinetic method that shares with photolysis the requirement that the protein-ligand complex is present in the sample under equilibrium conditions at the start of the experiment. Compared to photolysis this method has one great advantage and one great disadvantage. The advantage of T jump is that it can be applied to every system having a non-zero ΔH of binding, that is, to the large majority of protein-ligand complexes, whereas photolysis can be applied only to the tiny minority of systems in which photochemical breakage of the protein-ligand bond occurs. The great disadvantage is that this method can be applied only to systems in which the initial condition is close to the center of the ligand-binding isotherm (i.e., around $X_{1/2}$). This is because the principle of the method is to quickly raise the temperature of the sample by ~ 5–$10\,°C$, thus moving the system from a lower temperature isotherm to a higher temperature one, and to follow the system's relaxation to the new equilibrium condition. Thus, near the asymptotes of the equilibrium isotherm less signal change can be observed, as is evident from inspection of Figure 6.7.

The starting condition of a T-jump experiment can be decided only after extensive preliminary characterization of the system. Biological systems may present significant heat capacity changes: thus a system that yields a poor T-jump signal at a given starting temperature may give a larger one at a different temperature. This property may

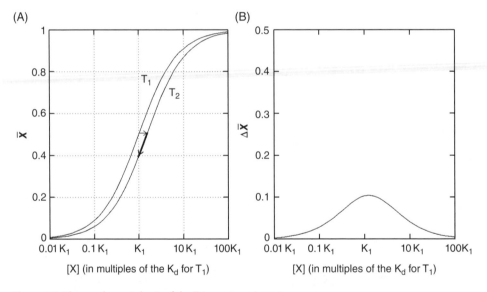

Figure 6.7 Thermodynamic basis of the T-jump experiment.
Panel A: Simulated equilibrium isotherms at two temperatures ($T_1 = 25\,°C$, $T_2 = 30\,°C$) for a non-cooperative protein with ΔH of 15 Kcal/mol. The dissociation reaction is endothermic and $K_2 = 1.52\,K_1$ for the simulated ΔT of $5\,°C$. The arrows indicate the reaction path started by a rapid T-jump of $5\,°C$ (orizontal arrow). Notice that while the ligand-binding isotherm shifts from T_1 to T_2, the ligand concentration does not change; hence the protein-ligand complex is suddenly transferred to a non-equilibrium condition. From this point the system relaxes to its new equilibrium condition along isotherm T_2, by dissociation of approximately one-fifth of the ligand bound at T_1 (oblique arrow). The signal change to be followed corresponds to the latter relaxation.
Panel B: expected ligand saturation changes after a T jump of $5\,°C$ as a function of ligand concentration, simulated using the same parameters as in panel A. As expected, the signal change is greatest if the starting saturation is close to 0.5 and decreases as it increases or decreases.

compensate the obvious limitation that biological systems are viable only between the temperatures at which the solvent freezes and that at which the protein is reasonably stable.

T jump allows the researcher to explore the kinetics of ligand binding or release for mixtures of ligation intermediates. Thus, it provides valuable information on cooperativity and its mechanisms (e.g., Kirschner *et al.*, 1966; Brunori *et al.*, 1981). Because the solution can be rapidly heated, but not rapidly cooled, the ΔH of the reaction dictates whether one will observe a process of ligand binding or release. However, in either case, the approach to equilibrium in the presence of excess free ligand will be governed by the sum of the (pseudo-first order) association and dissociation rate constants, as occurs in dilution experiments.

A further technique, pressure jump or P jump, provides results somewhat similar to T jump. In this case the equilibrium system is perturbed by a sudden increase or decrease in pressure, which forces the system to find a new equilibrium condition in which the volume is minimized or expanded. Large pressures are required because ligand-linked changes in volume of protein-ligand systems are usually quite small (e.g., Phillips *et al.*, 2008).

6.6 Integration of Equilibrium and Kinetic Data for Cooperative Systems

The study of ligand-binding kinetics of oligomeric proteins is especially valuable to clarify the mechanisms of positive homotropic cooperativity. However, as a consequence of the fact that autocatalysis offers only a limited amount of information about cooperativity (see Section 6.5), special experiments must be devised. Ideally, the researcher interested in a cooperative protein would like to determine the kinetic rate constants of ligand association and dissociation for each ligation intermediate of the protein. In practice, however, a more realistic goal would be the determination of the association and dissociation rate constants of the first and last binding site, that is, of the reactions:

$$P_n + X \Leftrightarrow P_n X \text{ governed by } k_{a,1} \text{ and } k_{d,1}$$

and

$$P_n X_{n-1} + X \Leftrightarrow P_n X_n \text{ governed by } k_{a,n} \text{ and } k_{d,n}$$

where n is the number of ligand binding sites of the macromolecule.

The initial rate constant of the association reaction observed after rapidly mixing the protein and the ligand yields the association rate constant for the first ligand molecule ($k_{a,1}$), and, in favorable cases, also the dissociation rate constant for the same step ($k_{d,1}$; see Section 2.2 and Figure 2.3). The initial rate constant is better estimated if the experimental data are fitted using a two-site Bateman scheme (eqns. 6.1 through 6.3), irrespective of the effective number of binding sites. Any successive rate constant one can derive from a rapid mixing experiment is unreliable. The limiting rate constant for ligand dissociation in a pulse experiment carried out at low concentrations of the ligand provides an alternative estimate of the dissociation constant for the first ligand ($k_{d,1}$; see the time course marked with asterisks in Figure 6.4). The ratio $k_{d,1}/k_{a,1}$ can be compared with the equilibrium dissociation constant of the first ligand molecule (K_1 or K_T).

Two experimental approaches are suitable for the determination of the rate constant for the dissociation reaction of the last ligand, that is, $k_{d,n}$: rapid chemical destruction of the free ligand or ligand replacement. In both cases the protein is fully liganded at the beginning of the experiment.

Rapid mixing of the ligand protein with a reactant or enzyme capable of destroying the ligand may yield autocatalytic time courses if kinetic cooperativity is significantly expressed in the ligand dissociation reaction, thus it is advisable to fit the experimental data using a two-step Bateman scheme. The rate constant of the initial kinetic process is the best estimate of $k_{d,n}$.

The typical time course of a replacement reaction is not autocatalytic and can usually be fitted to a single exponential, whose apparent rate constant contains information about the rate constants of the dissociation reactions of the two ligands used (see Section 2.6). Although the apparent rate constant of a ligand replacement reaction also contains information about the association rate constants, these cannot be reliably measured in this type of experiment. Moreover, one should be aware that the interpretation of a ligand replacement reaction requires an extensive set of experiments carried out at different concentrations of the two ligands. Thus, this approach is demanding and requires a substantial amount of protein.

As discussed in the preceding section, dilution experiments yield quite complex results, more so in the case of cooperative proteins because they are governed by both the association and dissociation rate constants for more than a single step. Thus, they are not ideally suited for the determination of $k_{d,n}$ of cooperative systems.

The rate constant for combination of the last ligand ($k_{a,n}$) may be measured by photolysis if the protein-ligand complex is photosensitive, or possibly by T jump.

The ratio of the two kinetic constants for association and dissociation of the last ligand equals the pertinent equilibrium constant $K_n = k_{d,n}/k_{a,n}$. Besides its relevance in the characterization of the reaction and its mechanism, the determination of $k_{a,n}$ and $k_{d,n}$ is particularly valuable for cooperative systems. Indeed, in many cooperative systems the equilibrium ligand-binding isotherm does not allow a precise estimate of K_n, because the pertinent ligation intermediate ($P_n X_{n-1}$) is poorly populated in general, and becomes prevalent only at very high levels of ligand saturation, where the signal is small and the experimental error is large. Thus, confirmation of the value of K_n measured in equilibrium experiments by kinetic measurements is desirable. For example, Gibson and Edelstein (1987) used kinetic methods to rule out the presence of quaternary enhancement in the binding of the fourth ligand to human hemoglobin.

6.7 Ligand-Binding Kinetics in the Framework of the Allosteric Model

The allosteric model of cooperativity has precise implications for the kinetics of ligand binding and release, but its application is quite complicated. The crucial problem for application of the allosteric model of cooperativity to the study of ligand-binding kinetics is determination of rate constants of the allosteric structural change of the various ligation intermediates. The direct determination of the rate constants for the conversion of R state to T state and vice versa is difficult even for experimental conditions in which only one ligation state is present, for example, for the unliganded derivative in the absence of the ligand. Moreover, while the values of the equilibrium allosteric constants L_1 through L_n can be derived from the parameters L_0, K_R and K_T (see eqn. 4.22, Section 4.6), the model makes no prediction on the corresponding kinetic rate constants except for the obvious property that the ratio of the forward and backward kinetic rate constants of the same reaction step equals the corresponding equilibrium constant. We shall consider in this Section the approximation proposed by Shulman, Hopfield, and Ogawa (Shulman *et al.*, 1972, 1975) to circumvent this problem.

Moreover, we shall describe two approaches that have led to the direct photochemical measurement of the rate of allosteric structural change in hemoglobin by Sawicki and Gibson (1976) and by Ferrone and Hopfield (1976). These approaches are very system-specific and difficult to generalize to other systems; however, they provide the important and general proof of concept that under favorable conditions the direct measurement of the kinetic rate constants of an allosteric structural change is possible.

The allosteric model of cooperativity described in Chapter 4.6 explains the ligand-binding isotherms of cooperative proteins using only three thermodynamic parameters K_R, K_T and L_0. At first sight it may seem that it should provide an easy and straightforward description of ligand-binding kinetics, especially under experimental conditions in which only one of the two possible reactions (association or dissociation) predominates. Indeed, in Section 6.6, we recognized the determination of the kinetic rate constants for ligand association and dissociation of the first and last ligand as a reasonable goal, and suggested the appropriate experimental approaches. The two-state model requires the rate constants of the T and R conformations, which one may approximate to those of the first and last binding step as follows: $k_{a,T} \approx k_{a,1}$, $k_{d,T} \approx k_{d,1}$, $k_{a,R} \approx k_{a,n}$, and $k_{d,R} \approx k_{d,n}$. However, this apparently straightforward plan is upset by the difficulty in determining the kinetic rate constants for the allosteric structural change (i.e., the kinetic counterparts of the allosteric constants L_0 through L_n). Thus, unless one can measure at least some of these rate constants directly, a crucial set of the model's parameters is beyond experimental reach.

At present the problem of a kinetic formulation of the two-state model has been satisfactorily solved only for the case of hemoglobin-CO (and, to a lesser extent, hemoglobin-O_2), to which we shall mainly refer in the remainder of this section. The reason for discussing such a specialized solution in a chapter aimed at more general subjects, rather than deferring it to the chapter dedicated to hemoglobin (Chapter 7), is that the problem is of general interest. Moreover, analysis of its solution for the very special case of hemoglobin illustrates the amount of information required, and is thus useful for a reader interested in repeating the same analysis on a different system. Suffice it to say here that, for the hemoglobin-CO system, $k_{a,R}$ is measured by photochemical methods at low levels of photolysis, and $k_{d,R}$ is measured by competition with microperoxidase or by replacement; $k_{a,T}$ is obtained by rapid mixing of unliganded hemoglobin and CO in the stopped flow; and, finally, $k_{d,T}$, the most difficult to measure, is estimated by competition with microperoxidase, starting from partially oxygenated derivatives (e.g., Hb$(O_2)_3$CO), and taking advantage of the fact that oxygen can be rapidly removed by dithionite (Sharma *et al.*, 1976). In the case of hemoglobin-O_2 $k_{a,R}$ is measured by photochemical methods, $k_{d,R}$ by rapid mixing with dithionite or by replacement with

CO, $k_{a,T}$ by photochemical methods applied to T-state mixed metal hybrid hemoglobins (Shibayama *et al.*, 1995), and $k_{d,T}$ by oxygen pulse.

Hopfield, Shulman, and Ogawa (1972, 1975) suggested an elegant way to circumvent the determination of the rate constants of the allosteric structural change. These authors remarked that in all experiments in which ligand association or dissociation is significantly slower than the allosteric structural change, one may assume that the population of the allosteric states approximates the equilibrium condition for any given ligand saturation, that is, for any point of the time course of the reaction. The equilibrium values of L_1 through L_n can be calculated from the parameters L_0 and c using eqn. 4.22. Thus, one can use information derived from equilibrium experiments to simulate the reaction time courses.

To simulate the time courses of ligand dissociation using the solution proposed by Hopfield and co-workers, one needs to calculate the n apparent kinetic rate constants for dissociation using $k_{d,T}$, $k_{d,R}$ and the T- and R-state distribution of each ligation intermediate:

$$k_{d,i} = \left(k_{d,R} + L_0 c^i k_{d,T}\right) / \left(1 + L_0 c^i\right)$$

The apparent rate constants thus obtained correspond to the n kinetic rate constants of a sequential reaction scheme.

The experimentally obtained time courses can then be simulated using analytical integration of n sequential steps using the equations derived by Bateman (eqns. 6.1 through 6.3).

Simulation of the time courses of ligand association can be performed using exactly the same method, but replacing the dissociation rate constants with the corresponding association rate constants.

This method has some limitations that should be taken into account. (i) Bateman's equations apply to a sequentially arranged series of irreversible first-order reactions; they work perfectly for pseudo-first order reactions, but the experimental conditions must be selected in order to assure that the reverse reaction is negligible (i.e., at the end point of the time course the population of either the liganded or unliganded species is negligible). (ii) The value of c (i.e., the ratio K_R/K_T) is ligand-dependent and must be determined for the ligand used in the kinetic experiment. (iii) The method assumes that equilibration of the T and R allosteric states of the macromolecule is faster than ligand binding or release. Observation of an autocatalytic reaction time course guarantees that the rate of the allosteric transition cannot be slower than that of ligand binding or release, but does not guarantee that it is faster, thus some kind of control experiment is appropriate. In the case of the ligand association reaction, the simplest control is provided by the effect of increasing the concentration of the ligand. If the allosteric transition is fast, and the ligand concentration satisfies the pseudo-first order requirement, an increase of ligand concentration accelerates the time course without changing its shape, that is, the plots of $\bar{X}_t$ as a function of the product $[X] \cdot t$ are superimposable. In the case of the dissociation reaction a similar test cannot be run, but the rate of the approach to equilibrium of the two allosteric states is the same in the two directions T→R and R→T, because it is the sum of the forward and backward rate constants. Thus, if the test has been run for the association reaction, and the dissociation reaction is not significantly faster, the assumption of rapid equilibration of the allosteric states is justified. Direct proof of the applicability of the equilibrium approximation has been obtained in the case of hemoglobin by measurement of the rate of the allosteric structural change of at least some ligation intermediates, but few systems allow this determination.

Direct determination of the rate constants of the allosteric transition is possible, but has demanding requirements that few systems meet. The first requirement is that the R and T states must differ in some observable feature, for example, absorbance or fluorescence; differences in the NMR proton resonances may be present, but the time resolution of NMR spectroscopy may be insufficient for the present scope. These differences must not depend on ligation, that is, the absorbance and/or fluorescence spectra of the unliganded T state protein should differ from those of unliganded R state. Ideally, the same should apply to the fully liganded protein. This requirement may be difficult to ascertain, but in principle should not be too uncommon because the two allosteric states are likely to differ in some structural feature, for example, specific contacts at the intersubunit interfaces. The best-studied case is that of hemoglobin, whose T and R states present characteristic absorbance differences (for a review see Bellelli and Brunori, 1994).

The second, and possibly more stringent, requirement is that one can change some condition of the system in order to perturb the equilibrium between the two allosteric states. The perturbation should require a much shorter time than the time course of the allosteric relaxation, if the rate constant of the process is to be measured. T jump is a possibility if the allosteric transition has a significant ΔH, but unfortunately, the technique works best when the two states are both significantly populated, and this usually occurs only for poorly represented ligation intermediates. We are not able to quote a pertinent example, but in the somewhat similar case of the interconversion of two valency isomers of the reduced state of the copper-containing amine oxidase from pea, Dooley and co-workers could use a T-jump instrument to determine the rate constant of the process (Turowski *et al.,* 1993). In this case, the two isomers of the protein present large spectroscopic differences, due to a reversible internal electron transfer step, and unrelated to the binding of substrates. Moreover, the two isomers are both significantly populated at room temperature.

A very effective technique for the measurement of the rate of the allosteric transition is photolysis of the protein-ligand complex. Because few protein-ligand complexes are photolabile, the application of this method cannot be generalized, and we shall consider here only the case of hemoglobin that, as already stated, is particularly favorable. Two derivatives of hemoglobin lent themselves to this type of measurement: unliganded and triply CO-liganded hemoglobin.

The case of unliganded hemoglobin is most straightforward. In 1976, Sawicki and Gibson used the 300 ns pulse of a dye laser to remove all the bound CO from fully liganded human carbon monoxy-hemoglobin. In the equilibrium HbCO complex the R state predominates over the T state by 10,000 times or more. The duration of the flash used to photochemically dissociate the complex is much shorter than the time required for the allosteric transition, thus at the end of the flash the unstable derivative unliganded R state Hb was obtained. This derivative decays to T state unliganded Hb before rebinding CO, with a rate constant of $6400\,s^{-1}$ at alkaline pH and T=20 °C (Sawicki and Gibson, 1976). The authors obtained similar results also using oxy-hemoglobin, and could demonstrate that the difference spectrum ^{R}Hb-^{T}Hb is independent of the ligand, as expected (Sawicki and Gibson, 1977). It is interesting to summarize the combination of favorable conditions that made this experiment successful. First, the authors took advantage of a relatively large and well-characterized absorbance difference between unliganded R state and unliganded T state hemoglobin. Second, they used a photosensitive R-state liganded complex, which could be rapidly and completely converted to the R-state unliganded derivative using a short but intense light pulse. Third, ligand rebinding after the photolysis pulse was too slow to compete with the decay of the fully unliganded protein from R to T state.

The determination of the rate of the allosteric transition for the triply liganded intermediate of HbCO is due to Ferrone and Hopfield (1976), who were able to induce fluctuations in the saturation of carbon monoxy-hemoglobin using a light source whose intensity was sinusoidally time-modulated. The absorbance signal could be deconvoluted to obtain in-phase and out-of-phase components, and the first out-of-phase component could be attributed to the $^{R}Hb(CO)_3$ <==> $^{T}Hb(CO)_3$ interconversion, which was assigned the rate constants of $780 \, s^{-1}$ in the direction R to T, and $2500 \, s^{-1}$ in the direction T to R, at pH 7 and 22 °C. The ratio between the two rate constants yields the equilibrium constant $L_3 = 0.31$, which agrees with estimates obtained from oxygen equilibrium measurements.

Taking advantage of the knowledge accumulated by these (and other) authors, W.A. Eaton and co-workers (Henry *et al.*, 1997) were able to successfully describe the evolution of time-resolved spectra of complex mixtures of hemoglobin-CO ligation intermediates in terms of the allosteric model. The computational problems of this project were significant, but the authors could use previous equilibrium and kinetic experiments to constrain many of the fitted parameters. However, the same cannot be said for oxygen whose rebinding kinetics subtly deviate from the prediction of the allosteric model (Gibson, 1999).

References

Bateman, H. (1910)The solution of a system of differential equations occurring in the theory of radio-active transformations. Proc Cambridge Phyl Soc, 15: 423–427.

Bellelli, A. and Brunori, M. (1994) Optical measurements of quaternary structural changes in hemoglobin. Methods Enzymol, 232: 56–71.

Brunori, M., Coletta, M. and Ilgenfritz, G. (1981) Temperature jump of hemoglobin. Methods Enzymol, 76: 681–704.

Ferrone, F.A. and Hopfield, J.J. (1976) Rate of quaternary structure change in hemoglobin measured by modulated excitation. Proc Natl Acad Sci USA, 73(12): 4497–4501.

Gibson, Q.H. and Roughton, F.J. (1956) The determination of the velocity constants of the four successive reactions of carbon monoxide with sheep haemoglobin. Proc R Soc Lond B Biol Sci, 146(923): 206–224.

Gibson, Q.H. (1959) The photochemical formation of a quickly reacting form of haemoglobin. Biochem J, 71(2): 293–303.

Gibson, Q.H. (1973) The contribution of the α and β chains to the kinetics of oxygen binding to and dissociation from hemoglobin. Proc Nat Acad Sci USA, 70: 1–4.

Gibson, Q.H. and Edelstein, S.J. (1987) Oxygen binding and subunit interaction of hemoglobin in relation to the two-state model. J Biol Chem, 262: 516–519.

Gibson, Q.H. (1999) Kinetics of oxygen binding to hemoglobin A. Biochemistry, 38(16): 5191–5199.

Henry, E.R., Jones, C.M., Hofrichter, J. and Eaton, W.A. (1997) Can a two-state MWC allosteric model explain hemoglobin kinetics? Biochemistry, 36(21): 6511–6528.

Hopfield, J.J., Shulman, R.G. and Ogawa, S. (1971) An allosteric model of hemoglobin. I. Kinetics. J Mol Biol, 61(2): 425–443.

Kirschner, K., Eigen, M., Bittman, R. and Voigt, B. (1966) The binding of nicotinamide-adenine dinucleotide to yeast d-glyceraldehyde-3-phosphate dehydrogenase: temperature-jump relaxation studies on the mechanism of an allosteric enzyme. Proc Natl Acad Sci USA, 56: 1661–1667.

Lambeth, D.O. and Palmer G. (1973) The kinetics and mechanism of reduction of electron transfer proteins and other compounds of biological interest by dithionite. J Biol Chem, 248: 6095–6103.

Phillips, R.S., Miles E.W., McPhie P., Marchal S., Georges C., Dupont Y. and Lange R. (2008) Pressure and temperature jump relaxation kinetics of the conformational change in Salmonella typhimurium tryptophan synthase l-serine complex: Large activation compressibility and heat capacity changes demonstrate the contribution of solvation. J Am Chem Soc, 130: 13580–13588.

Prinz, H. and Maelicke, A. (1983) Interaction of cholinergic ligands with the purified acetylcholine receptor protein. II. Kinetic studies. J Biol Chem, 258: 10273–10282.

Sawicki, C.A. and Gibson, Q.H. (1976) Quaternary conformational changes in human hemoglobin studied by laser photolysis of carboxyhemoglobin. J Biol Chem, 251: 1533–1542.

Sharma, V.S., Ranney, H.M., Geibel, J.F. and Traylor, T.G. (1975) A new method for the determination of ligand dissociation rate constant of carboxyhemoglobin. Biochem Biophys Res Commun, 66: 1301–1306.

Sharma, V.S., Schmidt, M.R. and Ranney, H.M. (1976) Dissociation of CO from carboxyhemoglobin. J Biol Chem, 251: 4267–4272.

Shibayama, N., Yonetani, T., Regan, R.M. and Gibson, Q.H. (1995) Mechanism of ligand binding to Ni(II)-Fe(II) hybrid hemoglobins. Biochemistry, 34: 14658–14667.

Shulman, R.G., Ogawa, S. and Hopfield, J.J. (1972) An allosteric model of hemoglobin. II. The assumption of independent binding. Arch Biochem Biophys, 151(1): 68–74.

Shulman, R.G., Hopfield, J.J. and Ogawa, S. (1975) Allosteric interpretation of haemoglobin properties. Q Rev Biophys, 8(3): 325–420.

Turowski, P.N., McGuirl, M.A. and Dooley, D.M. (1993) Intramolecular electron transfer rate between active-site copper and topa quinone in pea seedling amine oxidase. J Biol Chem, 268: 17680–17682.

Appendix 6.1 Kinetic Statistical Factors

Kinetic rate equations of symmetric oligomeric proteins require the application of statistical factors, as do equilibrium constants. Moreover, the ratio of the statistical factors to be applied to the dissociation and association kinetic rate constants equals the statistical factor to be applied to equilibrium constant.

In this appendix we demonstrate that application of the statistical factors to a symmetric, non-cooperative homodimer whose ligand association time course obeys eqns. 6.1

through 6.3 yields an exponential increase of the fractional ligand saturation $\bar{X}_t$. By definition, the microscopic association rate constants for the first and second site of the non-cooperative homodimer are identical ($k_{a,1} = k_{a,2} = k_a$), and the apparent pseudo-first order rate constants differ only because of the statistical factors ($k_{a,1}' = 2k_a[X]$; $k_{a,2}' = k_a[X]$; $k_a' = k_a[X]$). We substitute in eqns. 6.1 through 6.3 and obtain:

$$[P]_t = [P]_{tot}\, e^{-2k_a't}$$

$$[PX]_t = [P]_{tot}\, 2k_a' \left\{ e^{-2k_a't} / \left(k_a' - 2k_a' \right) + e^{-k_a't} / \left(2k_a' - k_a' \right) \right\}$$

$$= [P]_{tot}\, 2 \left(e^{-k_a't} - e^{-2k_a't} \right)$$

$$[PX_2]_t = [P]_{tot} \left(1 - 2e^{-k_a't} + e^{-2k_a't} \right)$$

and:

$$\bar{X}_t = \left([PX]_t + 2[PX_2]_t \right) / 2[P]_{tot} = 1 - e^{-k_a't}$$

7

Hemoglobin and its Ligands

The discovery of hemoglobin (Hb) is credited to Hunefeld in 1840, even though previous authors had already noticed the presence of iron and iron-containing substances in the blood; in 1866, F. Hoppe Seyler described the reversible combination of Hb with oxygen. To summarize over one and a half centuries of research in a single book chapter is clearly impossible, but describing ligand binding to hemoglobin without reference to its historical development would be misleading because many crucial features of ligand binding to this macromolecule were discovered early, misinterpreted, and then reinterpreted several times before reaching the present day, more or less coherent, description. Moreover, a historical perspective about the experiments carried out on this complex and finely regulated protein provides excellent guidelines for the study of other macromolecules and their ligands.

Hemoglobin, together with the structurally and evolutionarily related protein myoglobin (Mb), was the first protein whose structure was solved by x-ray crystallography, and consequently the first whose functional properties could be correlated with the three-dimensional structure. Hb is a very special example, whose study can be extremely rewarding. The aim of this chapter is to analyze Hb as a model system, whose structural and functional properties can exemplify many of, if not all, the subjects we dealt with from a theoretical viewpoint in Chapters 4 through 6; thus, we selected for discussion those properties and experimental approaches that better exemplify the subject matter discussed in the preceding chapters. We start with a description of the ligand-binding properties of Hb and its co-factor, the heme. Hb ligand-binding properties include homotropic and heterotropic interactions, allostery, and ligand-linked dissociation. We next describe the structure of liganded and unliganded Hb and, where available, that of its ligation intermediates. This separation between functional and structural data is somewhat arbitrary, because the majority of experiments were designed on the basis of both types of information, which beautifully evolved together and complemented each other; yet we acknowledge that it may be confusing to switch back and forth between structure and function.

An important reason for describing in some detail the ligand-binding properties of Hb in the context of this book is that the subjects we dealt with in the preceding chapters are extremely general and have been discussed without reference to their structural bases, which vary among different proteins. Thus, the very important subject of structure-function relationships has been essentially ignored. Hb is possibly the protein for which structure-function relationships have been studied to the deepest detail, thus it

Reversible Ligand Binding: Theory and Experiment, First Edition. Andrea Bellelli and Jannette Carey.
© 2018 John Wiley & Sons Ltd. Published 2018 by John Wiley & Sons Ltd.

seemed an appropriate example to discuss. The aim of this chapter is to describe how the general thermodynamic and kinetic phenomena described in Chapters 4 through 6 relate to the structural properties of at least one selected protein. The information provided in this chapter is therefore highly specific to Hb and cannot be generalized: other proteins may implement cooperativity, autocatalysis, and heterotropic effects using very different structural mechanisms. Indeed, in some cases we shall discuss events that occur on a very subtle scale: atom movements of a few tenths of an angstrom, rupture or formation of single H-bonds, and so on. We hope that the information provided in this chapter may serve as an inspiration even for readers who are not specifically interested in Hb, because, even though structural details are unique to each protein, ligand-linked structural changes occur in (almost) every protein and the experimental approaches one may use to investigate them are of general interest.

7.1 The Heme and Its Ligands

Oxygen binding to Hb and related proteins occurs via formation of a coordination bond with an iron ion at the center of the heme (Figure 7.1), the characteristic cofactor of these proteins. Under physiological conditions the iron is in the ferrous oxidation state, and oxygen binds only to this state. The iron may undergo oxidation to the ferric (and occasionally ferryl) state, either in vivo or in vitro. Ferric Hb binds non-physiological ligands and does not contribute to oxygen transport; nevertheless it deserves mention because important experiments were carried out on its derivatives.

The heme (iron protoporphyrin IX) is a tetrapyrrole aromatic ring that coordinates iron with its four nitrogens. Because iron ions form five- or six-coordinate complexes, two coordination positions of the heme iron are unoccupied and are able to bind other ligands. In this section we shall be concerned with a description of the most important heme ligands and the properties of their complexes with Hb and Mb. The study of the free heme, or of model iron-porphyrin compounds, has considerably contributed to our understanding of the properties of Hb and Mb. In the description of the heme-ligand complexes, we shall often compare Hb and Mb with heme model compounds to highlight important similarities and differences.

As a general rule it is quite difficult to isolate and characterize the *four-coordinate* isolated heme: besides its poor water solubility, which requires it to be dissolved in organic solvents or detergent micelles, the iron has high affinity for external ligands and in the absence of stronger ligands it coordinates water and, in the ferric state, OH^-. Thus the four-coordinate heme iron is only observed as a transiently populated species in the course of photochemical or pH-jump experiments more often carried out on hemoglobin or myoglobin than on the free heme itself (Giacometti *et al.*, 1977).

In hemoglobin and myoglobin one of the two axial coordination positions of the heme iron is occupied by a His residue provided by the protein moiety, called the proximal His, and the other is available for external ligands. The role of the proximal His is crucial both to anchor the heme to the protein, and to modulate the affinity for the sixth ligand. In the majority of heme model compounds the role of the proximal His is mimicked by imidazole or some other nitrogenous base. The *five-coordinate* derivative of the ferrous heme iron is characteristic of unliganded or deoxygenated Hb and Mb. This derivative can be obtained from the freshly prepared protein by equilibration under a flow of

(A)   (B)

(C)   (D)

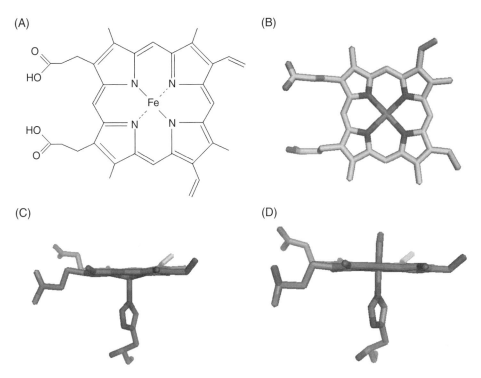

Figure 7.1 The heme.
   Panel A: the structural formula of heme (iron protoporphyrin IX). Panel B: top view of the three dimensional structure of unliganded heme, taken from the crystal structure of the α subunits in deoxyHb (PDB code 2dn2, see Park *et al.,* 2006), in atomic colors with green carbon atoms. Panel C: side view of unliganded heme and the proximal His from the α subunits of deoxyHb (PDB code 2dn2) showing the doming of the porphyrin and the out of plane position of the five-coordinated iron. Panel D: side view of the heme taken from the crystal structure of the α subunits of oxyHb (PDB code 2dn1), showing the proximal His and the bound oxygen, in atomic colors with cyan carbon atoms. The projection view is approximately the same as that of panel C to highlight the flattening of the heme and the in plane position of the six-coordinated iron. (*See insert for color representation of the figure.*)

nitrogen or argon under gentle agitation. Unliganded ferrous Hb and Mb have strong magnetic susceptibility with spin 4/2 (Pauling and Coryell, 1936), and their absorbance spectrum presents bands at 555 and 431 nm (for human Hb).

   In unliganded ferrous Hb and Mb the sixth coordination position is unoccupied, the heme is domed in the direction of the proximal His, and the iron is out of the heme nitrogen plane by ∼0.2–0.4 Å (Fermi *et al.,* 1987), as shown in Figure 7.1C. M.F. Perutz identified two main reasons for the domed structure of the five-coordinate heme (Perutz, 1979): (i) The five-coordinate ferrous iron adopts a position close to the center of the square-base pyramid described by its five ligands. (ii) The ionic radius of the unliganded, paramagnetic, and high-spin ferrous ion is too large to fit in the plane of the heme nitrogens. For hemoglobin and myoglobin (but not for free heme), Perutz added a third factor, namely (iii) the tertiary structure of the protein effectively pulls the proximal His, with the iron attached, thus contributing to the heme doming. The relevance of this factor depends on the structure of the protein, and may vary greatly among hemoproteins.

Oxygen binding, the physiological function of Hb and Mb, occurs via coordination of dioxygen to the vacant axial position of deoxyHb or deoxyMb. Thus, the heme iron becomes *six-coordinate*. Before describing some basic properties of HbO_2 and MbO_2, a general consideration on six-coordinate derivatives is in order that applies to all ligands of the ferrous heme iron (e.g., carbon monoxide, nitric oxide (NO), isocyanides, nitrosoaromatic compounds, and nitrogen heterocycles e.g., imidazole and pyridine and their derivatives), as well as most ligands of the ferric heme iron: cyanide, azide, fluoride, the hydroxyl ion, and so on: in six-coordinated derivatives of ferrous Hb and Mb the heme is flat or almost so and the iron ion lies in its plane (Figure 7.1D). Perutz proposed the following reasons to account for this fact: (i) The six-coordinated iron has bi-pyramidal geometry and the central position among its ligands lies in the heme nitrogen plane. (ii) The ionic radius of the iron ion shrinks as ligation causes the metal to become diamagnetic (Pauling and Coryell, 1936). The movement of the iron toward the heme plane upon ligation due to factors (i) and (ii) pulls the proximal His in the same direction, and causes a localized, but significant, rearrangement of the tertiary structure of the globin.

In principle, factors (i) and (ii) operate in the free heme as well as in Hb and Mb. In practice it is not easy to characterize the free heme in solution because the iron tends to auto-oxidize to the ferric state and to coordinate Lewis bases present in solution (e.g., water, chloride ions, etc.). Model compounds of the heme, that is, substituted iron protoporphyrins specifically designed to provide the fifth ligand and to protect the ferrous iron from the solvent, have been synthesized and they have usually confirmed Perutz's hypotheses. For example, Collman and co-workers (1978) synthesized iron and cobalt derivatives of the model compound meso-tetra pivalamidephenyl porphinato iron (this compound is nicknamed "picket fence" heme, because the phenyl groups are orthogonal to the heme plane) bearing either imidazole or 2-methyl imidazole covalently bound to the aromatic scaffold in a position suitable for metal coordination. Thus, the five-coordinate metal in these model compounds can become six-coordinate if a suitable external ligand is added (the authors used O_2 and CO). These authors demonstrated that the fifth ligand exerts an important control on the affinity of the sixth ligand. The oxygen affinity of the iron-imidazole heme iron is 60-fold higher than that of iron-2-methyl imidazole. The authors explained this result as a consequence of the steric hindrance of 2-methyl imidazole, which resists flattening of the heme. Both imidazole and 2-methyl imidazole are compatible with the domed structure of the five-coordinated heme, but only the former is fully compatible also with the planar structure of the six-coordinate heme.

Hb and Mb are usually prepared as their *oxygenated derivative* in which the two axial coordination positions of the ferrous iron are occupied by the proximal His and dioxygen. This derivative is diamagnetic (Pauling and Coryell, 1936), and has a bright red color, with absorption bands at 577, 541, and 415 nm (for human Hb). X-ray diffraction demonstrates that the heme is flat and that the Fe-O-O complex is bent, forming an angle of 156° in human HbO_2 (Shaanan, 1982) (Figure 7.1D). This is consistent with data from heme model compounds: for example, Collman and co-workers measured a Fe-O-O angle of approximately. 135° in the O_2 complex of meso-tetra pivalamidephenyl porphinato iron (Collman *et al.*, 1974). The comparison between HbO_2 (or MbO_2) and model compounds is important, because one may doubt whether the geometry observed in the hemoproteins may be enforced by the closely packed heme pocket.

Table 7.1 Partition constants for gaseous ligands of ferrous human Hb.

Ligands	Partition constant	Reference
O2 : CO	$K_p = K_{R\,O2} / K_{R\,CO} = 241$	Di Cera *et al.* (1987)
O2 : CO (Low saturation)	$K_p = K_{T\,O2} / K_{T\,CO} = 123$	Di Cera *et al.* (1987)
CO : NO (pH=9.1, sheep Hb)	$K_p = K_{R\,CO} / K_{R\,NO} = 1630$	Gibson and Roughton (1957)

The partition constants are defined as ratios between the equilibrium dissociation constants of the two ligands considered, expressed in mmHg.

X-ray crystallography and neutron diffraction studies revealed that in many Hbs and Mbs the bound O_2 forms a H-bond with a conserved His residue, called the distal His (Phillips and Schoenborn, 1981). Site-directed mutagenesis of the distal His has demonstrated that in Mb and in the α subunits of Hb this H-bond substantially increases the affinity for O_2 (Olson *et al.*, 1988).

On long incubation times HbO$_2$ and MbO$_2$ tend to undergo oxidation of the heme iron. This effect is due to the fact that in the iron-oxygen complex a partial electron transfer occurs, leading to a ferric-superoxide state as first suggested by Weiss (1964). Thus there is a finite, though small, probability that the dissociation of the complex yields ferric iron and superoxide ion, instead of ferrous iron and dioxygen. Under physiological conditions, this side effect is counteracted by enzymatic reducing systems that reduce the ferric heme iron, and scavenge superoxide ion.

Carbon monoxide is a stronger ligand than oxygen (Table 7.1); hence HbCO is easily prepared by addition of CO to solutions of oxy- or deoxy-Hb and Mb. Human HbCO is diamagnetic and presents absorption bands at 569, 540, and 419 nm. An important feature of the CO derivative of Hb and Mb is its high photosensitivity. The quantum yield of this derivative at the end of a short light pulse is unity and no rebinding occurs in the sub-ns time regime. This is not the case for other ligands (e.g., the O_2 and NO derivatives), which exhibit significant rebinding in the ps- and sub-ps time regimes. Removing CO from HbCO or MbCO is a lengthy operation, which requires flushing with O_2 under continuous illumination, preferably over an ice bath.

The Fe-CO complexes of Hb and Mb are bent, or at least the Fe-CO axis is not orthogonal to the heme plane (Moore *et al.*, 1988). By contrast in model compounds the Fe-CO complex is linear (i.e., the Fe-C-O angle is 180°) and orthogonal to the heme plane (Collman *et al.*, 1979). The difference in the geometry of the complex in Hb and heme model compounds is attributed to the fact that the linear binding geometry of CO would clash with the residues of the heme pocket. As a consequence of the forced distortion of the complex, the affinity of Hb and Mb for CO is significantly lower than that of unhindered heme model compounds, with typical partition constants (see Section 1.8) $K_{d,O2}/K_{d,CO} \simeq 20,000$ in sterically unhindered model compounds, *versus* only ~ 25 in sperm whale myoglobin and ~ 250 in human hemoglobin (Collman *et al.*, 1979). At least one other factor contributes to decrease the relative affinity for CO of Hb and Mb, that is, CO does not form a H-bond with the distal His, and the weakly polar environment created by this residue is unfavorable to CO. It has been suggested that these effects are physiologically important in view of the small but non-negligible amount of CO which is physiologically produced by the organism (Springer *et al.,* 1989].

A very interesting difference between the O_2 and CO complexes of Hb (or Mb) is that while O_2 tends to oxidize the heme iron, CO has the opposite effect. The reason CO acts as a reductant of the heme iron depends on the following reaction (Bickar *et al.,* 1984):

$$2\,HbFe^{+3} + CO + H_2O --> 2\,HbFe^{+2} + CO_2 + 2\,H^+$$

Because oxidation to the ferric state may lead to denaturation and/or heme loss, the CO derivative of hemoglobin and myoglobin is much more resistant to denaturation than the O_2 derivative, and may be stored for longer times without loss of its biological properties (the same applies to the unliganded derivative).

Nitric oxide has over 1500-fold higher affinity for the ferrous heme iron than CO and is poorly photosensitive (Table 7.1). The Fe-NO complex of Hb and Mb is bent as occurs in heme model compounds, with an angle similar to that observed in the Fe-O_2 complex (Yi *et al.,* 2014). NO is a radical that easily reacts with several molecules in common usage (e.g., oxygen, dithionite); thus the preparation of solutions of NO is a more delicate procedure than the subsequent preparation of HbNO or MbNO. The preparation of HbNO (or MbNO) should never start from HbO$_2$ (or MbO$_2$), because addition of NO to these derivatives leads to iron oxidation, with formation of nitrate (via a peroxynitrite intermediate) (Eich *et al.,* 1996):

$$HbFe^{+2}O_2 + NO --> HbFe^{+3} + NO_3^-$$

If the concentration of NO in the sample is high enough, the ferric NO complex is obtained ($HbFe^{+3}NO$), which slowly decays to the ferrous one by releasing one nitrosonium ion (NO^+) and binding a fresh molecule of NO; otherwise, aquo-met Hb is obtained ($HbFe^{+3}H_2O$).

The HbNO and MbNO complexes are paramagnetic, with spin = ½, thanks to the unpaired electron of the ligand; this makes HbNO and MbNO good candidates for EPR spectroscopy.

A very peculiar property of the NO derivative is that, unlike O_2 and CO, this ligand reduces the energy of the Fe-His bond on the proximal side (Traylor and Sharma, 1992). Under some experimental conditions, this may lead to breakage of the Fe-His bond, at least in the α subunits of Hb, leading to a five-coordinate complex domed toward the distal side (Szabo and Perutz, 1976); the transition between the six- and five-coordinate state is easily monitored by EPR.

Other ligands of ferrous Hb and Mb are isocyanides and nitroso aromatic compounds. These compounds have been characterized quite extensively, but we shall not deal with them any more and we refer the interested reader to the pertinent literature (Gibson, 1960; Reisberg and Olson, 1980). The reason for this choice is that isocyanides and nitrosoaromatic compounds are non-physiological ligands, whose interest is linked to the realization of very specialized experiments that are outside the scope of this book.

Ferric (or met) derivatives of Hb and Mb present significant variations from the above picture. Typical ligands of the ferric iron are anions: for example, cyanide, azide, fluoride, OH$^-$. It is uncommon for the same compound to bind to both oxidation states of the iron with reasonable affinity; this happens with NO, cyanide, imidazole, and pyridine derivatives (Antonini and Brunori, 1971). The ferric heme iron is usually six-coordinate, the proximal His and one external ligand occupying the axial coordination positions. In the absence of stronger ligands, a water molecule or hydroxyl ion is found at the sixth coordination position. However, the factors that force the doming of the heme and the out-of-plane position of the metal are much less powerful for the ferric than for the ferrous iron because the ferric ion has a shorter ionic radius than the high spin ferrous iron.

Under physiological conditions, only very limited amounts of ferric Hb (or Mb) are present, in the form of the H_2O or OH^- derivatives (called aquo-metHb and hydroxy-met Hb, respectively). These derivatives are reduced by intracellular enzymatic systems to unliganded ferrous Hb.

External ligands of ferric Hb (or Mb) other than water or OH^- are non-physiological, but may have relevance in toxicology studies or as tools to selectively explore some properties of Hb and Mb. We shall make occasional reference to experiments carried out on the cyanide derivative of ferric Hb. All derivatives of ferric Hb (or Mb) are paramagnetic with spin 1/2 or 5/2, thus they have been characterized extensively by EPR spectroscopy. At variance with the complexes of ferrous iron, those of ferric iron are non-photolabile, and cannot be studied by photochemical methods.

7.2 Reversible Ligand Binding and Cooperativity

In this section we shall present an overall description of ligand binding to ferrous Hb. The physiological role played by Hb is extremely demanding, and not always fully appreciated. A gross estimate of the O_2 consumption by humans is ~ 10 mmoles/min. This is a large amount of gas that Hb carries between two locations where the O_2 partial pressure is essentially fixed: the alveoli in the lung where $P_{O2} \simeq 100$ mmHg, and the tissue capillaries where $P_{O2} \approx 40$ mmHg (see Table 7.2). Thus, Hb delivers a large amount of oxygen in response to a small change in the partial pressure, barely above a factor of 2. To put this datum into proper context, we may recall that in Chapter 1 we considered the fraction of ligand bound or released as a consequence of a 10- or 100-fold change in the ligand concentration. Two factors contribute to the remarkable efficiency of Hb, namely the high degree of cooperativity of the O_2 binding isotherm, that causes the $\bar{X}$ versus log (P_{O2}) plot to be steep, and the presence in the red blood cell of heterotropic ligands some of them at variable concentration, that bind to sites other than the heme and lower the O_2 affinity. Thus, homotropic cooperativity and heterotropic effects will be the main subjects dealt with in the present section.

Hemoglobin provides a very special and fortunate example of reversible ligand binding because its relatively complex behavior can be directly compared with that of the isolated heme and heme model compounds and with that of the much simpler protein myoglobin. No other protein has been studied to the same level of detail, and for no other protein has a similar wealth of data been collected. Some of the properties that make Hb such a special case are: (i) the rich spectroscopic features of the heme, which provide easy access to quantitative measurements; (ii) the photosensitivity of the liganded derivatives of the ferrous heme, which allows photochemical methods to be applied; (iii) the possibility of dissociating the tetramer into fully functional subunits,

Table 7.2 Carbon dioxide and oxygen content of human blood.

	Arterial blood (mMol/L)	Venous blood (mMol/L)	Artero-venous difference
pH, plasma	7.40	7.37	
dissolved in plasma	0.68	0.78	0.10
as plasma bicarbonate	13.52	14.51	0.99
as plasma carbamates	0.30	0.30	0.01
Total, plasma	14.49	15.59	1.10
pH, erythrocyte cytoplasm	7.20	7.175	
dissolved in the erythrocytes	0.40	0.46	0.06
as bicarbonate in the erythrocytes	5.01	5.46	0.44
as hemoglobin carbamate	0.75	0.84	0.09
Total, erythrocytes	6.16	6.75	0.59
Total CO_2, blood	20.65	22.34	1.69
Total O_2, blood	8.4	6.4	2.0

Data for CO_2 from Geers and Gross (2000) in mMoles per liter of whole blood. The hemoglobin content of whole human blood is approx. 8.5 mMol/L (on a heme basis) leading to an estimated $HbCO_2$ saturation of 0.08 and 0.1 in arterial and venous blood respectively. The ratio $[CO_2]/[HCO_3^-]$ depends on the plasma or erythrocyte pH according the Henderson-Hasselbalch equation.

whose functional properties can be studied and compared with those of native tetramers; and (iv) the availability of a host of natural mutants. In addition, Hb and Mb were the first proteins whose structure was solved by x-ray diffraction, and among the first to be expressed heterologously and subjected to site-directed mutagenesis. Surprisingly, despite years of intense study a complete and fully satisfactory interpretation of cooperativity in Hb is still lacking. Agreement exists on some general hypotheses about hemoglobin's function, but on several details there is uncertainty and some experiments do not fit the general picture. One reason for this is that ligand-binding isotherms of Hb are compatible with several models of cooperativity and convey relatively little information on the intrinsic mechanism of homotropic interactions (Marden *et al.*, 1989). Although ligand binding has been interpreted in light of further independent evidence (e.g., from kinetics, x-ray crystallography), no unified model of the structural and thermodynamic basis for cooperativity has yet proven capable of describing all the available data.

Because the physiological ligand of Hb (and Mb) is dioxygen, and several other interesting heme ligands are gases, the binding isotherms are often expressed as functions of the gas partial pressure, rather than its concentration in solution; thus in the literature we often find $P_{1/2}$ (the gas partial pressure required to achieve half saturation) in place of $X_{1/2}$, log (P_{O2}) in place of log($[O_2]$), and so on.

The ligand-binding isotherms of ferrous Mb and of the isolated α and β subunits of Hb conform to simple hyperbolas, when plotted on a linear scale of ligand concentration, whereas those of tetrameric Hb have the sigmoidal shape characteristic of positive homotropic linkage (positive cooperativity). When plotted according to the Hill equation

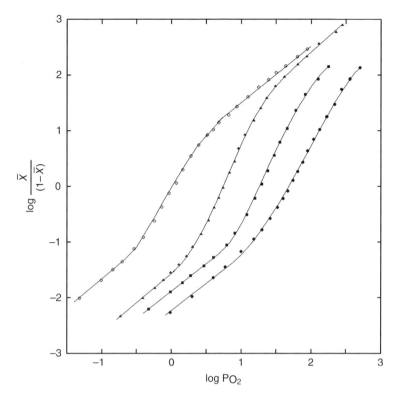

Figure 7.2 Oxygen-binding isotherms of human Hb in different experimental conditions (from Imai, 1982). Experimental conditions, all curves: T=25 °C; Hb 0.6 mM per heme; 0.05 M Tris or BisTris buffers. Symbols: open circles: pH=9.1, 2.6 mM Cl^-; closed triangles pH=7.4, 0.1 M Cl^-, 5% CO_2; closed circles: pH=7.4, 0.1 M Cl^-, 2 mM inositol hexakis phosphate (IHP); closed squares: pH=6.5, 0.1 M Cl^-.

(see Section 1.6), the ligand binding isotherms of Hb yield a slope significantly higher than one (usually in the range 2.5-3; see Figure 7.2). Many of the data available in the literature have been presented as Hill plots, and the results expressed in terms of overall affinity measured as $\log (P_{1/2})$ and cooperativity, estimated empirically from the Hill coefficient, n. These parameters are highly reproducible, as they are derived from the central region of the binding isotherm, between 10% and 90% saturation, where the distortion of the experimental error is small (see Figure 1.3). However, they have only an empirical significance and owe their persistent usage to the fact that a comprehensive and agreed upon model to explain Hb function is not available.

It seems worthwhile to briefly recapitulate Hill's hypothesis, in order that one is at least aware of its merits and its fallacy. As early as 1910, Archibald V. Hill (who was later awarded the Nobel Prize) explained cooperativity by suggesting that Hb is an oligomer whose partially liganded intermediates are populated only to a negligible extent (Hill 1910, 1913). He thus assumed that a reasonable simplification of the reversible reaction with ligands was as follows.

$$Hb_n + nO_2 \Leftrightarrow Hb_n \left(O_2\right)_n$$

The above reaction scheme completely abolishes partially liganded reaction intermediates, and leads to the following expression of the law of mass action, from which Hill derived the plot that bears his name.

$$K_d = [Hb_n][O_2]^n / \left[Hb_n (O_2)_n \right]$$

Hill was right on several crucial points: he correctly understood that Hb is an oligomer; that the reaction has stoichiometry higher than 1:1; and that partially saturated intermediates are poorly populated (see Figure 4.7B). However, he made the wrong assumption that the population of partially saturated intermediates was so low that they could be neglected altogether. This assumption led him to assume n as an estimate of the average degree of oligomerization of Hb in solution. That is, Hill assumed that human hemoglobin, having $n \approx 2.5-3$, is a mixture of species in which dimers and trimers predominate. Hill's hypothesis on cooperativity is clearly obsolete, hemoglobin being a stable tetramer, and the empirically measurable coefficient n has not the meaning Hill intended.

The modern usage of the Hill plot as a model-free empirical description of complex data sets misunderstands the value of Hill's original contribution, and neglects its obvious flaw. This is both unfair to Hill, and unnecessary, given that better model-free estimates of cooperativity are available (Section 4.6).

An important feature of the Hill plots of Hb lies in the upper and lower asymptotes with unitary slope. These extend over a quite limited range of ligand saturations (Figure 7.2) and graphically represent the non cooperative binding of the first and last (fourth) ligand. It is ironic that the asymptotes contradict the all-or-none binding of n ligands in a single reaction step implicit in Hill's hypothesis (Hill, 1913), but of course the precision of the measurements available at the time was insufficient to reveal their presence. Unfortunately, direct determination of K_1 and K_4 from the Hill plot is uncertain, because of the great distortion of the experimental error in the region of the asymptotes (see Figure 1.3). As already stated, the Hill plot should never be used for quantitative data analysis.

The oxygen affinity of Hb may vary over a very large range, because of heterotropic ligands that bind elsewhere than the heme. These include hydrogen ions, chloride, phosphate, and other inorganic anions, organic phosphates (e.g., 2,3 D-glycerate bisphosphate, DPG, and inositol hexakis phosphate, IHP), and carbon dioxide. Heterotropic interactions have been described in Sections 1.9 and 4.9–4.10. Because at least some of the heterotropic ligands of Hb are ubiquitous or nearly so, their description must precede any meaningful analysis of the oxygen affinity. All physiologically relevant heterotropic ligands present a negative linkage with heme ligands, that is, they lower the oxygen affinity, as one may verify from Figure 7.2 and Table 7.3. Positive heterotropic linkage is observed only in the non-physiological cases of the acidic Bohr effect (see below) and of covalent ligands of the residue Cys β93.

The binding sites of heterotropic ligands have been identified through detailed investigations carried out by means of chemical modifications, selective proteolytic digestion, analysis of natural mutants and, more recently, site-directed mutagenesis to generate artificial mutants. In many cases the heterotropic ligand bound to its site has been directly identified by means of x-ray crystallography of the Hb-ligand complex.

Table 7.3 Oxygen affinity of ferrous human hemoglobin. Experimental conditions as in Figure 7.2 (from Imai, 1982). O_2 affinities are given as $P_{1/2}$, the O_2 partial pressure required to half-saturate Hb under the given experimental condition. As a term of comparison, one may consider that PO_2 is $\sim$ 150 mmHg in air, $\sim$ 100 mmHg in the arterial blood, and $\sim$ 40 mmHg in mixed venous blood (for a resting subject).

Experimental conditions	oxygen affinity (as the $P_{1/2}$)
pH=9.1, 2.6 mM Cl⁻	1.07 mmHg
pH=9.1, 0.1 M Cl⁻	2.15 mmHg
pH=7.4, 7 mM Cl⁻	2.05 mmHg
pH=7.4, 0.1 M Cl⁻	5.32 mmHg
pH=7.4, 0.1 M Cl⁻, 5% CO_2	7.80 mmHg
pH=7.4, 0.1 M phosphate buffer	8.73 mmHg
pH=7.4, 0.1 M Cl⁻, 2 mM DPG	14.0 mmHg
pH=7.4, 0.1 M Cl⁻, 2 mM IHP	48.8 mmHg
pH=6.5, 0.1 M Cl⁻	18.8 mmHg
pH=6.5, 0.1 M Cl⁻, 2 mM IHP	136 mmHg

The effect of hydrogen ions is named after its discoverer, the Danish physiologist Christian Bohr. The effects of protonation in human Hb are two-fold because below pH = 6.5 protonation increases O_2 affinity of Hb (acidic Bohr effect, a case of positive heterotropic interaction), whereas above pH = 6.5 protonation decreases O_2 affinity (alkaline Bohr effect, a case of negative heterotropic interaction). Only the alkaline part of the Bohr effect is physiologically relevant in mammals because the blood pH never drops to the values required to express the acidic Bohr effect. The slope of the Bohr plot, that is, the graph log $P_{1/2}$ versus pH, peaks at approx. −0.5, indicating that the macromolecule releases half a mole of hydrogen ions per mole of oxygen bound (two hydrogen ions per tetramer). This is a functional estimate, which reflects convolution of the number of titratable protons and the changes of their pKa induced by oxygenation. Thus, one should look at the Bohr coefficient as a measure of the extent of the (negative) linkage, rather than the actual number of ionizable groups. Indeed, the Bohr effect is increased in the presence of other heterotropic effectors, because additional ionizable groups can be recruited (Riggs, 1971).

The identification of amino acid residues contributing Bohr protons (i.e., the titratable residues whose pKas change during oxygenation) has proven quite difficult. The classical view, painfully constructed by several authors who measured the Bohr effect of wild type human Hb and natural mutants (or other mammalian Hbs) in which each candidate residue is mutated, is as follows. The most important residues contributing to the *acidic* Bohr effect include His at position 143 of the two β subunits (Perutz *et al.*, 1980). The most important binding sites of the *alkaline* Bohr protons in human Hb are the amino terminal Val of the two α subunits and the C-terminal His of the two β subunits (His 146β) (Kilmartin and Rossi-Bernardi, 1969; Kilmartin and Wootton, 1970). Minor contributions arise from other residues whose pKa changes to a lesser extent during oxygenation, for example, Lys 82β (Perutz *et al.*, 1980). Several of the Bohr residues, in their protonated state, are also binding sites for anions (e.g., chloride for Val 1 of the α subunit, phosphate and chloride for His 146 of the β subunit, D-glycerate 2,3 bis-phosphate for Lys 82 of the β subunit), thus these residues are responsible for the anion-dependent part of the Bohr effect (Riggs, 1971; Kilmartin *et al.*, 1973). There is evidence from NMR that several other residues are involved (Ho and Russu, 1987), but selective removal of the C-terminal residues by carboxy-peptidase digestion and chemical modification of the amino terminus make it clear that Val 1α and His 146β account for a very large fraction of the Bohr effect (Kilmartin and Rossi-Bernardi, 1969; Kilmartin and Wootton, 1970); thus the contribution of other residues is comparatively minor.

Carbon dioxide forms a reversible carbamino adduct with the amino terminal residues of both subunits, thus in principle a Hb tetramer can transport four molecules of CO_2. In practice, however, under physiological conditions Hb is never CO_2 saturated, and accounts for the transport of less than one-tenth of the total CO_2 present in human blood, the rest being freely dissolved CO_2 (7%) or bicarbonate (85%) (Table 7.2). The fractional saturation of Hb with CO_2 ranges between 8% (in arterial blood) and 10% (in venous blood). For comparison, over 98% of the total O_2 present in human blood is bound to Hb and less than 2% is freely dissolved. Binding of CO_2 decreases the affinity for O_2, thus carbon dioxide is a negative heterotropic effector; moreover, overlapping of the binding sites of CO_2 and hydrogen ion at the amino group of Val 1α causes CO_2 to interfere also with the Bohr effect.

Other physiologically relevant (negative) heterotropic effectors that control the O_2 affinity of Hb are chloride and DPG. Chloride (and other anions) binds to several amino acid residues, one of which is the already quoted Val 1α. Other residues responsible for chloride binding overlap with those of the DPG binding site. DPG and other organic phosphates (e.g., inositol hexakis phosphate, ATP, NADP) bind to a cavity between the two β subunits and establish ionic interactions with the positively charged residues that line this cavity, namely His 2β, Lys 82β, and His 143β (Arnone, 1972).

7.3 The Structure of Hemoglobin

Heme is poorly soluble in water, and tends to stack by π interactions and precipitate; moreover, a physiologically effective control of the oxidation state of the heme iron cannot be achieved if heme is freely dissolved in water in the presence of oxygen, because it is immediately oxidized to the ferric state. The heme in biological systems avoids these effects because it is always bound to proteins that keep it in solution, prevent π-stacking, and control the redox state of the iron. In hemoglobin and myoglobin, the heme is wedged into a crevice of the protein molecule that shields it from the solvent and prevents oxidation of the ferrous iron. The protein moiety is a 150 residue polypeptide chain called the *globin* (the exact length varies in different hemoglobin subunits and myoglobins), which adopts a characteristic three-dimensional structure, the *globin fold*. Myoglobins are usually monomeric, and vertebrate hemoglobins are made up of four subunits of two types: α-like and β-like, each carrying one heme. The functional assembly of tetrameric Hb requires two α-like and two β-like subunits; only under artificial or pathological conditions (or in some non-vertebrate Hbs) are other types of assembly observed (e.g., the $β_4$ homotetramer, HbH, in α thalassemia). In healthy adult humans, some 97% of the circulating Hb is of the type $\alpha_2\beta_2$ (called HbA$_1$), the remaining being accounted for by HbA$_2$, in which the β subunits are substituted by the δ ones ($\alpha_2\delta_2$), and possibly by fetal Hb (HbF, $\alpha_2\gamma_2$), in which the β subunits are substituted by the γ ones. Post translationally modified forms of these exist (e.g., glycosylated HbA$_1$, HbA$_1$c) and usually account for only a few percent of the total.

The globin fold is made up by eight α-helices, named A through H, separated by short interhelical random coil segments named after the helices they connect, for example, CD or FG. Overall, the shape of the macromolecule is an irregular ovoid or disc. It is very convenient to number the residues of the polypeptide chain with reference to the helix or inter helical segment they belong to; thus, one often finds notations of the type "Thr 38α (C3)" or "Thr C3, 38α" to indicate the 38th residue of the α subunit which (in human Hb) is a Thr and occupies the third position of helix C. The advantage of the topological notation (C3 in this example) is that the positions in the helices are more invariant than the absolute positions in the polypeptide chain. To give another example, in essentially all known globins the eighth position of helix F (F8) is occupied by the conserved proximal His residue that coordinates the heme iron. The absolute position of His F8 is 87 in the α subunits of human Hb, 93 in the β subunits, and 93 in sperm whale Mb. Obviously, the absolute positions are influenced by insertions and deletions that occurred during evolution of the globins, whereas the topological positions depend on the highly conserved structure of the globin fold and are relatively constant.

As often happens in globular proteins the C and N termini of the globin fold lie on the same side of the ovoid and not far apart; thus the first and last helices (A, B, G, and H) are close to each other on the side opposite to the heme-binding crevice, which is lined by helices E and F. In the β subunits of Hb and in Mb, helix D and the CD corner close the heme crevice in the region where helices E and F diverge. In the α subunits helix D is lacking and its function is played by an elongated CE corner. The residue at position CD1 (CE1 in the α subunits) is a highly conserved Phe that is implicated in heme affinity. Overall, the heme crevice is a short triangular prism, having helices E, F, and C/D as the three sides, and helices B, G, and H as one of the ends; the remaining end is lacking, and as a consequence the heme propionates reach the surface of the macromolecule, where they interact with the solvent. The globin fold of sperm whale MbCO is represented in Figure 7.3.

In hemoglobin two α and two β subunits, each having the globin fold, form a hetero-tetramer that is conveniently described as a homodimer of $\alpha\beta$ heterodimers. At concentrations in the low μM range liganded hemoglobin dissociates into $\alpha\beta$ heterodimers, which constitute real and physiologically relevant substructures. By contrast, reversible dissociation of Hb into subunits occurs only at extreme dilution (Mrabet *et al.*, 1986). The intersubunit interface of the $\alpha\beta$ heterodimer resembles the type that Monod called isologous, that is, symmetric and made up by identical structural features, in this case helices B, G, and H, with the contacts αB–βH, αG–βG, and αH–βB. Obviously, this interface is pseudo-symmetric rather than truly symmetric, because the amino acid sequences of α and β subunits are different. The structure of the dimer, showing its interface, is depicted in Figure 7.4.

Two identical $\alpha\beta$ dimers assemble into a tetramer, via isologous contacts at the contact regions between the C helices and FG corners of the α and β subunits. In order to distinguish the tetramer's contacts from those of the dimer, the subunits of the tetramer are named α1, α2, β1, and β2; thus the tetramer is composed by two identical dimers

(A)   (B)

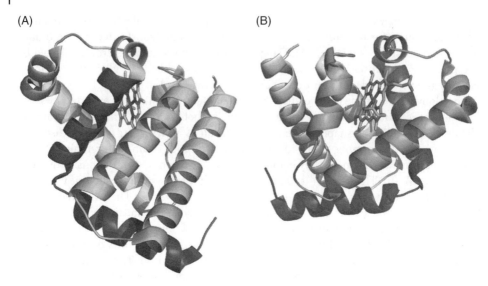

Figure 7.3 Ribbon representation of the backbone of sperm whale myoglobin.
α-helices A and B in red; C,D,E, and F in green; G and H in yellow; the heme is represented as sticks in atomic colors with green carbon atoms.

Panel A shows the macromolecule from the side farthest from the heme, and panel B from the opposite side, where the heme propionates interact with solvent. The bottom of the heme pocket is formed mostly by helices B and G, whereas its sides are formed by helices C, E, and F. (*See insert for color representation of the figure.*)

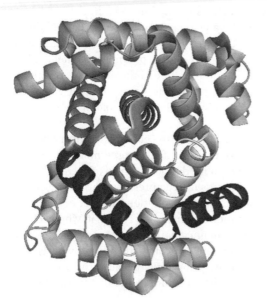

Figure 7.4 The αβ dimer of human deoxyhemoglobin.

The secondary structures of the α subunit are shown in red (helices A and B), green (helices C through F) and yellow (helices G and H); the corresponding secondary structures of the β subunit are shown in magenta, green, and orange. The figure shows the interface contacts αB-βH (red-orange), αG-βG (yellow-orange), and αH-βB (yellow-magenta). The heme-binding regions (green in both subunits) are relatively far from the αβ intersubunit interface. (*See insert for color representation of the figure.*)

α1β1 and α2β2, and the intersubunit interfaces between dimers occur at the contact regions α1C-β2FG (with the perfectly symmetric α2C-β1FG) and α1FG-β2C (with the perfectly symmetric α2FG-β1C). It is important to note that the α1β2 dimer can be

Figure 7.5 The human deoxy-hemoglobin tetramer and the a1b2 intersubunit interface. α subunits are shown in green, β subunits in pink. The FG corner and C helices of each subunit, which constitute the α1β2 interface are highlighted using stronger colors. (*See insert for color representation of the figure.*)

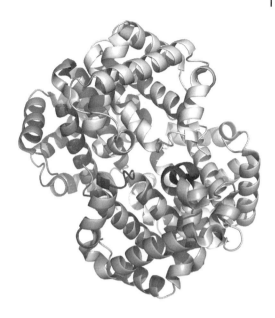

identified structurally within the tetramer but is never populated in solution and that the tetramer dissociates into α1β1 type dimers, that is, dimers in which the interface contacts occur at helices B, G, and H, as demonstrated by the structure of the hemoglobin-haptoglobin complex (Lane-Serff *et al.*, 2014). Thus, references in the literature to the α1β2 dimer actually describe the α1β2 interface of the tetramer. The α1β2 interface of human deoxy hemoglobin is depicted in Figure 7.5.

7.4 Ligation-Dependent Structural Changes

Small ligation-dependent structural changes are observed in the tertiary structures of the subunits, and larger changes are observed in the quaternary assembly of the tetramer, as judged by comparison of crystal structures obtained under different conditions (Perutz, 1970; Baldwin and Chothia, 1979). Ligation-dependent structural changes are minimal in Mb. Ligation causes a flattening of the heme with movement of the iron toward the nitrogen plane (Figure 7.1). The proximal His and the entire F helix to which it belongs move slightly as a consequence of the movements of the iron and the structural changes of the heme. These structural changes are essentially limited to the F helix and the FG corner, and to a lesser extent to other heme contacts that constitute what Gelin *et al.* (1983) called the "allosteric core" of the subunits (see Figure 7.6).

The FG corner and the C helix constitute the inter-dimeric α1β2 interface of the tetramer, and the tertiary structural changes induced by ligation are amplified at this interface. Indeed, while the α1β1 interface is essentially insensitive to the ligation state of the heme, because helices B, G, and H contact the periphery of the heme but are quite removed from the iron, the opposite is true for the α1β2 interface, which is much closer to the iron and its ligand, the FG corner being only few residues away from the proximal His F8.

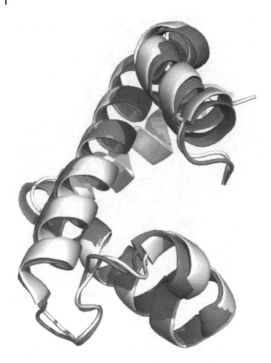

Figure 7.6 Superposition of the α subunit of deoxy-hemoglobin (green) and carbon monoxy-hemoglobin (cyan). The figure shows only residues 1-93 (helices A through E) in order to better visualize the C-helix (bottom) and F-helix and FG corner (top). Superposition was optimized for the region where ligand-linked structural changes are minimal, corresponding to the α1β1 contact region (helices B, G and H, residues 2-35 and 94-138), and the orientation emphasizes the region where the ligand-linked structural changes are largest, corresponding to the α1β2 contact region. (*See insert for color representation of the figure.*)

In Hb, the ligation-dependent movement of the F helix and FG corner exerts a powerful effect on the interdimeric interfaces of the tetramer (α1β2 and α2β1), which is evident if one compares the structure of fully unliganded and fully liganded Hb. The most obvious structural change is observed at the α1C-β2FG contact (and the symmetric α2C-β1FG), where His β2 97(FG4) packs between Thr α1 41(C6) and Pro α1 44(CD2) in deoxy Hb, and between Thr α1 38(C3) and Thr α1 41(C6) in HbO$_2$ (Figure 7.7). Thus, ligand binding causes His β2 FG4 to move over its contact with the α1 C helix by one full helical turn (Baldwin and Chothia, 1979). Other structural changes accompany this switch, for example, a hydrogen bond between Asp β2 99(G1) and Thr α1 38(C3), present in deoxy Hb, is broken upon oxygenation. At other contacts of the same interface, the structural changes are equally marked but less well defined. The α1FG-β2C contact region was called the "flexible joint" by Baldwin and Chothia to note this fact. As a consequence of these ligand-linked structural changes, the inter-dimeric interface undergoes a significant rearrangement, which causes one α1β1 dimer to slide by over 1 Å, and to rotate by over 14° with respect to the other.

The picture presented above, which may be referred to as the "classical" one, has been greatly enriched and substantially confirmed by the structural study of partially liganded derivatives (Section 7.7).

Although the quaternary structure changes are very large, they affect the O$_2$ affinity because of the way they affect the tertiary structure of the subunits *via* the specific arrangement of the α1β2 interface. Thus, in the course of oxygenation of Hb, subtle tertiary structure changes in the liganded subunits trigger the large quaternary structure change of the tetramer. The quaternary structure change of the tetramer, in turn, causes subtle tertiary structure changes in the yet unliganded subunits, increasing their

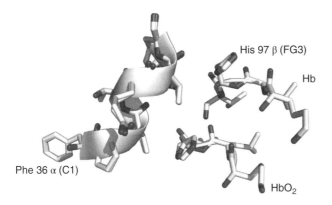

His 97 β (FG3)

Hb

Phe 36 α (C1)

HbO₂

Figure 7.7 Ligand-dependent quaternary structural change at the αC-βFG contact region of the α1β2 interface of human hemoglobin (detail). Upon oxygenation the FG corner of the β subunit slides by one helical turn over the C helix of the α subunit. The α C helices of the structures of HbO₂ (light blue, PDB file 2dn1) and Hb (green, PDB file 2dn2) were superimposed to make evident the relative movement of the β FG corner. (*See insert for color representation of the figure.*)

oxygen affinity. *The structural bases of Hb cooperativity should therefore be looked at in this interplay between the tertiary and quaternary levels of structural organization of the macromolecule.* Table 7.4 summarizes some of these structural changes as distances between the C helix and FG corner of the same subunit, or between the FG corners between the α and β subunits of the same α1β1 dimer.

Two other regions of the subunits experience significant ligation-dependent structural changes, namely the last two C-terminal residues and the residues that come in close contact with the bound ligand on the distal side of the heme (i.e., around the distal His, at topological position E7), especially in the β subunit.

In unliganded Hb, the penultimate residue of each subunit (Tyr HC2) packs in a cavity located at the α1β2 interface, and interacts with the residue Val FG4 of the same subunit. The last residues of the α (Arg HC3) and β subunits (His HC3) form the ionic interactions listed in Table 7.5. By contrast in liganded Hb the last two C-terminal residues of all subunits are disordered and the interactions described in Table 7.5 are broken. These interactions are required to stabilize the structure of unliganded Hb, and deletion of the last two C-terminal residues by limited proteolytic digestion or genetic removal abolishes cooperativity.

The reader may notice that the salt bridges of the C-terminal residues involve the two groups which are most relevant to the alkaline Bohr effect (Val 1α and His 146β): indeed the molecular mechanism of the Bohr effect is strongly linked to the fact that deprotonation of the Bohr residues interrupts these interactions, and thus favors the ligand-linked quaternary structure change (Perutz, 1970). Accordingly, removal of the last two residues of the subunits by carboxypeptidase digestion abolishes cooperativity, increases the oxygen affinity, and diminishes or abolishes the Bohr effect (Antonini *et al.*, 1961). Allosteric effectors may partly restore these properties (Bonaventura *et al.*, 1972). The same effects are observed in the naturally occurring mutant Hb McKees Rocks, which bears the genetic deletion of the last two residues of the β subunit (Winslow *et al.*, 1976). Removal of only the last residue of each subunit, yielding the so-called des-Arg and des-His Hbs (which lack the C-terminal Arg of the α subunits or the C-terminal His of

Table 7.4 Interatomic distances as markers of the ligand-dependent structural changes in hemoglobin.

Hb derivative	PDB code	α1C6-α1FG4	β1C6-β1FG4	α1FG4-β1FG4	β1Fe-β2Fe	P$_{1/2}$ (mmHg)
Human deoxy (T)	2HHB	9.73	9.66	31.74	39.5	41.5
Human deoxy (T)	2DN2	9.71	9.66	31.76	39.5	41.5
Human αNi βFeCO (T)	1NIH	9.88	9.73	31.67	40.0	44
Human oxy (T)	1GZX	9.70	9.84	31.78	39.8	150
Trout Hb1 deoxy (T)	1OUT	9.88	9.53	32.73	39.1	148
Tuna deoxy (T)	1V4W	9.68	9.15	33.12	38.99	
Tuna deoxy (T)	1V4X	9.44	9.19	32.95	38.88	
Human oxy (R)	2DN1	10.38	9.91	29.05	34.9	0.3
Human CO (R)	2DN3	10.38	10.04	29.25	35.0	
Human CO (R2)	1BBB	10.28	10.49	28.58	34.33	
Human CO (R3)	1YZI	10.4	10.94	27.49	30.7	
Horse deoxy HbBME (R)	1IBE	10.11	10.27	29.91	34.3	R-like
Horse CO BZF (R)	1IWH	10.61	9.86	29.79	34.7	
Trout Hb1 CO (R)	1OUU	9.92	9.96	31.05	33.3	
Tuna Hb CO (R)	1V4U	9.75	10.22	28.9	34.39	
Sperm whale deoxy Mb	1A6N	10.80				
Sperm whale CO Mb	1MBC	10.69				

The table shows that interatomic distances at the Cα of aminoacids occupying selected topological positions neatly groups into two classes, one for the T state and one for the R state. Unliganded R state and liganded T-state group with the corresponding thermodynamic and structural state, rather than with the corresponding ligation state (from Bellelli and Brunori, 2011). Myoglobin presents an interesting term of comparison, because it lacks the quaternary constraint, and it cannot access the T-state tertiarty structure markers, even when unliganded. The reader may want to compare the αC-βFG distances in this Table to the schematic depiction of the allosteric model in Figure 4.8.

Table 7.5 Ionic interactions of the C-terminal residues in unliganded Hb.

α-carboxylate of Arg HC3 α1	---	α-amino group of Val NA1 α2
guanidinium group of Arg HC3 α1	---	γ-carboxylate of Asp H9 α2
α-carboxylate of His HC3 β1	---	ε-amino group of Lys C5 α2
imidazole group of His HC3 β1	---	γ-carboxylate of Asp FG1 β1

the β subunits, but not the penultimate Tyr residues) exerts milder effects, with reduction of the Bohr effect, and increase of oxygen affinity (Moffat *et al.*, 1973; Kilmartin *et al.*, 1980).

The residues on the distal side of the heme pocket, where the heme ligands bind, undergo a ligand-dependent structural rearrangement that is different in the α and β subunits. In the unliganded α subunits the distal cavity is partially occupied by a water molecule that is expelled by oxygen (or other heme ligands); moreover, the seventh residue of the E helix, the distal His, forms a H-bond with the iron-bound oxygen. His E7 is optimally placed to interact with the distal atom of a diatomic ligand that binds in a bent conformation, like O_2 and NO (see above, Section 7.1). By contrast, this residue clashes with a diatomic ligand binding in a straight conformation like CO. As a consequence of these interactions, His E7 favors binding of O_2 (and presumably NO) and decreases that of CO. This effect is important to reduce the risk of CO intoxication, because small amounts of CO are produced physiologically and the gas may be absorbed as an environmental pollutant.

In unliganded Hb, the distal pocket of the β subunits does not contain a water molecule, but is collapsed and allows no room for oxygen or other heme ligands. Thus, a more extensive structural rearrangement is required during ligation. Moreover, the H-bond between O_2 and His E7, if it forms at all, is weaker and less relevant to O_2 affinity than in the α subunits, as demonstrated by the effects of site directed mutation His E7 -> Gly (Olson *et al.*, 1988).

7.5   Quaternary Constraint

The α and β subunits of hemoglobin can be separated by prolonged incubation with the thiol reagent p-hydroxy mercuryl-benzoate (PMB) at acidic pH (Bucci and Fronticelli, 1965). The reagent labels all Cys residues of the macromolecule (one on the α subunits at position 104,G11 and two on the β subunits at positions 93,F9 and 112,G14) and since two of these lie at the G-G contact region of the $\alpha1\beta1$ intersubunit interface, their covalent modification with such a bulky reagent promotes dissociation of the tetramer. The α^{PMB} and β^{PMB} subunits can then be separated by ion exchange chromatography, thanks to their large difference in surface charge. PMB can be removed by incubation with a thiol reagent (e.g., β-mercaptoethanol or dodecanthiol). The native subunits, after removal of PMB, tend to aggregate: the αs form a weakly bound dimer, whereas the βs form a stable tetramer called HbH. The intersubunit contacts of HbH are topologically very similar to those of Hb (Borgstahl *et al.*, 1994a, 1994b). In all cases the isolated subunits or their oligomers bind heme ligands non-cooperatively and with affinity similar to that of R-state Hb. Thus, Hb is a case of pure quaternary constraint, in which the native tetramer has lower affinity than its isolated monomeric constituents. Importantly, upon mixing equimolar amounts of the α and β subunits, after removal of PMB, fully functional, low-affinity, cooperative Hb tetramers are obtained, indistinguishable from native Hb.

Quaternary enhancement, that is, an increase of ligand affinity for the last binding step of the tetramer over the affinity of the isolated subunits, has been proposed by some authors but has been essentially ruled out by photochemical kinetic measurements that can significantly populate the pertinent ligation intermediate (Gibson and Edelstein, 1987; Philo and Lary, 1990). Indeed it was experimentally found that the rate constants of oxygen binding and dissociation for the fourth binding step do not appreciably differ from those of the isolated subunits. Thus, if quaternary enhancement exists at all it must be extremely small indeed.

One may question whether the $\alpha\beta$ dimers formed upon dissociation of the Hb tetramer are cooperative, like Hb, or non-cooperative, like the isolated α and β subunits in whatever aggregation state. Solving this problem has required a large amount of work, but the final answer is unequivocal: the heterodimers bind ligands non-cooperatively (Edelstein *et al.*, 1970). This result is fully consistent with the observation, reported above, that ligand-dependent structural changes are essentially limited to the interdimeric α1β2 interface.

The structural bases of the quaternary constraint have been studied by site-directed mutagenesis and x-ray crystallography, but have not been completely unraveled yet. The soundest hypothesis was formulated by M.F. Perutz (Perutz, 1970): quaternary constraint would be due to the fact that the α1β2 interface contacts of deoxyHb force a position of the F helix in which the His F8-Fe bond is stretched. This stretch is reflected in the N-Fe bond distance and hinders the movement of the iron toward the heme plane caused by ligand binding thus constraining the affinity. Perutz's hypothesis has been largely confirmed for the α subunits of human Hb in experiments in which the proximal His was replaced by Gly by site-directed mutagenesis (Barrick *et al.*, 1997). In this site-directed Hb mutant imidazole was added and found to bind to the heme iron at the proximal side, whereas external ligands can bind at the distal side. The tetramer containing the mutated α subunits and wild type β subunits ($\alpha_2^{H87G}\beta_2$) presents high ligand affinity and essentially no cooperativity, as predicted by Perutz's hypothesis. In the case of the β subunits, the results of the same experiment are not as clear-cut, because the $\alpha_2\beta_2^{H93G}$ mutant presents some residual cooperativity. Thus, in the case of the β subunits the main determinant of the quaternary constraint is probably not the position of the F helix. Still, experiments carried out by site directed mutagenesis of Trp 37β (C3), a residue located at the α1β2 interface, reveal that the quaternary constraint is due to the structural arrangement of that interface, even though how it is transduced to the heme is unclear (Noble *et al.*, 2001).

7.6 Structural Aspects of Cooperativity: Allostery

Several hypotheses have been advanced to explain the cooperativity of hemoglobin at the molecular level. The most influential of these has been the allosteric model proposed by J. Monod, J. Wyman, and J.P. Changeux in 1965. In spite of some obvious shortcomings, there is general consensus that this model effectively captures some fundamental features of Hb and that more accurate models should be viewed as its evolution (Eaton *et al.*, 2007; Bellelli, 2010). A general description of the two-state model was given in Sections 4.6 and 4.7; here, we shall review the structural bases of allostery in hemoglobin.

A minimal summary of the allosteric model of cooperativity is as follows. The model hypothesizes that the macromolecule can sample two structural conformations, called T (for tense) and R (for relaxed). The two structures are true thermodynamic states because they are postulated to have different ligand affinities. Moreover, the T structural conformation is more populated in the absence of the ligand and has lower ligand affinity; the R structural conformation is more populated in the presence of saturating ligand concentration and has higher ligand affinity. Each allosteric state binds ligands non-cooperatively; however, given the above premises, increases of ligand concentration and degree of saturation cause a decrease of the population of T state and an increase of the population of R state, thus promoting an increase of the apparent ligand affinity of the system that is the weighted average of the affinities of the two states.

Some properties of hemoglobin conform very well to the above hypotheses. In particular, we may assume, with some caution, that the structure of T-state Hb (^{T}Hb) is similar to that of unliganded Hb, and the structure of R-state (^{R}HB) is similar to that of liganded Hb (Table 7.4). That is, in the case of Hb thermodynamic states and three-dimensional quaternary structures may correspond quite closely, and one is justified in speaking of T state and T structure, as well as R state and R structure. This correspondence between states and structures cannot be assumed as general and other allosteric systems may present more complex structure-function relationships. We consider the state-structure correspondence in Hb as a sort of basic finding, and present here some supporting experimental evidence. We shall next consider how this basic finding shapes the implementation of the allosteric model to Hb. Finally, in the next section, we shall discuss some more refined experiments that in some cases provide further support, and in other cases do not support, the basic relationship between states and structures.

The allosteric model postulates that the T and R states are both populated at any degree of ligand saturation, and that ligand binding to either state in the absence of the allosteric structural change is non-cooperative. Unliganded Hb has been crystallized, and its three-dimensional structure has been solved (Fermi *et al.*, 1984; Park *et al.*, 2006); some details of its structure have been described in Section 7.3. One may reasonably expect that this structure corresponds to the T state, because model fitting of the ligand binding isotherms suggests that in the absence of the ligand the T state predominates over the R state by 10^4- to 10^6-fold. Experimental evidence confirms this expectation: under selected experimental conditions, the crystal lattice forces may prevent the ligand-induced allosteric structure change, which otherwise would break the crystals, thus allowing ligand binding to ^{T}Hb within the crystal to be recorded. This experimental approach allowed Mozzarelli and co-workers to record oxygen-binding isotherms of ^{T}Hb in crystals. These authors found that (i) oxygen binding to ^{T}Hb crystals is non-cooperative, and (ii) it occurs with the same affinity recorded for the T state in solution, in the presence of strong heterotropic effectors (Mozzarelli *et al.*, 1991, 1997; Rivetti *et al.*, 1993). Thus we may confidently assume that the structure of unliganded Hb corresponds to the T state.

There is further evidence in favor of the above identification of T structure and T state. Taking advantage of the experimental conditions that allow the oxygenation of crystals of ^{T}Hb, Paoli and co-workers were able to obtain crystals of ^{T}Hb(O$_2$)$_4$ and to solve the 3D structure of the macromolecule, thus obtaining the structure of ^{T}Hb$_4$ (Paoli *et al.*, 1996). The structure obtained by these authors is nearly identical to that of ^{T}Hb, and the interface contacts are exactly the same, as are the C-terminal salt bridges. Flattening of the six-coordinate heme is incomplete, a feature that has been interpreted as a strain at the level of the heme, caused by the unliganded-like position of the proximal His and the F helix (Paoli *et al.*, 1997).

Crystals of fully liganded Hb can be easily obtained and the structures of several liganded derivatives including ^{R}Hb(O$_2$)$_4$, and ^{R}Hb(CO)$_4$ have been solved (Shaanan, 1983; Baldwin, 1980; Park *et al.*, 2006). Some details of these structures, including the arrangement of the interface contacts, have been described in Section 7.3 and compared to those of unliganded ^{T}Hb$_0$. Assigning the structure of fully liganded Hb to the R state is less straightforward than assigning the structure of unliganded Hb to the T state, because no experimental condition has been found to record the deoxygenation isotherms of crystals of Hb(O$_2$)$_4$: as observed by F. Haurowitz in 1938, deoxygenation dissolves the crystals of liganded Hb. However, a direct determination of the reactivity of crystals of ^{R}Hb(CO)$_4$ was obtained by Parkhurst and Gibson who recorded the time course of CO rebinding to crystals of horse ^{R}Hb(CO)$_4$ after partial photolysis. In this case, the deligation is only transient and incomplete, and Hb rebinds CO before the crystals break (Parkhurst and Gibson, 1967). These experiments confirmed that Hb frozen in the structure obtained by crystallization of the fully liganded CO derivative has R-like reactivity. Thus, we are justified to identify an R state and an R structure.

The 3D structure of unliganded ^{R}Hb could be solved using chemically modified derivatives unable to switch to the T structure when unliganded. The best model of unliganded ^{R}Hb is provided by horse Hb reacted with bis(N-maleimidoethyl) ether (BME) [Wilson *et al.*, 1996). DeoxyHbBME has the interface contacts characteristic of fully liganded ^{R}Hb, and the doming of the heme is much less pronounced than in unliganded ^{T}Hb. Two other structures that may be relevant to the present discussion are those of liganded and unliganded HbH (β_4) (Borgstahl, 1994a, 1994b), whose interface contacts closely mimic those of Hb. HbH is interesting because it is a non-cooperative, high oxygen affinity tetramer, whose structure is very similar to that of fully liganded ^{R}Hb irrespective of HbH being liganded or unliganded; thus, in this case no doubt is possible on the identification between structure and reactivity.

The two-state model postulates perfect functional symmetry of the protein oligomer, and Hb comes as close as possible to a perfectly symmetric tetramer, given that it is a symmetric homodimer of pseudo-symmetric $\alpha\beta$ dimers (it is manifestly impossible for the subunits of a tetramer to form four identical isologous interfaces, and indeed in Hb two different types of intersubunit interfaces exist, $\alpha1\beta1$ and $\alpha1\beta2$); moreover, the α and β subunits have essentially the same ligand affinity in both liganded and unliganded Hb (Mims *et al.*, 1983). From the above discussion on the identification of structures and states, we feel entitled to say that symmetry applies to both states, because neither ^{T}Hb nor ^{R}Hb admit differences in the interfaces of the same type (i.e., in no structure does the $\alpha1\beta2$ differ from $\alpha2\beta1$, nor $\alpha1\beta1$ differ from $\alpha2\beta2$).

We may write the binding polynomial and fractional ligand saturation for a symmetric tetramer that obeys the two-state model as follows.

$$[Hb]_{tot} = {}^R Hb\left[\left(1+[x]/K_R\right)^4 + L_0\left(1+[x]/K_T\right)^4\right] \qquad \text{(eqn. 7.1)}$$

$$\bar{X} = \frac{[X]/K_R\left(1+[X]/K_R\right)^3 + L_0[X]/K_T\left(1+[X]/K_T\right)^3}{\left(1+[X]/K_R\right)^4 + L_0\left(1+[X]/K_T\right)^4} \qquad \text{(eqn. 7.2)}$$

As stated in the above Section 7.2, the O$_2$ affinity of Hb cannot be discussed without reference to heterotropic ligands. In the original formulation of the model, the effect of heterotropic ligands was explained by assuming preferential binding to the T state. Because almost all effectors present negative heterotropic linkage, the hypothesis of preferential binding to the R state, though possible, did not seem practically relevant. Under this assumption, one should substitute L_0 in eqns. 7.1 and 7.2 with $L_{0, \text{app}}$, defined as:

$$L_{0,app} = L_0 \frac{1+[Y]/{}^T K_Y}{1+[Y]/{}^R K_Y} \qquad \text{(eqn. 7.3)}$$

where ^TK_Y and ^RK_Y represent the dissociation equilibrium constants of the effector Y from ^THb and ^RHb respectively. This formula presumes a stoichiometry of 1 molecule of the effector per tetramer, and applies to DPG or IHP; in the case of effectors binding with higher stoichiometry, for example, protons, it would be more complex. Notice that in this formulation it is assumed that the effector Y does not change K_T or K_R (the dissociation equilibrium constants of the heme ligand X from ^THb and ^RHb respectively). Thus, the original model of Monod, Wyman, and Changeux did not envisage direct heterotropic linkage between X and Y in either ^THb or in ^RHb.

This assumption soon proved an Achilles heel of the two-state model, because most allosteric effectors do indeed change the value of K_T (see Figure 7.2). The greater resistance of K_R to allosteric effectors is only apparent and is explained by the lower affinity of ^RHb for heterotropic ligands, which under common experimental conditions bind scarcely to ^RHb, if at all. It has been consistently demonstrated by several authors that if the concentration of allosteric effectors is increased enough, these also bind to ^RHb and change the value of K_R (Antonini *et al.*, 1982; Imai, 1983; Yonetani, 2002). As pointed out by Yonetani, the dependence of K_R and K_T on heterotropic ligands has the physiologically important effect of increasing the cooperativity of Hb over and above what one would expect from the allosteric structural change alone (see Sections 4.9 and 4.10).

A consequence of the fact that the affinity of (at least) ^THb depends on heterotropic effectors, some of which are always present in solution (e.g., hydrogen ions), is that, as demonstrated by eqn. 4.30, *^THb may present a form of statistical cooperativity due to the oxygen-dependent release of heterotropic effectors.* This consideration may explain why intrinsic T-state cooperativity has been suggested by some experimental results obtained in solution (Di Cera *et al.*, 1987; Gibson, 1999; Ackers and Holt, 2006), and has not been observed in crystals (Mozzarelli *et al.*, 1991), where heterotropic effectors cannot be released during oxygenation.

The allosteric model may be corrected to accommodate the dependence of all three parameters on allosteric effectors. This, however, leads to a much more complex binding polynomial, on whose formulation there is no general agreement; at least two variants exist, the three-state model of Minton and Imai (1974) and the so called "global allostery model" of Yonetani and co-workers (2002). We shall not describe these models, which are quite specialized, but we remark that the elegance of the two-state model resides in its simplicity and in its general predictive ability, rather than in the strict adherence to the experimental detail: one can always increase the latter characteristic at the expense of the former two, but this is not always a good choice. Moreover, combination of the two-state model with the effect of heterotropic linkage with protons and anions within the T and R states, would probably provide a better and conceptually simpler explanation of the experimental data.

Several experimental data agree well with the predictions of the allosteric model. We shall consider in the following sections: (i) the 3D structures of partially liganded mixed metal hybrid Hbs (see Section 7.9); (ii) the direct determination of kinetic rate constants for allosteric structural change at different degrees of ligation (Section 7.8); (iii) the effect of site directed mutagenesis or proteolytic digestion of residues involved relevant to allosteric conformational changes above (Section 7.11); and (iv) the interpretation of ligand-linked dissociation (Section 7.10).

7.7   Structure and Energy Degeneracy

The two-state allosteric model hypothesizes two thermodynamic states, characterized by different affinity (i.e., binding free energy). The simplest possible realization of the model occurs when each state is associated to a structure (i.e., a conformational isomer of the protein), and the allosteric constant describes the equilibrium between the two isomers. This is clearly the picture that Monod and co-workers had in mind when they imagined the model, and it would be very difficult to interpret otherwise some basic premises like the requirement of symmetry or the role of isologous interfaces (Monod *et al.*, 1965). Hb comes close to realizing these premises, perhaps closer than any other protein, but still has deviations and peculiarities that deserve consideration, if anything to evaluate whether they require adaptations of the model.

An important property of Hb, that was unforeseen when Monod, Wyman, and Changeux conceived the model, is the degeneracy of the R state. More than a single structure is known for the high-affinity liganded R state of human Hb, depending on the crystallization conditions. At least three structures of liganded Hb can be obtained, called R, R2, and R3, and possibly some intermediate ones (Safo and Abraham, 2005). Thus, the R state presents *structural degeneracy*. It is our opinion that structural degeneracy of the R state is not incompatible with the allosteric model and does not necessarily require any extension of the model. Indeed, Hb can have access to an unlimited number of unconstrained structures, which are all functionally equivalent: the isolated subunits, the $\alpha\beta$ heterodimers, and the β_4 homotetramers are all variants of the high-affinity R state. The R state in solution may well be a mixture of structural isomers having the same ligand affinity, because they are all unconstrained. The relative population of each isomer would be governed by an equilibrium constant analogous to L_0, but since all isomers have the same ligand affinity their relative populations would not change upon ligation and they would behave as a single thermodynamic state.

On the other hand, more than one structure would be quite unlikely for the constrained T state. In this case, a very specific structural arrangement is required, identified in the conformation of the F helix and FG corner imposed by the T-state structure of the $\alpha1\beta2$ interface, and it is implausible that a different structural arrangement could produce the same constraint. Consistent with this premise, structural degeneracy of ^{T}Hb has not been observed in the same Hb crystallized under different conditions: rather a family of closely related T-like structures has been collected for a family of site-directed interface mutants, each protein apparently having only one T-like structure (Kavanaugh, 2005). We suggest that if two different constrained structures exist, the protein would become a three-state system, as in Minton and Imai's model of cooperativity (Minton and Imai, 1973): that is, in the case of the constrained state, structure degeneracy would probably be associated to a degeneracy of the ligand binding energies (see below). However, structural degeneracy is not required to explain the clear-cut *degeneracy of the ligand binding energies of ^{T}Hb* caused by heterotropic effectors (Section 7.2). We may summarize the point as follows: the unconstrained state is compatible with different structural arrangements, all having the same ligand binding energy. The constrained state has basically a single, very specific structure, very subtle changes of which, caused by the binding of allosteric effectors, are associated to large changes in the ligand binding energy. The available evidence on ^{T}Hb is insufficient to decide whether it should be considered an equilibrium mixture of two states, in which case the appropriate description of Hb would be provided by the three-state model, or rather a case of heterotropic linkage in which the effectors act by an induced fit mechanism.

There is one point in the analysis of structural and energy degeneracy reported in this section that remains problematic, namely the energy degeneracy of ^{R}Hb observed in the presence of allosteric effectors. We observed in Section 7.6 that heterotropic effectors are known that lower the oxygen affinity of ^{R}Hb as well as ^{T}Hb. This observation conflicts with the identification of the R state and the R structures as unconstrained, and indeed Table 7.4 shows that the interatomic distances at the α1β2 interface of ^{R}HbCO in complex with the effector bezafibrate (BZF) resemble those of the unconstrained R state, with the possible exception of the β subunits. It has been suggested that these effectors perturb the protein structure at sites different from the cooperative α1β2 interface, possibly at the distal site. In any case, the structural bases of negative heterotropic effects in ^{R}Hb remain uncertain.

The simplest global description of Hb cooperativity at present seems to be the following. Hb is basically a two state system, presenting a low-affinity T state clearly associated with a well defined structure, and a structurally degenerate high-affinity R state. Each state presents negative heterotropic linkage phenomena between heme and non-heme ligands. Heterotropic linkage causes changes of ligand affinity of the T and R state and increases the homotropic cooperativity of heme ligands. Thus, in Hb the effects described in Sections 4.6 and 4.9 are present simultaneously.

7.8 Kinetics of Ligand Binding

The study of the kinetics of ligand binding to Hb has provided important information on its reaction mechanisms; however, hemoglobin binding kinetics was unraveled while the instrumentation was being developed and the structure of the protein was being solved: thus, the original experiments have often required extensive re-interpretation.

As a general rule the time courses of ligand binding to or release from hemoglobin present autocatalysis, that is, the reaction becomes faster and faster as it proceeds. Autocatalysis is the kinetic counterpart of equilibrium cooperativity, and is absent under conditions in which ligand binding is non-cooperative (e.g., because of chemical modification or genetic mutation). However, autocatalysis is not equally expressed in the time courses of ligand association and dissociation, and at times it may be quite minor and difficult to detect. For example, in the case of oxygen it is mainly expressed in the dissociation reaction, whereas the association reaction is scarcely autocatalytic, if at all. The opposite is true for CO, which presents strongly autocatalytic combination time courses, and weakly autocatalytic dissociation time courses. As pointed out by several authors (Szabo, 1978; Moffat *et al.,* 1979), this behavior is consistent with transition state theory and suggests that: (i) for ligands whose association rate constants are high (i.e., close to the diffusion limit; this applies especially to O_2 and NO) autocatalysis is predominantly expressed in the dissociation reaction, whereas the opposite holds for ligands whose combination rate constants are low (e.g., CO and isocyanides); and (ii) rapidly binding ligands, whose autocatalysis is predominantly expressed in the dissociation reaction, should have structurally reactant-like transition states, whereas

Table 7.6 Overall kinetic rate constants for the gaseous ligands of ferrous Hb and Mb.

Protein	$k_{d,O2}$ (s^{-1})	$k_{a,O2}$ (M^{-1}s^{-1})	$k_{d,CO}$ (s^{-1})	$k_{a,CO}$ (M^{-1}s^{-1})	$k_{d,NO}$ (s^{-1})	$k_{a,NO}$ (M^{-1}s^{-1})
^{R}Hb α	12	5.9×10^7	0.01	6.5×10^6	(1×10^{-5})	(2.5×10^7)
^{R}Hb β	21	5.9×10^7	0.01	6.5×10^6	(1×10^{-5})	(2.5×10^7)
^{T}Hb α	2000	2.9×10^6	0.1	1×10^5	(1×10^{-3})	(2.5×10^7)
^{T}Hb β	2000	1.2×10^7	0.1	1×10^5	(1×10^{-3})	(2.5×10^7)
isolated α subunits	28	5×10^7	0.013	4×10^6	4.6×10^{-5}	2.4×10^7
isolated β subunits	16	6×10^7	0.008	4.5×10^6	2.2×10^{-5}	2.4×10^7
Sperm Whale Mb	10	1.5×10^7	0.015	5×10^5	1.2×10^{-4}	1.7×10^7

Data for O_2 and CO are from Mims *et al.*, 1983 and Shibayama et al., 1995; for NO from Cassoly and Gibson, 1975 and from Moore and Gibson, 1976. Notice that in the case of NO the values for R- and T- state Hb are not resolved for the two subunits; hence, they are in parentheses.

slowly binding ligands, whose autocatalysis is predominantly expressed in the association reaction should have product-like transition states.

Q.H. Gibson suggested long ago that the diffusion properties of the diatomic gaseous ligands are essentially the same both in the bulk and inside the protein matrix, whereas activation energy at the heme iron or inside the heme pocket may vary greatly, and explains the observed differences in the kinetic rate constants of the association and dissociation reactions of common ligands (Table 7.6).

The discovery of autocatalysis should probably be ascribed to F.J.W. Roughton, among the first students to attempt the quantitative interpretation of the time courses of ligand binding to hemoglobin (Bateman and Roughton, 1935). However, over two decades of further work and the development of the stopped flow apparatus were required before the more complete study "The determination of the velocity constants of the four successive reactions of carbon monoxide with sheep hemoglobin" (Gibson and Roughton, 1956) could be completed.

The hemoglobin concentration employed at the time was high, due to the poor sensitivity of the detector, and the concentration of CO was higher than that of Hb by only three-fold due to the time resolution of the instrument. As a consequence, the reaction could not be approximated as a pseudo-first order reaction, and numerical integration was required, which was a remarkable accomplishment in view of the limited computational tools available at the time. As demonstrated in Section 6.5 the apparent rate constant of a sequence of reactions is limited by the rate constant of the first one; thus, Gibson and Roughton relied on inference to estimate the progression of the rate constants beyond $k_{a,1}$.

The direct determination of the rate constant for binding to the fourth site of Hb was achieved only after the introduction of photochemical methods. The discovery of "The photochemical formation of a quickly reacting form of haemoglobin" is again due to Q.H. Gibson (1959), but was quite puzzling at the time, since the fraction of the quickly reacting form was strongly dependent on pH and hemoglobin concentration.

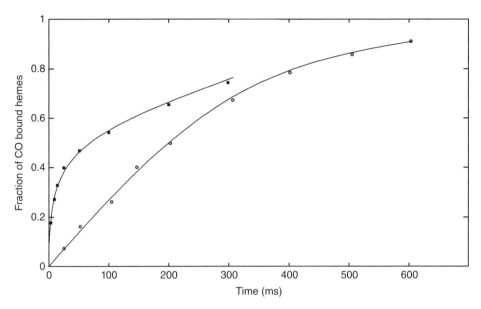

Figure 7.8 Time courses of CO combination to human hemoglobin by stopped flow (open circles) and flash photolysis (closed squares). Experimental conditions: T = 20 °C; buffer 0.1 M Bistris plus 0.1 M NaCl pH = 7; Hb concentration 2.8 µM (on a heme basis); CO concentration 25 µM. The time course obtained by rapid mixing of unliganded Hb and CO (open circles) is autocatalytic; that obtained by photolysis is described by two exponentials, one for the quickly reacting dimers, the other for the slowly reacting tetramers. Redrawn from Bellelli et al. 1987.

It took eight years before Gibson and E. Antonini, using a newly developed instrument that coupled a stopped flow and a flash photolysis apparatus, could tentatively suggest that the quickly reacting form "may be due to the dissociation of ligand bound hemoglobin" (Gibson and Antonini, 1967). This interpretation, though essentially correct, did not take into account the fact that two very different states of the macromolecule, both quickly reacting, may coexist in solution in this type of experiment, that is, R-like dimers and partially liganded R-state tetramers (Fig. 7.8). Kinetic data can be interpreted quite accurately within the framework of the two-state model (Henry *et al.*, 1997) and the coupling of stopped flow and photolysis has allowed the researcher to obtain a comprehensive and consistent set of measurements, summarized in Table 7.6.

With the advent of pulsed lasers of great brightness and very short duration, photochemical methods gave access to faster time regimes. In 1976, Sawicki and Gibson were able to follow the time course of the allosteric structural change, taking advantage of subtle absorbance differences between the high- and low-affinity states of the macromolecule. These authors used a 1 µs laser pulse of sufficient intensity to completely photolyze the HbCO complex and followed the time course of the relaxation from unliganded ^{R}Hb to unliganded ^{T}Hb, which occurs much faster than the rebinding of the ligand (Sawicki and Gibson, 1976). The reaction scheme is as follows:

$$^R\text{HbCO} \rightarrow {}^R\text{Hb} + \text{CO} \rightarrow {}^T\text{Hb} + \text{CO} \rightarrow {}^R\text{HbCO}$$

The first reaction is the photochemically induced dissociation of CO and occurs during the light pulse, CO dissociation in the dark being negligible under these experimental conditions. The next reaction is the conversion of unliganded R-state Hb to T state, whose signal is provided by the different absorbance of these two forms, with a peak at 430 nm (Bellelli and Brunori, 1994). The last reaction is the bimolecular rebinding of CO, coupled to the allosteric structural change. The two last processes are not resolved in time, and their coupling makes the reaction autocatalytic. Depending on the concentration of the protein, some R-like dimers may be present in solution. Their reaction scheme is as follows:

$$^R\alpha\beta CO \rightarrow ^R\alpha\beta + CO \rightarrow ^R\alpha\beta CO$$

The unliganded dimers cannot convert to T state, hence their rebinding occurs at a faster rate than that of tetramers (Table 7.6).

Thus, in this simple experiment one observes (and measures the rate constant of): (i) the allosteric structural change ^{R}Hb --> ^{T}Hb; (ii) the bimolecular rebinding of CO to R-state dimers; and (iii) the bimolecular and autocatalytic rebinding of CO to T-state tetramers. Reduction of the intensity of the laser pulse makes removal of CO incomplete. The less CO is removed, the smaller the amount of unliganded Hb that undergoes the T --> R transition and the greater the amount of partially liganded Hb that remains in the R state and rebinds CO over the same time regime as the dimers. Comparison of the effects of pulses of different intensity helps one to correctly interpret the absorbance changes and improves precision of the measurement of $k_{a,R}$.

Photochemical methods may be applied to the study of the rebinding of the geminate pair in Hb (see Section 2.5). This approach, which requires shorter and brighter pulses than those used to study the ^{R}Hb → ^{T}Hb transition, probes the pre-flash state of the protein because rebinding of the geminate pair occurs over a shorter time scale than any structural rearrangement. An interesting observation can be made in the case of the hemoglobin-CO complex, because the second geminate pair exhibits significant geminate recombination (up to 50%) if the starting derivative is R state and no recombination if it is T state.

In order to better characterize the second geminate pair, it is convenient to use laser pulses for photolysis in the order of 1-5 ns, because, at least in the case of NO and O_2, shorter pulses yield the proximate pair, whose rebinding prevents the accumulation of the second pair. If the light pulse lasts longer than the time course of rebinding of the proximate pair, the same molecule can be photodissociated over and over again, effectively building up the population of the second pair. Thus, the duration of the photolysis pulse effectively selects the photochemical intermediate one obtains. As a general rule, the yield of the geminate recombination of the second geminate pair is less than that of the first, and a significant amount of the ligand photodissociated by a ns laser pulse escapes to the solvent, from which rebinding is bimolecular. Even though a ns flash is not optimal for the study of the allosteric structural change of hemoglobin because of geminate rebinding, it is probably a good compromise. Eaton and co-workers obtained from this type of experiment a very complete description of the rebinding of CO to human hemoglobin (Table 7.7) (Hofrichter *et al.*, 1983).

Table 7.7 Kinetic processes observed in carbonmonoxy-hemoglobin after a 10 ns laser pulse capable of inducing full photochemical breakdown of the complex (from Hofrichter *et al.*, 1983).

Process	1/e time (rate constant)
I: geminate rebinding to ^{R}Hb	40–50 ns ($\sim 2.5 \times 10^7$ s^{-1})
II: tertiary relaxation of subunits where geminate rebinding did not occur	0.7–1 µs ($\sim 1 \times 10^6$ s^{-1})
III: quaternary R -> T structural change	18–20 µs ($\sim 5 \times 10^4$ s^{-1})
IV: CO bimolecular rebinding to ^{R}Hb	ligand dependent ($\sim 5 \times 10^6$ M^{-1}s^{-1})
V: CO bimolecular rebinding to ^{T}Hb	ligand dependent ($\sim 2.5 \times 10^5$ M^{-1}s^{-1})

7.9 Ligation Intermediates: Measurement and Structure

A fundamental prediction of the allosteric model is that ligation intermediates are either T state or R state, and have either the T or the R structure: no intermediate ligand affinity state or quaternary structure is predicted to exist. One can imagine a variant of the allosteric model having three (or more) states (Minton and Imai, 1974), but also in that model all states are postulated to be populated independently of the presence of the ligand, thus none is an obligate intermediate between the other two. Moreover all states must be symmetric, at least in a functional sense, that is, the ligand affinity of the subunits within each state should be exactly the same.

Thus, a model of cooperativity based on the classical formulation of allostery would be contradicted if the protein presented a ligation intermediate in which high- and low-affinity subunits could coexist, or one whose existence requires a specific number of bound ligands. Unfortunately, the characteristic feature of positive cooperativity is that ligation intermediates are poorly populated compared to the fully liganded and fully unliganded states (Figure 4.7B), making the study of ligation intermediates a difficult task. The MWC model predicts the existence of at least ten ligation intermediates, five (from fully unliganded to fully liganded) for ^{T}Hb and five for ^{R}Hb; more if one assumes that the distribution of ligands within the same intermediate is to be considered (e.g., if $(\alpha O_2)_2\beta_2$ is to be distinguished from $\alpha\alpha O_2\beta\beta O_2$).

Ingenious experimental approaches have been devised to prepare and study intermediates with two heme ligands bound per tetramer. To prepare doubly liganded intermediates, a sample of Hb is freed from the heme to obtain the globin. The globin is then reconstituted with a metal substituted protoporphyrin IX (e.g., Ni-PPIX or Mn-PPIX) to obtain the corresponding metal substituted Hb (NiHb or MnHb), whose subunits are then separated using the standard method, based on p-chloro mercuryl benzoate reaction (PMB). Tetrameric hemoglobin is then reconstituted by mixing the metal substituted subunit with the natural partner subunit (obtained by PMB splitting of native Hb) (Ikeda-Saito *et al.*, 1977). At the end of the procedure one obtains the mixed-metal hybrid Hbs like $\alpha_2^{Ni}\beta_2^{Fe}$ (or $\alpha_2^{Mn}\beta_2^{Fe}$) and $\alpha_2^{Fe}\beta_2^{Ni}$ (or $\alpha_2^{Fe}\beta_2^{Mn}$). Because the metal used to replace iron is chosen from those that form stable five-coordinate complexes and do not bind the ligands of iron, it mimics unliganded iron. Therefore, upon addition of a suitable ligand (usually CO) one obtains the mixed metal hybrid Hbs $\alpha_2^{Ni}\beta_2^{FeCo}$ (Luisi *et al.*, 1990) and $\alpha_2^{FeCo}\beta_2^{Ni}$ (Luisi and Shibayama, 1989) liganded on only two of the four subunits. In the case of manganese only the hybrid $\alpha_2^{FeCo}\beta_2^{Mn}$ has been structurally characterized (Arnone *et al.*, 1986). Assuming that the non-iron metal can be considered a good mimic for unliganded, five-coordinate iron, the results of these studies are unequivocal: mixed metal hybrids with two bound ligands preferentially adopt the T structure (consistent with the fact that this structure is prevalent in human Hb until the third ligand binds). These hybrids have never been crystallized in any intermediate structure, neither R nor T. Strain is observed at the heme and at the interdimeric interface. More recently, however, Shibayama and co-workers were able to describe a doubly liganded Hb derivative presenting a structure intermediate between T and R (Shibayama *et al.*, 2014). As yet, the relevance of this result is incompletely assessed.

The experiments described above have been criticized because the crystal packing forces may stabilize forms that may not be present or predominant in solution, and may select the least soluble species, which is not necessarily the more abundant in solution. However, positive evidence for the presence of other types of tertiary or quaternary structures, neither R nor T, in solution is scarce. In view of the ongoing work on this point we give only a quick overview of the experimental results that seem to conflict with the two-state model. Besides the above reported hemoglobin derivative with structure and ligand affinity intermediate between those of ^{R}Hb and ^{T}Hb by Shibayama and co-workers (2014), we quote the results by G.K. Ackers and co-workers, who studied the ligand linked dissociation of Hb and identified a ligation intermediate having R-like reactivity in spite of the fact that it is supposed to have T-like structure (see Section 7.10). We also quote Q.H. Gibson's finding that the rebinding of O_2 after photolysis is incompatible with a simple two state model and suggests that a fraction of the macromolecule may present a hybrid structure that includes T-like and R-like subunits (Gibson, 1999). Finally, Eaton and co-workers detected R-like reactivity in ^{T}Hb trapped in silica gels (Viappiani *et al.*, 2004).

7.10   Ligand-Linked Dissociation Into Dimers

Hemoglobin presents the phenomenon of ligand-linked dissociation. At neutral pH and low ionic strength the dimer-tetramer equilibrium constant for unliganded Hb is in the order of 10^{-12} M (Thomas and Edelstein, 1972), and that of liganded Hb is in the order of 10^{-7} M (Edelstein *et al.*, 1970) at neutral pH and low ionic strength. Both values are strongly dependent on solution conditions, and increase with increasing ionic strength. As a consequence, at concentrations in the order of a few μM or less, upon oxygenation the low-affinity tetramer reversibly dissociates into high-affinity dimers, and this causes an apparent increase in ligand affinity (Figure 7.9).

Quantitative analysis of the oxygen-binding isotherms at concentrations where this phenomenon is significant is complex because, as explained in detail below, tetramers bearing the same number of ligands may dissociate into different dimers, and thus cannot be considered functionally equivalent, as they are in the majority of cooperativity models. For example, doubly liganded tetramers, depending on the distribution of the two ligands among the four subunits, may dissociate into two singly liganded dimers or into a doubly liganded dimer and an unliganded one. In practice, the quantitative study of ligand-linked dissociation yielded meaningful results only if the experiments were

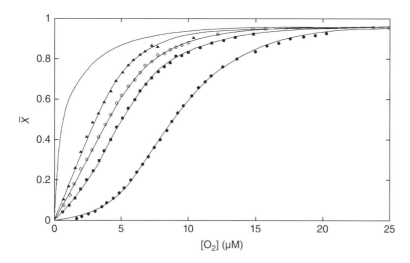

Figure 7.9 Effect of concentration on the oxygen affinity of human hemoglobin (modified from Mills *et al.*, 1976). Experimental conditions: 0.1 M Tris-HCl buffer pH = 7.4 containing 0.1 M NaCl and 1 mM EDTA; T = 21.5 °C; hemoglobin concentrations, on a heme basis (from right): 382 μM, 270 nM, 80 nM, and 40 nM. The leftmost curve indicates the calculated binding isotherm for pure αβ dimers. The dependence of ligand affinity on target concentration is the hallmark of ligand-linked dissociation (see Chapter 5).

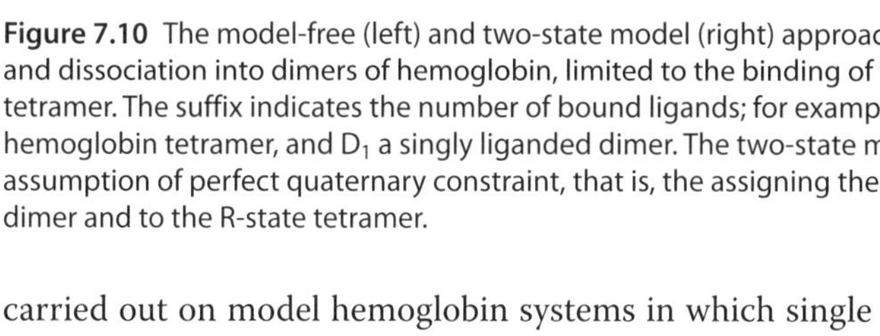

Figure 7.10 The model-free (left) and two-state model (right) approach to describe ligand binding and dissociation into dimers of hemoglobin, limited to the binding of the first ligand molecule to the tetramer. The suffix indicates the number of bound ligands; for example, Hb_0 indicates an unliganded hemoglobin tetramer, and D_1 a singly liganded dimer. The two-state model is written under the assumption of perfect quaternary constraint, that is, the assigning the same ligand affinity to the dimer and to the R-state tetramer.

carried out on model hemoglobin systems in which single specific ligation intermediates could be isolated (Section 7.9).

We outline below the model-free and the two-state model approach to ligand-linked dissociation for the binding of the first oxygen molecule to Hb (Figure 7.10).

The quantitative treatment for the model-free approach is straighforward. We define:

$$^D K_0 = [D_0]^2 / [Hb_0]$$
$$^D K_1 = [D_1][D_0] / [Hb_1]$$
$$K_{Hb1} = [Hb_0][X] / [Hb_1]$$
$$K_D = [D_0][X] / [D_1]$$

and we easily obtain:

$$^D K_1 = {}^D K_0 K_{Hb1} / K_D$$

or, generalizing:

$$^D K_i = {}^D K_{i-1} K_{Hbi} / K_D \tag{eqn. 7.4}$$

Eqn. 7.4 shows that the dimer dissociation constant increases by the factor K_{Hbi}/K_D (the ratio of the ligand affinity constants of the tetramer and the dimer, recalling that $K_{Hbi} > K_D$ because of the quaternary constraint) after the binding of each successive ligand molecule to the tetramer, and that the ratio of the dissociation constants of two successive ligation intermediates of the tetramer ($^D K_i / {}^D K_{i-1}$) equals the ratio of the ligand affinities (K_{Hbi}/K_D). This is a requirement of the thermodynamic linkage of subunit dissociation and ligand binding. In the absence of intrinsic cooperativity of the tetramer, one would observe K_{Hbi}/K_D = constant and the apparent cooperativity of the oxygen binding isotherm would be solely due to the change of $^D K_i$ (see Section 5.3). However, in almost all vertebrate Hbs, the tetramer is intrinsically cooperative and the ratio K_{Hbi}/K_D approaches unity as i approaches 4.

One may define a cooperativity factor as the ligand-linked change in the tendency of the tetramer to dissociate into dimers, with respect to any arbitrarily chosen reference intermediate. For example, if one takes as a reference state the unliganded tetramer, one obtains:

$$^D K_i / {}^D K_0 = \prod \mathbf{K}_{Hbi} / K_D{}^i$$

which would reduce to $(K_{Hb}/K_D)^i$ in the case of a (hypothetical) non cooperative tetramer.

An important aspect of this type of experiment and analysis lies in the fact that the relationship between ligation and dissociation of protein oligomers cannot ignore the geometry of assembly of the protein subunits. Hemoglobin is assembled via pseudo-isologous or isologous intersubunit interfaces. As a consequence, even though ligation intermediates having the same number of bound ligands may be functionally equivalent, they will behave differently in this type of experiment. For example, the four doubly liganded intermediates of Hb $\alpha_1 X \alpha_2 \beta_1 X \beta_2$, $\alpha_1 X \alpha_2 \beta_1 \beta_2 X$, $\alpha_1 X \alpha_2 X \beta_1 \beta_2$ and $\alpha_1 \alpha_2 \beta_1 X \beta_2 X$ dissociate into different dimers ($\alpha X \beta X + \alpha \beta$; $\alpha X \beta + \alpha \beta X$; $2\alpha X \beta$; and $2\alpha \beta X$ respectively). As a consequence, no two intermediates are fully equivalent in this type of measurement, but each must be considered in itself, and the pertinent equilibrium constant should not include a statistical factor, except in the case of perfectly symmetric intermediates. Thus this approach demands experiments capable of resolving each individual intermediate, beyond the mere number of ligands bound (see below).

Measurements of the constants for dissociation into dimers of ligation intermediates Hb_i and Hb_{i-1} may lead to three possible outcomes, according to eqn. 7.4:

i) the dimer dissociation constant increases as one more ligand binds to Hb_{i-1} to form Hb_i. This is the expected result for a cooperative Hb presenting the phenomenon of quaternary constraint, as it implies $K_{Hbi} > K_D$.

ii) The dimer dissociation constant remains unchanged upon binding of one more ligand. This implies $K_{Hbi} = K_D$, that is, ligation intermediate Hb_i has as high an affinity for the ligand as the dimer. This condition in vertebrate Hbs can occur for the last binding step(s).

iii) The dimer dissociation constant decreases as one more ligand binds to Hb_{i-1}. This implies that $K_{Hbi} < K_D$, that is, it implies that intermediate Hb_i presents quaternary enhancement, rather than quaternary constraint.

The majority of cooperativity models for Hb suggest a more or less even increase of ligand affinity as ligation proceeds; moreover, as stated above, there is no doubt that Hb cooperativity is essentially based on quaternary constraint and that quaternary enhancement is very minor, if present at all. Thus plausible expectations are case (i), and, at high degrees of ligation, where the ligand binding isotherm approaches its asymptote (see Figure 7.2), case (ii).

We now turn to the description of ligand-linked dissociation according to the MWC two-state model. A simplified two-state reaction scheme that takes into account the dimer-tetramer equilibrium for the first ligation step is depicted in Figure 7.10. If one assumes that Hb cooperativity is solely due to a perfect case of quaternary constraint, then the equilibrium constant for oxygen dissociation from the dimer and from the R-state tetramer is the same, and the dimer-tetramer equilibrium constant for ^RHb is independent of ligation. We may indicate this parameter as ^DK_R without any suffix to indicate the ligation state. The same does not apply to ^THb: as Figure 7.10 makes clear, the following relationship applies:

$$^DK_{T,1} = {}^DK_{T,0} K_T / K_R$$

and, generalizing:

$$^DK_{T,i} = {}^DK_{T,0} \left(K_T / K_R \right)^i$$

that is, the dimer-tetramer equilibrium constant of ^THb increases as ligation proceeds (Edelstein and Edsall, 1986).

We further observe that there is a relationship between the allosteric constant L_0 and the ratio of the dimer-tetramer equilibrium constants for the two allosteric states. This is demonstrated by the fact that there are two possible paths from $^RHb_0$ to $^THb_0$, that is, the direct conversion, governed by L_0, or the dissociation of $^RHb_0$ to two unliganded dimers and the reassociation of two unliganded dimers to $^THb_0$; this implies that the allosteric constant L_0 equals the ratio of the dimer-tetramer equilibrium constants of the two allosteric conformations of Hb in the absence of the ligand, that is:

$$L_0 = {}^DK_R / {}^DK_{T,0}$$

and, generalizing:

$$L_i = {}^DK_R / {}^DK_{T,i} = \left({}^DK_R / {}^DK_{T,0} \right) \left(K_R / K_T \right)^i$$

As a consequence of the above relationships, one may explore hemoglobin cooperativity by studying the dimer-tetramer equilibria of its ligation intermediates or mimics thereof. G.K. Ackers pioneered this approach by devising clever experiments that make use of mixed metal hybrid Hbs (Section 7.9) or mixed valency hybrid Hbs (i.e., Hb tetramers in which one type of subunit is present as the cyano-met derivative, the other type as the ferrous derivative; mixed valency hybrids may be prepared using a procedure similar to that used for mixed-metal hybrids). These hybrid Hbs mimic doubly liganded states, and can be mixed with unliganded or fully liganded tetramers in order to obtain, by recombination, mimics of the singly liganded and triply liganded species. Ackers then measured the rate constants of tetramer dissociation into dimers.

The experiment takes advantage of the serum protein haptoglobin (Hp), whose physiological function is to sequester with high affinity hemoglobin dimers released by intravascular hemolysis. Because the heme is an extremely powerful quencher of the intrinsic fluorescence of proteins, the formation of the Hp-Hb complex is conveniently followed by quenching of Hp fluorescence. Thus, by rapidly mixing the Hb solutions with Hp one may follow the time course of formation of the complex Hp-$\alpha\beta$, which is rate limited by the rate constant of dissociation of Hb tetramers into dimers (Alfsen *et al.*, 1970).

Devising experiments to measure the rate constant of reassociation of Hb dimers is much more difficult, and can be done for only some Hb derivatives: for example, rapid deoxygenation of HbO_2 (which at μM concentrations is significantly dissociated into dimers) by mixing with sodium dithionite is followed by the slow bimolecular process of reassociation of unliganded dimers to form deoxyHb (whose dimer dissociation constant is much smaller than that of HbO_2). This process is associated with the $^RHb_0$-$^THb_0$ difference spectrum, and can be recorded by absorbance spectroscopy using a stopped flow apparatus (Bellelli and Brunori, 1994). Unfortunately, ligation intermediates cannot be subjected to the same experiment, thus, in order to infer the dimer-tetramer equilibrium constant from kinetic measurements, one has to assume that the second order rate constant of the reaction of dimer association is ligand-independent. Because of the above problems, inferring the equilibrium constant of the dimer-tetramer equilibrium of ligation intermediates rests on the (unproven) assumption that the dimer association rate constants are independent of ligation.

We can schematically describe Ackers' experiments on ligand-linked dissociation of Hb as follows.

i) The kinetic rate constant of dissociation of unliganded tetramers ($^Dk_0$) is measured by mixing with haptoglobin; this experiment is perfectly unequivocal.
ii) The kinetic rate constant of dissociation of fully liganded tetramers ($^Dk_4$) or its mimics (e.g., $HbFe^{+3}CN^-$) is measured by mixing with haptoglobin; this experiment is again unequivocal.
iii) Mixed valency hybrids mimicking the doubly liganded tetramers may be obtained by mixing unliganded β subunits with the cyano-met derivative of α subunits (or vice-versa); these hybrid Hbs allow the researcher to measure two different $^Dk_2$ values for ligation intermediates $\alpha_2^{CN}\beta_2$ and $\alpha_2\beta_2^{CN}$.
iv) By mixing the parental species $\alpha_2^{CN}\beta_2$ with $\alpha_2\beta_2$ and allowing time for equilibration, one obtains three species characterized by different tetramer-dimer dissociation rate constants, that is, $\alpha_2\beta_2$, $\alpha\alpha^{CN}\beta_2$ and $\alpha_2^{CN}\beta_2$. Since the first and third have already been characterized, this experiment allows the researcher to determine the tetramer dissociation constant of a mimic of a singly liganded tetramer bearing the ligand only on the α subunits.
v) Experiments analogous to (iv) are repeated with different parental species, in order to resolve the ^Dk of the mimics of all other ligation intermediates.

The results of this beautiful series of experiments are summarized in Table 7.8. Ackers' data have often been misinterpreted because of the innovative experimental method, and

Table 7.8 Cooperative free energies for the 10 ligation intermediates of tetrameric Hb, determined using three different mimics of the liganded heme.

	Tetramer	Parent Species	Free Energy $Fe^{+2}/Fe^{+3}CN^-$	Free Energy Fe^{+2}/Mn^{+3}	Free Energy $Co^{+2}/Fe^{+3}CO$
0:1	$\alpha_1^0\beta_2^0$ $\beta_1^0\alpha_2^0$	none	0 kcal/mol	0	0
1:1	$\alpha_1^x\beta_2^0$ $\beta_1^0\alpha_2^0$	$\alpha_1^x\beta_2^0 + \alpha_1^0\beta_2^0$ $\beta_1^0\alpha_2^x$ $\beta_1^0\alpha_2^0$	3.1	2.9	1.5
1:2	$\alpha_1^0\beta_2^0$ $\beta_1^x\alpha_2^0$	$\alpha_1^0\beta_2^x + \alpha_1^0\beta_2^0$ $\beta_1^x\alpha_2^0$ $\beta_1^0\alpha_2^0$	3.3	3.7	2.0
2:1	$\alpha_1^x\beta_2^0$ $\beta_1^x\alpha_2^0$	$\alpha_1^x\beta_2^x + \alpha_1^0\beta_2^0$ $\beta_1^x\alpha_2^x$ $\beta_1^0\alpha_2^0$	3.1	3.4	2.1
2.2	$\alpha_1^x\beta_2^x$ $\beta_1^0\alpha_2^0$	$\alpha_1^x\beta_2^0 + \alpha_1^0\beta_2^x$ $\beta_1^0\alpha_2^x$ $\beta_1^x\alpha_2^0$	6.4	6.6	3.0
2:3	$\alpha_1^x\beta_2^0$ $\beta_1^0\alpha_2^x$	none	6.1	6.8	3.1
2:4	$\alpha_1^0\beta_2^x$ $\beta_1^x\alpha_2^0$	none	6.4	6.2	3.2
3:1	$\alpha_1^x\beta_2^x$ $\beta_1^x\alpha_2^0$	$\alpha_1^x\beta_2^x + \alpha_1^0\beta_2^x$ $\beta_1^x\alpha_2^x$ $\beta_1^x\alpha_2^0$	6.3	6.5	3.0
3.2	$\alpha_1^x\beta_2^0$ $\beta_1^x\alpha_2^x$	$\alpha_1^x\beta_2^x + \alpha_1^x\beta_2^0$ $\beta_1^x\alpha_2^x$ $\beta_1^0\alpha_2^x$	6.2	6.5	3.1
4.1	$\alpha_1^x\beta_2^x$ $\beta_1^x\alpha_2^x$	none	6.0	6.9	2.5

Source: From Ackers and Holt (2006); see also Bellelli (2010).

the unconventional representation of the results. Thus, we shall make an effort to clarify the data reported in Table 7.8 in stepwise fashion. We observe that the values of the cooperative free energies neatly group into three classes: zero, 3 and 6 kcal/mol. We do not include the case of Co metal hybrids, which have lower but equally spaced cooperative free energy changes; the reader may apply the reasoning that follows to that case as well. Because of the experimental setup and the definition of the equilibrium constants these values are expressed per mol of dimer, which is confusing because the cooperative free energy is conventionally expressed per mol of heme. If we convert to kcal per mol of heme, the resulting values are zero, 1.5, and 3. Note that 3 kcal/mol of heme is a good estimate of the free energy difference for the binding of the first and fourth molecule of oxygen, that is, an estimate of $RT \ln(K_1/K_4)$ or, in the MWC model, $RT \ln(K_T/K_R)$.

If we read the columns of Table 7.8 pertinent to the $Fe^{+2}/Fe^{+3}CN^-$ hybrids or the Fe^{+2}/Mn^{+3} hybrids *from the bottom*, we observe that:

i) the mimic of the fully liganded tetramer (species 4:1) presents the same free energy of dissociation into dimers as the triply liganded tetramer (species 3:2 and 3:1). This implies that the last ligation step occurs with the same affinity as ligation to the dimer, that is, $K_4 = K_D$, consistent with the proven quaternary constraint of hemoglobin.

ii) Moving one ligation step up we find that three of the four doubly liganded species (2:2; 2:3; and 2:4) have the same stability against dimerization as the triply liganded species. This again implies that binding of one ligand to 2:2, 2:3, or 2:4 to form 3:1 or 3:2 occurs with the same affinity as binding to the dimer, that is, in this case $K_3 = K_D$. One doubly liganded species, 2:1, bearing the two ligands on the same dimer within the tetramer differs from the other three and represents the anomaly in the series because it is a more stable tetramer with respect to dissociation into dimers by 3 kcal/mol of dimer (or 1.5 kcal/mol of heme) than all other doubly liganded species. When this species binds one more ligand to convert to either 3:1 or 3:2 its tendency to dissociate into dimers increases and eqn. 7.4 tells us that in this case K_3 is 12-fold lower than K_D.

iii) The singly liganded intermediates (1:1 and 1:2) are less stable toward dissociation into dimers by 3 kcal/mol of dimer than the doubly liganded species 2:2, 2:3 and 2.4. Eqn. 7.4 tells us the ligation step leading from 1:1 or 1:2 to 2:2, 2:3 and 2.4 occurs with K_2 that is 12-fold lower than K_D. By contrast the singly liganded species have the same stability as the anomalous doubly liganded one (2:1). This implies that the ligation step leading from 1:1 or 1:2 to 2:1 occurs with $K_2 = K_D$, that is, it is not associated to any change in the tendency of the tetramer to dimerize (Section 5.6). This (if true) is an instance of R-like reactivity within ^{T}Hb, and is thus incompatible with the original MWC model.

iv) Finally, the unliganded species is more stable than the singly liganded ones by 3 kcal/mol of dimer, and the fully liganded by 6 kcal/mol. This implies that the first ligand binds with 150-fold lower affinity to the unliganded tetramer than to the dimer.

Essentially all the ligation intermediates conform to the prediction of the MWC model, with the exception of the doubly liganded species bearing two ligands in one α1β1 dimer (species 2:1). Ackers claimed that this result proves the existence of a "molecular code" of Hb cooperativity, which, simply stated, says that the allosteric structural change occurs preferentially when both dimers bear at least one ligand, whereas ^{T}Hb is compatible with one and even two ligands in the same dimer. Strictly speaking, this finding is incompatible with the original formulation of the two-state allosteric model, since the model does not attribute relevance to the position of the ligands within each ligation intermediate, but only to their number. This deviation, if real, may reflect the imperfect symmetry of Hb, which is a dimer of heterodimers rather than a homotetramer as envisioned in the MWC model. The unusual type of experiment and complex presentation of results may obscure the significance of Ackers' findings. Ackers found R-like (or "more than T") reactivity within ^{T}Hb, for the 1:1 or 1:2 to 2.1 ligation step, which implies strong cooperativity within the T state. He also found T-like (or "less than R") reactivity within ^{R}Hb, for the ligation step 2:1 to 3:1 or 3:2. These findings challenge not only the two-state model, but almost all we know about the structure of Hb. Ackers is not the only author to suggest, on the basis of experimental findings, that non allosteric cooperativity may exist within T-state Hb. For example, the same hypothesis was put forward, on the basis of different experimental data sets, by Di Cera and co-workers (Di Cera et al., 1987). However, cooperativity within T-state Hb is to be expected as a consequence of the ligand-dependent release of Bohr protons and other heterotropic effectors (see Section 4.9), a factor that apparently escaped consideration by Ackers and Di Cera. Thus, the discrepancy between Ackers findings and the MWC model may require re-evaluation, and may partially be explained by the already mentioned dependence of K_T on allosteric effectors (Sections 7.2 and 7.7).

Ackers' data have been criticized for several reasons. (i) They do not seem to be consistent with other experiments on ligation intermediates, for example, with the experimental determination of the concentrations of ligation intermediates by cryoelectrofocusing by Perrella and co-workers (Perrella and Di Cera, 1999), or with structural data on doubly liganded intermediates by Luisi and co-workers (Luisi and Shibayama, 1989; Luisi *et al.*, 1990), which crystallize in the T-conformation even though they bear ligands on both dimers, thus violating the molecular code of cooperativity. (ii) Rearrangement of mixed valency or metal hybrids during the long incubation time of the parental species would lead to a redistribution of modified and unmodified hemes, thus confusing the identification of the intermediate of interest (Shibayama *et al.*, 1997). (iii) Simulation of oxygen-binding isotherms using the parameters of Table 7.8 yields much lower cooperativity than experimentally observed (Edelstein, 1996). (iv) No clear structural basis exists for the molecular code of cooperativity. Ackers's reply to these criticisms was that the extent of the anomaly of ligation intermediate 2:1 depends on the mimic of the liganded heme and that in the case of O_2 and CO it may be less than in the case of cyano-met and other mimics.

7.11   Non-Human Hemoglobins and Human Hemoglobin Mutants

All vertebrates possess Hb genes, and all of them, except some fishes living in the Antarctic ocean, synthesize Hbs. Thus, one may compare the functional and structural properties of a very large number of naturally occurring variants of the same protein. The hemoglobins from mammals are very similar to each other. From the functional point of view, they have been classified into two broad groups: those having an intrinsically low oxygen affinity and high cooperativity, which respond poorly to heterotropic effectors, and those having an intrinsically high oxygen affinity, which respond strongly to allosteric effectors. Prototypes of the first group are the Hbs from ruminants, of the second group the Hbs from man and carnivora. The structural bases of these functional differences are subtle and incompletely clarified.

The importance of comparative studies on Hbs from different organisms is two-fold. Before the widespread use of site-directed mutagenesis, Hbs from different animals and naturally occurring mutants of human hemoglobin constituted the preferred or unique model systems to study the functional role of selected amino acid residues. Moreover, Hb is capable of a remarkable functional plasticity that allows wide variations in its functional properties, within an essentially constant structural arrangement; thus, its study provides extremely instructive examples of molecular adaptation. We list below some of the most clear-cut cases.

The α subunits of Hb from the opossum, *Didelphis marsupialis*, present the unusual substitution of the distal His E7 with Gln (Sharma *et al.*, 1982). The same substitution occurs in the Mb of the Asian elephant, *Elephas maximus* (Romero-Herrera *et al.*, 1981). Gln is the only residue that can functionally replace His at position E7, since it has similar bulk and forms a H-bond with dioxygen; thus, the observation that it is also the only known substituent of His at position E7 (in vertebrates) points to the functional relevance of distal effects in Hbs.

Cold adaptation in mammalian Hbs has attracted great interest. The extremities of cold living mammals (e.g., the arctic reindeer) may experience a significant reduction in temperature with respect to the rest of the body, and given the large negative ΔH of oxygenation of Hb, this would lead to hypoxia. In these mammals the ΔH of oxygenation of ^{T}Hb is essentially the same as that of other mammmals, whereas that of ^{R}Hb is close to zero or possibly slightly positive due to entropy-enthalpy compensation and coupling with the heat of the allosteric transition and that of release of allosteric effectors (Giardina *et al.*, 1989; Brix *et al.*, 1990]. As a consequence oxygen affinity and cooperativity exhibit a very unusual temperature dependence, which allows normal oxygen delivery at low temperature. This effect is remarkable since the ΔH of oxygen binding to the Hbs of non-cold adapted mammals is very close to that of the free heme (-14 kcal/mol according to Traylor and Berzinis, 1980): thus, in the arctic reindeer the protein moiety "corrects" this unfavorable thermodynamic property of its cofactor.

Some fish hemoglobins present a very special property, discovered by R.W. Root (1931) and named after him. The Root effect is a drastic reduction of oxygen affinity induced by acidic pH and leading to Hb becoming incompletely oxygenated in air or even under several atmospheres of pure oxygen. The essential features of the Root effect (misinterpreted by Root) are: (i) the stabilization of the T-allosteric conformation even in the fully liganded derivative; and (ii) a marked functional heterogeneity of one type of subunit in the T-allosteric conformation. The functional relevance of the Root effect is related to the necessity of releasing high partial pressures of oxygen to some special target organs, namely the swim bladder and the choroid rete in the eye. The case of the swim bladder is most straightforward: an "acid gland" releases lactic acid in the vessels of the swim bladder; this, in turn, causes Hb to release oxygen and to fill the swim bladder against the physical pressure of the surrounding water, which may amount to several atmospheres. Root effect Hbs are interesting not only as refined physiological devices, but also because they make possible some conceptually relevant experiments. For example Brunori and co-workers used trout Hb IV to record the spectroscopic signal assigned to the ^{T}Hb(CO)$_4$ - ^{R}Hb(CO)$_4$ structural transition (Giardina *et al.*, 1975). This experiment takes advantage of pH to induce the allosteric structural change, and of the high affinity for CO that prevents the gas from being released. Interestingly, the ^{T}Hb$_4$-^{R}Hb$_4$ difference spectrum recorded in this experiment exactly matches the ^{T}Hb$_3$-^{R}Hb$_3$ difference spectrum recorded by Ferrone and Hopfield using the modulated photoexcitation method (see Bellelli and Brunori, 1994 for a comparison).

The hemoglobin from lamprey is of special interest: it binds oxygen cooperatively thanks to a pure effect of ligand linked dissociation. The protein is an equilibrium mixture of homotetramers and homodimers when unliganded, and becomes monomeric upon oxygenation (Andersen, 1971). The oxygen affinity of the oligomeric assemblies is negligible and essentially the protein conforms to the description provided in Section 5.4 (except that only the dimeric and not the tetrameric assembly was considered in that section).

As in every protein, amino acid residues of Hb and Mb play very special roles, and their mutation is expected to significantly change the structural and functional properties of the macromolecule. In the past the only possible approaches were provided by natural mutants, often causing hemoglobinopathies, and by hemoglobins from other species. These approaches have obvious limitations, first of all, because the more important the role of the residue of interest, the less likely to find its variants. Moreover, comparing hemoglobin from different species involved comparing several amino acid substitutions at once. Site-directed mutagenesis drastically changed the experimental approach to this problem and allowed the researcher to selectively change single amino acid residues (most often in human Hb and in sperm whale Mb, which confirmed their role as prototypes of the respective class).

The pertinent literature is huge and we shall consider only some residues and topological positions of very special interest; we refer the reader to specialized reviews for a broader treatment (e.g., Bellelli *et al.*, 2006). The site-directed mutants discussed in this section are: (i) those of the proximal His F8; (ii) those of the distal His E7; and (iii) those of the "gate" residue at position E11.

The functional role of the proximal His was investigated by site-directed mutagenesis by Barrick and co-workers (Barrick *et al.*, 1997). These authors replaced His F8 with Gly in sperm whale Mb and in the α and β subunits of human Hb and added free imidazole as the fifth ligand. As a consequence their site-directed mutants resembled the corresponding wild type proteins, except that they lacked any chemical connection between the protein moiety and the fifth heme ligand. The most clear-cut case is that of a recombinant Hb constituted by the mutated α subunits and wild type β subunits ($\alpha_2^{F8His \rightarrow Gly} \beta_2$). In this mutant the α and β subunits present a very marked difference in affinity: the αs have very high affinity, as expected, because the quaternary constraint mediated by the protein is removed; the βs have very low affinity because the quaternary structure is T state, and ligand binding to the α subunits does not promote the allosteric structural change. The opposite mutant ($\alpha_2 \beta_2^{F8His \rightarrow Gly}$), however, binds oxygen with high affinity, suggesting that the mutated β subunits, which bind first, maintain the ability to promote the allosteric structural change. This result has been taken as an indication that in the β subunits distal effects have a greater relevance than in the αs in causing the allosteric structural change, and is qualitatively consistent with Perutz's observation that the heme pocket of the β subunits in unliganded Hb is collapsed and undergoes a significant structural rearrangement upon ligation.

Substitution of the distal His E7 with Gly (or other residues unable to form a H-bond with O_2) causes a significant decrease of the oxygen affinity, mainly due to an increase of the dissociation rate, and a large increase of the CO/O_2 partition constant, consistent with the putative role of the residue in stabilizing bound oxygen and destabilizing bound CO. In addition to the distal His, the function of a large number of other residues lining the heme pocket has been investigated by site-directed mutagenesis (Olson *et al.*, 1988). Two residues of the heme pocket deserve specific mention: Val E11, the residue one helical turn away from the distal His that gates the access to the heme pocket, whose mutation increases the rate of ligand binding and also of autoxidation; and Leu B10 that contacts the heme and presumably finely adjusts its position in the pocket, whose mutation may significantly affect the affinity for heme ligands in both directions (reviewed in Bellelli *et al.*, 2006).

Table 7.9 A selection of well-characterized naturally occurring or site-directed point mutations at the $\alpha 1 \beta 2$ interface (from Bellelli *et al.*, 2006).

Mutation	name (if any)	remarks
Thr αC3 $\rightarrow$ Trp		
Tyr αC7 $\rightarrow$ His		
Arg αFG3 $\rightarrow$ Leu	Hb Chesapeake	high affinity (decreased L_0)
Val αG3 $\rightarrow$ Trp		low affinity (increased L_0)
Tyr βC1 $\rightarrow$ Phe	Hb Philly	non-cooperative
Trp βC3 $\rightarrow$ Glu		stable a_1b_1 dimer
Phe βC7 $\rightarrow$ Tyr		low O_2 affinity
Asn βG10 $\rightarrow$ Lys	Hb Presbyterian	low O_2 affinity
Asn βG10 $\rightarrow$ Gln		low O_2 affinity

   Several mutants of the intersubunit interfaces, notably of the $\alpha 1 \beta 2$ interface have been designed and synthesized in order to explore the molecular mechanisms of cooperativity and quaternary constraints. Relevant residues explored by site directed mutagenesis or corresponding to naturally occurring hemoglobinopathies are listed in Table 7.9.

References

Ackers, G.K. Holt, J.M. (2006) Asymmetric cooperativity in a symmetric tetramer: human hemoglobin. J Biol Chem, 281: 11441–11443.

Alfsen, A., Chiancone, E., Wyman, J. and Antonini, E. (1970) Studies on the reaction of haptoglobin with hemoglobin and hemoglobin chains. II. Kinetics of complex formation. Biochim Biophys Acta, 200: 76–80.

Andersen, M.E. (1971) Sedimentation equilibrium experiments on the self-association of hemoglobin from the lamprey *Petromyzon marinus*. A model for oxygen transport in the lamprey. J Biol Chem, 246: 4800–4806.

Antonini, E., Wyman, J., Zito, R., Rossi-Fanelli, A. and Caputo, A. (1961) Studies on carboxypeptidase digests of human hemoglobin. J Biol Chem, 236: PC60-PC63.

Antonini, E., Condò, S.G., Giardina, B., Ioppolo, C. and Bertollini A. (1982) The effect of pH and D-glycerate 2,3-bisphosphate on the O2 equilibrium of normal and SH(β 93)-modified human hemoglobin. Eur J Biochem, 121: 325–328.

Arnone, A. (1972) X-ray diffraction study of binding of 2,3- diphosphoglycerate to human deoxyhaemoglobin. Nature, 237: 146–149.

Arnone, A, Rogers, P., Blough, N.V., McGourty, J.L. and Hoffman, B.M. (1986) X-ray diffraction studies of a partially liganded hemoglobin [αFeCOβMn]2. J Mol Biol, 188: 693–706.

Baldwin, J.M. (1980) The structure of human carbonmonoxy haemoglobin at 2.7 Å resolution. J Mol Biol, 136: 103–128.

Barrick, D., Ho, N.T., Simplaceanu, V., Dahlquist, F.W. and Ho, C. (1997) A test of the role of the proximal histidines in the Perutz model for cooperativity in haemoglobin. Nature Struct Biol, 4: 78–83.

Bellelli, A. and Brunori M. (1994) Optical measurements of quaternary structural changes in hemoglobin. Methods Enzymol, 232: 56–71.

Bellelli, A., Brunori, M., Miele, A.E., Panetta, G. and Vallone, B. (2006) The allosteric properties of hemoglobin: insights from natural and site directed mutants. Curr Protein Pept Sci, 7: 17–45.

Bellelli, A. (2010) Hemoglobin and Cooperativity: Experiments and Theories. Curr Protein and Pept Sci, 11: 2–36.

Bellelli A. and Brunori M. (2011) Hemoglobin allostery: variations on the theme. Biochim. Biophys. Acta (Bioenergetics), 1807: 1262–1272.

Bickar, D., Bonaventura, C. and Bonaventura J. (1984) Carbon monoxide-driven reduction of ferric heme and heme proteins. J Biol Chem, 259: 10777–10783.

Bonaventura, J., Bonaventura, C., Giardina, B., Antonini, E., Brunori, M. and Wyman, J. (1972) Partial restoration of normal functional properties in carboxypeptidase A-digested hemoglobin. Proc Natl Acad Sci USA, 69: 2174–2178.

Borgstahl, G.E., Rogers, P.H. and Arnone, A. (1994a) The 1.8 Å structure of carbonmonoxy-β4 hemoglobin. Analysis of a homotetramer with the R quaternary structure of liganded α2β2 hemoglobin. J Mol Biol, 236: 817–830.

Borgstahl, G.E., Rogers, P.H. and Arnone, A. (1994b) The 1.9 Å structure of deoxyβ4 hemoglobin. Analysis of the partitioning of quaternary-associated and ligand-induced changes in tertiary structure. J Mol Biol, 236: 831–843.

Bucci, E. and Fronticelli, C. (1965) A new method for the preparation of the α and β subunits of human hemoglobin. J Biol Chem, 240: PC551–PC552.

Cassoly, R. and Gibson, Q.H. (1975) Conformation, co-operativity and ligand binding in human hemoglobin J Mol Biol, 91: 301–313.

Collman, J.P., Gagne, R.R., Reed, C.A., Robinson, T.W. and Rodley, G.A. (1974) Structure of an iron(II) dioxygen complex: A model for oxygen-carrying hemeproteins. Proc Natl Acad Sci USA, 71: 1326–1329.

Collman, J.P., Brauman, J.I., Doxsee, K.M., Halbert, T.R. and Suslick, K.S. (1978) Model compounds for the T state of hemoglobin. Proc Natl Acad Sci USA, 75: 564–568.

Collman, J.P., Brauman, J.I. and Doxsee, K.M. (1979) Carbon monoxide binding to iron porphyrins. Proc Natl Acad Sci USA, 76: 6035–6039.

Di Cera, E., Robert, C.H. and Gill, S.J. (1987) Allosteric interpretation of the oxygen-binding reaction of human hemoglobin tetramers. Biochemistry, 26: 4003–4008.

Eaton, W.A., Henry, E.R., Hofrichter, J., Bettati, S., Viappiani, C. and Mozzarelli, A. (2007) Evolution of allosteric models for hemoglobin. IUBMB Life, 59: 586–599.

Edelstein, S.J., Rehmar, M.J., Olson, J.S. and Gibson, Q.H. (1970) Functional aspects of the subunit association-dissociation equilibria of hemoglobin. J Biol Chem, 245: 4372–81.

Edelstein, S.J. and Edsall, J.T. (1986) Linkage between ligand binding and the dimer-tetramer equilibrium in the Monod-Wyman-Changeux model of hemoglobin. Proc Natl Acad Sci USA, 83: 3796–3800.

Edelstein, S.J. (1996) An allosteric theory for hemoglobin incorporating asymmetric states to test the putative molecular code for cooperativity. J Mol Biol, 257: 737–744.

Eich, R.F.1, Li, T., Lemon, D.D., Doherty, D.H., Curry, S.R., Aitken, J.F., Mathews, A.J., Johnson, K.A., Smith, R.D., Phillips, G.N., Jr. and Olson, J.S. (1996) Mechanism of NO-induced oxidation of myoglobin and hemoglobin. Biochemistry, 35: 6976–6983.

Fermi, G., Perutz, M.F., Shaanan, B. and Fourme, R. (1984) The crystal structure of human deoxyhaemoglobin at 1.74 Å resolution. J Mol Biol, 175: 159–174.

Fermi, G., Perutz, M.F. and Shulman, R.G. (1987) Iron distances in hemoglobin: Comparison of x-ray crystallographic and extended x-ray absorption fine structure studies. Proc Nati Acad Sci USA, 84: 6167–6168.

Geers, C. and Gros, G. (2000) Carbon dioxide transport and carbonic anhydrase in blood and muscle. Physiol Reviews, 80: 681–715.

Gelin, B.R., Lee, A.W. and Karplus, M. (1983) Hemoglobin tertiary structural change on ligand binding. Its role in the co-operative mechanism. J Mol Biol, 171: 489–559.

Giacometti, G.M., Traylor, T.G., Ascenzi, P., Brunori, M. and Antonini, E. (1977) Reactivity of ferrous myoglobin at low pH. J Biol Chem, 252: 7447–7448.

Giardina, B., Ascoli, F. and Brunori, M. (1975) Spectral changes and allosteric transition in trout haemoglobin. Nature, 256: 761–762.

Giardina, B., Brix, O., Nuutinen, M., el Sherbini, S., Bardgard, A., Lazzarino, G. and Condò, S.G. (1989) Arctic adaptation in reindeer. The energy saving of a hemoglobin. FEBS Lett, 247: 135–138.

Gibson, Q.H. (1960) The reactions of some aromatic C-nitroso compounds with haemoglobin. Biochem J, 77: 519–526.

Gibson, Q.H. and Edelstein, S.J. (1987) Oxygen binding and subunit interaction of hemoglobin in relation to the two-state model. J Biol Chem, 262: 516–519.

Gibson, Q.H. (1999) Kinetics of oxygen binding to hemoglobin A. Biochemistry, 38: 5191–599.

Gibson, Q.H. and Roughton, F.J. (1956) The determination of the velocity constants of the four successive reactions of carbon monoxide with sheep haemoglobin. Proc R Soc Lond B Biol Sci, 146(923): 206–224.

Henry, E.R., Eaton, W.A. and Hochstrasser, R.M. (1986) Molecular dynamics simulations of cooling in laser-excited heme proteins. Proc Natl Acad Sci USA, 83: 8982–8986.

Henry, E.R., Jones, C.M., Hofrichter, J. and Eaton, W.A. (1997) Can a two-state MWC allosteric model explain hemoglobin kinetics? Biochemistry; 36: 6511–6528.

Hill, A.V. (1910) The possible effects of the aggregation of the molecule of haemoglobin on its dissociation curves. J Physiol, 40 (suppl.): iv–vii.

Hill, A.V. XLVII. (1913) The combinations of haemoglobin with oxygen and carbon monoxide. Biochem J, 7: 471–480.

Ho, C. and Russu, I.M. (1987) How much do we know about the Bohr effect of hemoglobin? Biochemistry, 26: 6299–6305.

Hofrichter, J., Sommer, J.H., Henry, E.R. and Eaton, WA. (1983) Nanosecond absorption spectroscopy of hemoglobin: elementary processes in kinetic cooperativity. Proc Natl Acad Sci USAm 80: 2235–2239.

Ikeda-Saito, M., Yamamoto, H., Imai, K., Kayne, F.J. and Yonetani, T. (1977) Studies on cobalt myoglobins and hemoglobins. Preparation of isolated chains containing cobaltous protoporphyrin IX and characterization of their equilibrium and kinetic properties of oxygenation and EPR spectra. J Biol Chem, 252: 620–624.

Imai, K. (1982) Allosteric Effects in Hemoglobin. Cambridge University Press, Cambridge, UK.

Kavanaugh, J.S., Rogers, P.H. and Arnone, A. (2005) Crystallographic evidence for a new ensemble of ligand-induced allosteric transitions in hemoglobin: the T-to-T(high) quaternary transitions. Biochemistry, 44: 6101–6121.

Kilmartin, J.V. and Rossi-Bernardi, L. (1969) Inhibition of CO2 combination and reduction of the Bohr effect in haemoglobin chemically modified at its α-amino groups. Nature, 222(5200): 1243–1246.

Kilmartin, J.V. and Wootton, J.F. (1970) Inhibition of Bohr effect after removal of C-terminal histidines from haemoglobin β-chains. Nature, 228(5273): 766–767.

Kilmartin, J.V., Breen, J.J., Roberts, G.C.K. and Ho, C. (1973) Direct measurements of the pK values of an alkaline Bohr group in human hemoglobin. Proc Natl Acad Sci USA, 70: 1246–1249.

Kilmartin, J.V., Fogg, J.H. and Perutz, M.F. (1980) Role of C-terminal histidine in the alkaline Bohr effect of human hemoglobin. Biochemistry, 19: 3189–3183.

Luisi, B. and Shibayama, N. (1989) Structure of haemoglobin in the deoxy quaternary state with ligand bound at the α haems. J Mol Biol, 206: 723–736.

Luisi, B., Liddington, B., Fermi, G. and Shibayama, N. (1990) Structure of deoxy-quaternary haemoglobin with liganded β subunits. J Mol Biol, 214: 7–14.

Marden, M.C., Kister, J., Poyart, C. and Edelstein, S.J. (1989) Analysis of hemoglobin oxygen equilibrium curves. Are unique solutions possible? J Mol Biol, 208: 341–345.

Mills, F.C., Johnson, M.L. and Ackers, G.K. (1976) Oxygenation-linked subunit interactions in human hemoglobin: Experimental studies on the concentration dependence of oxygenation curves. Biochemistry, 15: 5350–5362.

Mims, M.P., Porras, A.G., Olson, J.S., Noble, R.W. and Peterson, J.A. (1983) Ligand binding to heme proteins. An evaluation of distal effects. J Biol Chem, 258: 14219–14232.

Minton, A.P. and Imai, K. (1974) The three-state model: a minimal allosteric description of homotropic and heterotropic effects in the binding of ligands to hemoglobin. Proc Natl Acad Sci USA, 71: 1418–1421.

Moffat, K., Olson, J.S., Gibson, Q.H. and Kilmartin, J.V. (1973) The ligand-binding properties of desHis (146β) hemoglobin. J Biol Chem, 248: 6387–6393.

Moffat, K., Deatherage, J.F. and Seybert, D.W. (1979) A structural model for the kinetic behavior of hemoglobin. Science, 206: 1035–1042.

Moore, E.G. and Gibson, Q.H. (1976) Cooperativity in the dissociation of nitric oxide from hemoglobin. J Biol Chem, 251: 2788–2794.

Moore, J.N., Hansen, P.A. and Hochstrasser, R.M. (1988) Iron-carbonyl bond geometries of carboxymyoglobin and carboxyhemoglobin in solution determined by picosecond time-resolved infrared spectroscopy. Proc Natl Acad Sci USA, 85: 5062–5066.

Mozzarelli, A., Rivetti, C., Rossi, G.L., Henry, E.R. and Eaton, W.A. (1991) Crystals of haemoglobin with the T quaternary structure bind oxygen noncooperatively with no Bohr effect. Nature, 351: 416–419.

Mozzarelli, A., Rivetti, C., Rossi, G.L., Eaton, W.A. and Henry, E.R. (1997) Allosteric effectors do not alter the oxygen affinity of hemoglobin crystals. Protein Sci, 6: 484–489.

Mrabet, N.T., Shaeffer, J.R., McDonald, M.J. and Bunn, H.F. (1986) Dissociation of dimers of human hemoglobins A and F into monomers. J Biol Chem, 261: 1111–1115.

Noble, R.W., Hui, H.L., Kwiatkowski, L.D., Paily, P., DeYoung, A., Wierzba, A., Colby, J.E., Bruno, S. and Mozzarelli, A. (2001) Mutational effects at the subunit interfaces of human hemoglobin: evidence for a unique sensitivity of the T quaternary state to changes in the hinge region of the α1β2 interface. Biochemistry 40: 12357–12368.

Olson, J.S., Mathews, A.J., Rohlfs, R.J., Springer, B.A., Egeberg, K.D., Sligar, S.G., Tame, J., Renaud, J.P. and Nagai, K. (1988) The role of the distal histidine in myoglobin and haemoglobin. Nature, 336: 265–266.

Paoli, M., Liddington, R., Tame, J., Wilkinson, A. and Dodson, G. (1996) Crystal structure of T state haemoglobin with oxygen bound at all four haems. J Mol Biol, 256: 775–792.

Paoli, M., Dodson, G., Liddington, R.C. and Wilkinson, A.J. (1997) Tension in haemoglobin revealed by Fe-His(F8) bond rupture in the fully liganded T-state. J Mol Biol, 271: 161–167.

Park, S.Y., Yokoyama, T., Shibayama, N., Shiro, Y. and Tame, J.R. (2006) 1.25 Å resolution crystal structures of human haemoglobin in the oxy, deoxy and carbonmonoxy forms. J Mol Biol, 360: 690–701.

Parkhurst, L.J. and Gibson, Q.H. (1967) The reaction of carbon monoxide with horse hemoglobin in solution, in erythrocytes, and in crystals. J Biol Chem, 242: 5762–2570.

Pauling, L. and Coryell, C.D. (1936) The magnetic properties and structure of hemoglobin, oxyhemoglobin and carbonmonoxyhemoglobin. Proc Natl Acad Sci USA, 22: 210–216.

Perrella, M. and Di Cera, E. (1999) CO ligation intermediates and the mechanism of hemoglobin cooperativity. J Biol Chem, 274: 2605–2608.

Perutz, M.F. (1979) Regulation of oxygen affinity of hemoglobin: influence of structure of the globin on the heme iron. Annu Rev Biochem, 48: 327–386.

Perutz, M.F., Kilmartin, J.V., Nishikura, K., Fogg, J.H. and Butler, P.J.G. (1980) Identification of residues contributing to the Bohr effect of human haemoglobin. J Mol Biol, 138: 649–668.

Phillips, S.E. and Schoenborn, B.P. (1981) Neutron diffraction reveals oxygen-histidine hydrogen bond in oxymyoglobin. Nature, 292: 81–82.

Philo, J.S. and Lary, J.W. (1990) Kinetic investigations of the quaternary enhancement effect and α/β differences in binding the last oxygen to hemoglobin tetramers and dimers. J Biol Chem, 265: 139–143.

Reisberg, P.I. and Olson, J.S. (1980) Equilibrium binding of alkyl isocyanides to human hemoglobin. J Biol Chem, 255: 4144–4150.

Riggs, A. (1971) Mechanism of the enhancement of the Bohr effect in mammalian hemoglobins by diphosphoglycerate. Proc Natl Acad Sci USA, 68: 2062–2065.

Rivetti, C., Mozzarelli, A., Rossi, G.L., Henry, E.R. and Eaton, W.A. (1993) Oxygen binding by single crystals of hemoglobin. Biochemistry, 32: 2888–2906.

Romero-Herrera, A.E., Goodman, M., Dene, H., Bartnicki, D.E. and Mizukami H. (1981) An exceptional amino acid replacement on the distal side of the iron atom in proboscidean myoglobin. J Mol Evol, 17: 140–147.

Root, R.W. (1931) The respiratory function of the blood of marine fishes. Biol Bull, 61: 427–456.

Safo, M.K. and Abraham, D.J. (2005) The enigma of the liganded hemoglobin end state: a novel quaternary structure of human carbonmonoxy hemoglobin. Biochemistry, 44: 8347–8359.

Sawicki, C.A. and Gibson, Q.H. (1976) Quaternary conformational changes in human hemoglobin studied by laser photolysis of carboxyhemoglobin. J Biol Chem, 251: 1533–1542.

Shaanan, B. (1982) The iron–oxygen bond in human oxyhaemoglobin. Nature, 296: 683–684.

Shaanan, B. (1983) Structure of human oxyhaemoglobin at 2.1 Å resolution. J Mol Biol, 171: 31–59.

Sharma, V.S., John, M.E. and Waterman, M.R. (1982) Functional studies on hemoglobin opossum. Conclusions drawn regarding the role of the distal histidine. J Biol Chem, 257: 11887–11892.

Shibayama, N., Yonetani, T., Regan R.M. and Gibson, Q.H. (1995) Mechanism of ligand binding to Ni-Fe hybrid hemoglobins. Biochemistry, 34: 14658–14667.

Shibayama, N., Morimoto, H. and Saigo S. (1997) Reexamination of the hyper thermodynamic stability of asymmetric cyanomet valency hybrid hemoglobin, (α + CN-β + CN-)($\alpha\beta$): no preferentially populating asymmetric hybrid at equilibrium. Biochemistry, 36: 4375–4381.

Shibayama, N., Sugiyama, K., Tame, J.R. and Park, S.Y. (2014) Capturing the hemoglobin allosteric transition in a single crystal form. J Am Chem Soc, 136: 5097–5105. doi: 10.1021/ja500380e.

Springer, B.A., Egeberg, K.D., Sligar, S.G., Rohlfs, R.J., Mathews, A.J. and Olson, JS. (1989) Discrimination between oxygen and carbon monoxide and inhibition of autooxidation by myoglobin. Site-directed mutagenesis of the distal histidine. J Biol Chem, 264: 3057–3060.

Szabo, A. and Perutz, M.F. (1976) Equilibrium between six- and five-coordinated hemes in nitrosylhemoglobin: interpretation of electron spin resonance spectra. Biochemistry, 15: 4427–4428.

Szabo, A. (1978) Kinetics of hemoglobin and transition state theory. Proc Natl Acad Sci USA, 75: 2108–2111.

Thomas, J.O. and Edelstein, S.J. (1972) Observation of the dissociation of unliganded hemoglobin. J Biol Chem, 247: 7870–7874.

Traylor, T.G. and Berzinis, A.P. (1980) Binding of O_2 and CO to hemes and hemoproteins. Proc Natl Acad Sci USA, 77: 3171–3175.

Traylor, T.G. and Sharma, V.S. (1992) Why NO? Biochemistry, 31: 2847–2849.

Viappiani, C., Bettati, S., Bruno, S., Ronda, L., Abbruzzetti, S., Mozzarelli, A. and Eaton, W.A. (2004) New insights into allosteric mechanisms from trapping unstable protein conformations in silica gels. Proc Natl Acad Sci USA, 101: 14414–14419.

Weiss J.J. (1964) Nature of the iron-oxygen bond in oxyhaemoglobin. Nature; 202: 83–84.

Wilson J., Phillips K. and Luisi B (1996) The crystal structure of horse deoxyhaemoglobin trapped in the high-affinity (R) state. J. Mol. Biol.; 264: 743–756.

Winslow R.M., Swenberg M.L., Gross E., Chervenick PA, Buchman RR, Anderson WF. (1976) Hemoglobin McKees Rocks (α2β2 145Tyr -> term.) A human nonsense mutation leading to a shortened β–chain. J. Clin. Invest; 57: 772–781.

Yi, J., Soares, A.S. and Richter-Addo, G.B. (2014) Crystallographic characterization of the nitric oxide derivative of R-state human hemoglobin. Nitric Oxide, 39: 46–50.

Yonetani T., Park SI., Tsuneshige A., Imai, K. and Kanaori K. (2002) Global allostery model of hemoglobin. Modulation of O2 affinity, cooperativity, and Bohr effect by heterotropic allosteric effectors. J. Biol. Chem; 277: 34508–34520.

Part III

Enzymes: A Special Case of Ligand-Binding Proteins

8

Single-Substrate Enzymes and their Inhibitors

Enzymes qualify as ligand-binding proteins, and have only one fundamental and unique property that distinguishes them from simple ligand-binding proteins: *one or more of their ligands, the substrate(s), can dissociate either in unmodified form or in a chemically modified form, the product(s).* Other enzyme ligands, the inhibitors and effectors, behave as described in the preceding chapters.

From a practical point of view, *the signal of enzymatic reactions is most often due to conversion of the substrate(s) into product(s),* and this makes enzymology a technically specialized field, but the thermodynamic relationships we examined in the preceding chapters remain fully valid. In the majority of cases the conversion of substrate(s) into product(s) is studied under experimental conditions and over time windows where it is practically irreversible. As a consequence, *it cannot be taken for granted that during an enzyme assay the complexes of the enzyme with its substrate(s), product(s), effectors, or inhibitors attain the population one would expect if they could reach their equilibrium condition.* One should be aware that the standard equations for inhibitors and effectors are usually derived under the equilibrium assumption (or, for the sake of precision, the pseudo-equilibrium assumption), which may or may not be adequate to the case of interest.

In the present and the following chapters we shall be concerned with the theoretical and practical aspects of ligand binding to enzymes as measured by assay of their catalytic activity. This is a limited viewpoint, which, though consistent with the scope of this book, leaves out several important aspects of the discipline of enzymology. We strongly suggest that the interested reader complements the information provided in this book with that contained in dedicated enzymology treatises (e.g., Cornish Bowden, 1995).

8.1 Enzymes, Substrates, and Inhibitors: A Special Case of Ligand Binding

Enzymes are proteins that bind one or more substrate ligands and catalyze their chemical transformation into one or more products. Under most experimental conditions the rate of product formation, called the reaction velocity V, is a hyperbolic function of the substrate concentration, [S] (Figure 8.1). This curve can be described empirically by two parameters: the upper asymptote, which defines the maximum velocity (V_{max}) of the catalyzed reaction, and the free substrate concentration required to obtain half the V_{max},

Reversible Ligand Binding: Theory and Experiment, First Edition. Andrea Bellelli and Jannette Carey.
© 2018 John Wiley & Sons Ltd. Published 2018 by John Wiley & Sons Ltd.

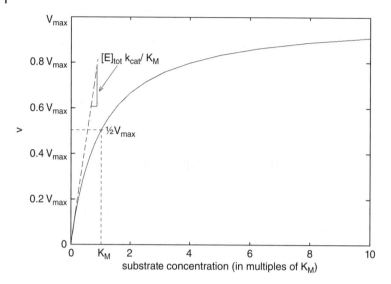

Figure 8.1 The Michaelis-Menten plot, showing the hyperbolic dependence of V on [S] in a typical enzyme-catalyzed reaction.

defined as the Michaelis constant (K_M). K_M is independent of enzyme concentration, whereas V_{max} is proportional to it; thus, it is often normalized to enzyme concentration to obtain the concentration independent parameter k_{cat}: $k_{cat}=V_{max}/[E]_{tot}$. The k_{cat}, also called *catalytic constant* or *turnover number* of the enzyme, indicates the maximum number of molecules of substrate that can be transformed by one molecule of enzyme in one time unit; its units are time^{-1} (usually s^{-1} or min^{-1}).

It was noticed since the beginning of the twentieth century that the dependence of V on [S] strictly resembles that of $\bar{X}$ on [X] (Section 1.4, Figure 1.1A). Explanation of this resemblance in terms of a plausible chemical model that uses the empirical parameters K_M and V_{max} was reached thanks to the contributions from several authors. The basic equation describing the relationship between K_M and V_{max} is named after the German enzymologist Leonor Michaelis and his Canadian co-worker Maude L. Menten. The simplest case, which unfortunately is not the most common although it is discussed in every text of elementary biochemistry, is that of a single-substrate enzyme that catalyzes a practically irreversible chemical reaction in which substrate S is converted to product P by way of a short-lived intermediate, the enzyme-substrate complex ES:

$$E+S \Leftrightarrow ES \rightarrow E+P \qquad \text{(eqn. 8.1)}$$

This reaction scheme was hypothesized because of the evident similarity between Figures 8.1 and 1.1A. Note that the back reaction, conversion of product to substrate, is neglected. The rate of formation or decay of the enzyme-substrate complex is:

$$\delta[ES]/\delta t = [E][S]k_1 - [ES](k_2 + k_3)$$

where k_1 is the second-order rate constant for forming the ES complex, k_2 is the first-order rate constant for dissociation of the ES complex, and k_3 is the first-order rate constant for conversion of ES to E + P.

The reaction described by eqn. 8.1 is irreversible and cannot reach a true equilibrium condition. However, it can reach the condition of steady state, where the concentrations of enzyme species E and ES are constant because the rates of formation and decay of the ES complex are equal. Thus, for the steady-state condition we posit $\delta[ES]/\delta t=0$ and write:

$$[E][S]/[ES]=(k_2+k_3)/k_1$$

This equation allows us to define the binding polynomial of the enzyme at steady state, using the substrate-free species E as a reference:

$$[E]_{tot}=[E]+[ES]=[E]\{1+[S]k_1/(k_2+k_3)\}$$

from which the fraction of the enzyme present in the form of the catalytically active ES complex can be derived:

$$[ES]/[E]_{tot}=[S]/\{(k_2+k_3)/k_1+[S]\}$$

The rate of conversion of the substrate into the product is:

$$V=\delta[P]/\delta t=[ES]k_3$$

Hence, for the reaction scheme of eqn. 8.1, the equation that defines the relationship between the rate of product formation at steady-state and the concentration of enzyme and substrate is:

$$V=[E]_{tot}k_3[S]/\{(k_2+k_3)/k_1+[S]\} \qquad \text{(eqn. 8.2)}$$

The graphical representation of eqn. 8.2 is a hyperbola, from which one can derive the analytical expression of the empirical parameters defined in Figure 8.1. V_{max} is obtained by assuming $[S]\to\infty$. Under this assumption one cancels all terms of eqn. 8.2 that do not contain $[S]$ and simplifies the remaining ones, to obtain:

$$V_{max}=[E]_{tot}k_3 \qquad \text{(eqn. 8.3)}$$

K_M equals the substrate concentration required to achieve $V=\frac{1}{2}V_{max}$. Thus one replaces $[S]$ with K_M and equates $2V=V_{max}$, to obtain:

$$[E]_{tot}k_3=2[E]_{tot}k_3K_M/\{(k_2+k_3)/k_1+K_M\}$$

that yields:

$$K_M=(k_2+k_3)/k_1 \qquad \text{(eqn. 8.4)}$$

More complex catalytic mechanisms yield more complex expressions, but the basic procedure to calculate V_{max} and K_M from the velocity is invariant: V_{max} is V at $[S]\to\infty$; K_M equals the substrate concentration required to achieve $V=\frac{1}{2}V_{max}$. k_{cat}, the normalized maximum velocity is defined as $k_{cat}=V_{max}/[E]_{tot}$, and in this case equals k_3. In more complex reaction schemes it may correspond to an equation containing constants and variables.

Using the above defined parameters, eqn. 8.2 may be rewritten as:

$$V = k_{cat}[E]_{tot}[S]/(K_M+[S])$$ (eqn. 8.5)

Eqn. 8.5 is called the Michaelis-Menten equation and its graphical representation is a hyperbola.

The substrate concentration [S] appearing in all the above equations refers to the free substrate, rather than the total substrate, the relation between the two being:

$$[S]_{tot} = [S]+[ES]$$

This is exactly analogous to the meaning of [X] in the equations used for ligand binding equilibrium (Section 1.3). In enzymology experiments, the total enzyme concentration is usually much smaller than the total substrate concentration, thus the approximation $[S]\approx[S]_{tot}$ is almost always justified.

Under appropriate experimental conditions, eqn. 8.5 is general, that is, it may apply to reaction schemes more complex than that described by eqn. 8.1. In these cases, of which we shall have several examples in the following sections, K_M, k_{cat}, and V_{max} should be considered empirical parameters rather than thermodynamic or kinetic constants. The fact that a vast majority of the known enzymes under appropriate experimental conditions present hyperbolic V versus [S] plots similar to the one reported in Figure 8.1 greatly simplifies the experimental design. However, it is extremely important that the researcher is aware that k_{cat} and K_M may correspond to more complex convolutions of parameters (and variables!) than those described for the reaction scheme in eqn. 8.1.

The parameters k_{cat} and K_M are empirically defined as the normalized maximum velocity and the free substrate concentration required to achieve half the maximum rate, respectively. An important objective of the characterization of any enzyme's catalytic mechanism is to define the dependence of these empirical parameters on the real kinetic constants of the chosen reaction scheme (see below). If we had the luxury to rename parameters whose names have been in use for a century, we would rename K_M the $S_{1/2}$ of the catalyzed reaction, for consistency with the $X_{1/2}$ defined as the half-saturating free-ligand concentration in the preceding chapters. Even if such a renaming is impossible, we invite the reader to consider K_M as analogous to $X_{1/2}$.

The affinity of the enzyme for its substrate under conditions where catalysis does not occur (which may or may not be experimentally accessible) is usually defined as $K_S = k_2/k_1$. Note that K_S is defined as a dissociation constant, with units of concentration. When this parameter may be experimentally determined it is of interest to compare it with K_M. In the cases where this comparison can be made, one finds that K_S is the lower limit of K_M under conditions in which the catalytic event is much slower than dissociation of the substrate from the enzyme (i.e., $k_3<<k_2$).

A physiologically important parameter of the reaction is the ratio k_{cat}/K_M, which corresponds to the slope of the tangent to the hyperbola at its origin (Figure 8.1), and is called the *specificity constant* or *kinetic efficiency* of the enzyme. If [S] << K_M we approximate:

$$V \approx [E]_{tot}[S]k_{cat}/K_M$$

One may rearrange the terms of the ratio k_{cat}/K_M to demonstrate that it depends on k_1, the rate constant for complex formation, and on the ratio $k_3/(k_2+k_3)$, which expresses the catalytic efficiency, that is, how many turnovers occur per binding event. Because in this book we used the term specificity to indicate the ability of a protein to discriminate between two ligands (Section 1.8), in this and the following Chapters we shall prefer the name *kinetic efficiency* over that of specificity constant.

Unfortunately, the Michaelis-Menten equation is frequently misunderstood in the literature. Indeed, eqn. 8.1 applies to a minority of enzymes, those that use only one substrate, but one may find experimental conditions in which almost any enzyme can produce graphs indistinguishable from Figure 8.1. In these cases, the apparent parameters k_{cat}, K_M and k_{cat}/K_M are easily determined but their relationships with the intrinsic parameters of the reaction are much more complex than eqns. 8.2 through 8.5, and may include variables as well as constants. In these cases the reproducibility of k_{cat} and K_M strongly depends on the experimental conditions. In this chapter we focus our attention on one-substrate enzymes; the more complex case of two-substrate enzymes will be dealt with in Chapter 9.

8.2 Importance of Initial Velocity Studies: Zero Order Kinetics

Because the substrate is being transformed during the reaction catalyzed by the enzyme, the steady-state condition, unlike the chemical equilibrium, is not stable, and the rate of product formation decreases as the substrate is consumed. Over short time intervals at low $[E]_{tot}$, the decrease in substrate concentration is small and the reaction velocity is nearly constant, approximating a *zero order* reaction. In a reaction that obeys zero order kinetics the rate of reagent consumption is independent of its concentration, that is, the reagent concentration decreases linearly over time. Zero order kinetics is only observed in catalyzed reactions, and always requires a complex reaction mechanism. Indeed, it is paradoxical, as it apparently implies that the probability of the reagent to decay increases as its concentration decreases. As one can easily verify from eqn. 8.5, zero order is a consequence of the fact that over short time intervals the concentration of the enzyme substrate complex, whose decay is first order, is approximately constant. Thus no single chemical species actually decays obeying a true zero order kinetics. Accordingly, the zero order law breaks down if the reaction is followed for long times, and/or at high enzyme concentrations because substrate consumption leads to a detectable decrease of the concentration of the complex ES. This effect may be prominent in the determinations carried out at low substrate concentrations, where the zero order kinetics portion of the time course may be short-lived.

If one tries to follow the reaction over long time intervals, another complication occurs. Indeed, in parallel with the decrease of substrate concentration, the product, which was absent at the beginning of the experiment, starts accumulating. The presence of the product can considerably complicate analysis of the experiment, for the following reasons. (i) If the conversion of S to P is reversible, the contribution of the backward reaction to the observed signal must be taken into account. (ii) Even in reactions that are practically irreversible under physiological conditions, the product may bind to the active site of the enzyme and inhibit its catalytic activity causing *product inhibition*. Because of the above considerations, classical enzymology experiments are designed to determine *initial velocities* only, and the sample is discarded after the measurement.

The discussion of enzyme reactions thus far has carried the tacit assumption that the velocities in Figure 8.1 and in eqns. 8.2 and 8.5 are initial velocities, recorded in the absence of the product, consistent with eqn. 8.1, which precludes product binding to the enzyme.

Unfortunately, there is a practical limitation to initial velocity studies. Indeed, from the considerations given above, it is clear that an initial velocity determination is not limited by time itself, but by substrate consumption and product formation. Thus, it should be properly defined as a determination carried out over a time span where substrate consumption is negligible. However, in the great majority of cases the signal is due to substrate consumption, and should be large enough to allow a precise measurement. An ideal substrate should yield large signal to concentration ratios for example, if the signal is absorbance, a large absorbance difference between the substrate and the product is desirable. This is not always the case. If a clear zero order process cannot be detected, it is customary to estimate the tangent of the initial velocity, but the greater the deviation of the tangent from the actual time course the greater the potential errors in the determination. While not solving the problem, a careful choice of enzyme concentration (which contributes to the determination of the velocity, see eqn. 8.5) may help in optimizing the accuracy of the measurement.

8.3   Linearizations of the Michaelis-Menten Hyperbola

As discussed in Chapter 1, transformation of hyperbolic curves into straight lines causes a distortion of experimental errors and is not advisable for quantitative analysis. However, before the widespread availability of personal computers, conversion of experimental data to a linear form was widely practiced, and it is impossible to read a paper on the subject without a proper understanding of the Lineweaver-Burk plot, to name only the most commonly used linearization. We shall therefore illustrate at least this linearization of the Michaelis-Menten hyperbola, with the recommendation that it can be used for illustrative purposes only and should not be relied upon for quantitative analysis of experimental data, for the same reasons as the Hill or Scatchard plots should not be used to quantitatively analyze ligand binding data.

Linearization of the Michaelis-Menten hyperbola obtained by plotting the reciprocals of [S] and V is due to Lineweaver and Burk and named after them. Use of this plot has been so widespread that no work in enzymology can be understood by anyone who is not familiar with it. However, the Lineweaver-Burk plot, in addition to severely distorting the experimental errors, greatly overweights data points collected at very low substrate concentration, which are often the least reliable ones (Section 8.2). The transformation of eqn. 8.5 is as follows:

$$1/V = K_M/V_{max}[S] + 1/V_{max}$$

This transformation yields a straight line with slope K_M/V_{max} and intercept $1/V_{max}$, as in Figure 8.2.

The field of enzymology is crowded with other types of linearizations that are less frequently used, but we shall not deal with all of them, for two reasons. Their use cannot be encouraged except for representative purposes, and as for the ligand-binding case,

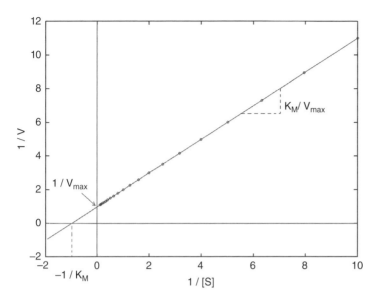

Figure 8.2 Lineweaver-Burk plot of the data from Figure 8.1.

direct fitting of experimental data in the hyperbolic form is preferable and easily achievable with modern computing. Furthermore, our main scope in this chapter is to compare the equations developed in enzymology with those developed for ligand binding (Chapters 1 through 7). Thus we feel that graphical representations specific to the enzyme field are out of scope. We refer the interested reader to specialized enzymology treatises (e.g., Cornish Bowden, 1995; Palmer, 1985).

8.4 Enzymatic Catalysis of Reversible Reactions

The reactions catalyzed by enzymes can often be reversible, and tend to an equilibrium condition where both the reagent(s) and the product(s) are significantly populated. In these cases *the enzyme accelerates progress to the same equilibrium condition the system would reach at a much lower rate in its absence.* Study of this type of reaction is of great theoretical interest, but rarely has practical relevance, because the experiments are preferentially designed to determine initial velocities, under conditions far removed from equilibrium. We shall consider here the rate equation for enzyme-catalyzed reversible reactions, and the case of one-way enzymes.

A simple case of a reversible reaction catalyzed by enzymes is isomerization. For example, the reader may be interested in revisiting the classical characterization of the forward and backward reactions of triose-phosphate isomerase by Plaut and Knowles (1972). The simplest reaction scheme is as follows:

$$\mathrm{E+S} \underset{k_2}{\overset{k_1}{\rightleftharpoons}} [\mathrm{ES} \rightleftharpoons \mathrm{EP}] \underset{k_4}{\overset{k_3}{\rightleftharpoons}} \mathrm{E+P}$$

where the number of internal steps (in square brackets) may be variable. An isomerization reaction catalyzed by an enzyme can proceed by a quite complex mechanism, with several intermediate species. For this example we shall consider only the simplest possible mechanism, in which a single intermediate is considered; thus, we group the terms [ES <==> EP] of the above scheme under the fictitious species ES*.

This system is described by four kinetic constants, k_1 through k_4, of which k_1 and k_4 are second order and k_2 and k_3 are first order. The differential rate equations that define this system are:

$$\delta[ES^*]/\delta t = [E](k_1[S]+k_4[P]) - [ES^*](k_2+k_3)$$
$$\delta[S]/\delta t = [ES^*]k_2 - [E][S]k_1$$
$$\delta[P]/\delta t = [ES^*]k_3 - [E][P]k_4$$

Under steady-state conditions there is no net change in [ES*], thus $\delta[ES^*]/\delta t=0$, and one obtains:

$$[E](k_1[S]+k_4[P]) = [ES^*](k_2+k_3)$$
$$[ES^*] = [E](k_1[S]+k_4[P])/(k_2+k_3)$$
$$[ES^*]/[E]_{tot} = (k_1[S]+k_4[P])/(k_1[S]+k_2+k_3+k_4[P])$$

(eqn. 8.6)

The rate of product formation is the net difference between the forward ($[ES^*]k_3$) and the backward ($[ES^*]k_2$) rates (Cornish-Bowden, 1995):

$$V = \delta[P]/\delta t = [E]_{tot}([S]k_1k_3 - [P]k_2k_4)/(k_1[S]+k_2+k_3+k_4[P])$$

(eqn. 8.7)

We define K_M and k_{cat} for the forward and backward rates, as they would be measured in initial velocity studies, as follows:

$$K_{MS} = (k_2+k_3)/k_1$$
$$k_{cat\,S} = k_3$$
$$K_{MP} = (k_2+k_3)/k_4$$
$$k_{cat\,P} = k_2$$

This allows us to rewrite eqn. 8.7 in a form that is equivalent, but a little more friendly:

$$V = [E]_{tot}([S]k_{cat\,S}/K_{MS} - [P]k_{cat\,P}/K_{MP})/(1+[S]/K_{MS}+[P]/K_{MP})$$

(eqn. 8.8)

Eqn. 8.8 has the important theoretical consequence that the equilibrium condition of a reversible reaction catalyzed by an enzyme, achieved when V=0, implies that *the equilibrium constant of the reaction equals the ratio of the kinetic efficiencies of the product and substrate, but not the ratio of k_{cat} or K_M values:*

$$[P]_{eq}/[S]_{eq} = K_{eq} = k_{cat\,S}K_{MP}/k_{cat\,P}K_{MS}$$

This explains the apparently paradoxical existence of the so-called *one-way* enzymes, that under the usual experimental conditions catalyze much more effectively the forward than the backward reaction or vice versa, apparently violating the expectation that an enzyme should catalyze the forward and backward reaction to the same extent in order to leave unaltered the equilibrium constant. The explanation of this puzzling observation is that an enzyme accelerates to the same extent the forward and backward reaction only when the substrate and product concentrations are both equal to their equilibrium values. At concentrations of substrate and product far from their equilibrium values, that is, in initial velocity experiments, the catalytic activity of the enzyme may be more pronounced on the forward than the backward reaction, or *vice versa*, without violation of thermodynamic principles (Cornish-Bowden, 1995).

Although eqn. 8.8 has theoretical relevance, it is hard to see how one may use it in practice. This is because in order to take advantage of this equation, one should prepare a solution containing a mixture of the substrate and the product, both at significant concentration but far from their equilibrium ratio, add the enzyme, and follow conversion of substrate into product or vice versa, until equilibrium is reached. This experiment is rarely carried out. The more usual case is to study either the forward or backward reaction separately, that is, the researcher adds the enzyme either to a solution of substrate in the absence of product, or to a solution of product in the absence of substrate, and records the initial velocity. Doing so amounts to equating either [S] or [P] to zero and effectively reduces eqn. 8.8 to eqn. 8.5. An additional reason to separately study the forward and backward reactions is that isomerizations rarely provide a good signal, though other kinds of reactions may not have this limitation. As a consequence isomerizations are often studied by taking advantage of a coupled reaction that provides the signal by chemically or enzymatically destroying the reaction product (at a faster rate than that of its formation), thus preventing the possibility of reaching the equilibrium condition.

8.5 The Study of Enzyme Inhibitors Under the Pseudo-Equilibrium Approximation

Inhibitors are chemical compounds that combine with an enzyme and prevent or impair catalysis. The simplest case is that of reversible competitive inhibitors, which combine with the enzyme at the same site as the substrate and inhibit the binding and transformation of the latter. Competitive inhibition is thus an instance of identical linkage (Section 1.8).

The textbook treatment of enzyme inhibitors entails an assumption that is so engrained in our understanding of the matter as to go essentially unnoticed. In the absence of inhibitors the reaction is analyzed using a *kinetic* approach, that is, we define the K_M as a function of kinetic rate constants (eqn. 8.4). The binding of an inhibitor, however, is treated under the assumption that it equilibrates with the enzyme, according to an equilibrium dissociation constant (K_I). That is, the textbook analysis does not consider or require the kinetic rate constants of association and dissociation of the inhibitor, $k_{a,I}$ and $k_{d,I}$. There is nothing wrong in the *pseudo-equilibrium* approximation, but it is important to define the conditions it must satisfy.

Before going any further we shall discuss the pseudo-equilibrium approximation and its great explanatory power, together with its shortcomings; then, in the following sections, we shall present the standard treatment of rapidly equilibrating inhibitors.

The treatment of slowly equilibrating competitive inhibitors, which defy the pseudo-equilibrium approximation, will be presented in Chapter 11.

The pseudo-equilibrium approximation consists of treating all the reactions of the catalytic cycle of the enzyme as if they were under equilibrium conditions, with the exception of the catalytic event itself, which therefore would be the only reaction requiring a kinetic rate constant. The catalytic cycle resolves into a set of chemical equilibria, and the conversion of substrate to product would occur on the equilibrium population of the pertinent intermediate. One might expect that under the pseudo-equilibrium approximation K_M should be replaced by K_S. However, this is unnecessary because k_2 and k_3 always appear as summed together in the pertinent kinetic equations, and the reaction they govern both produce the same species, E. *The pseudo-equilibrium approximation is simple and powerful, but requires that the rate constant of the catalytic event* (k_3) *is the lowest among all the first-order or pseudo-first order kinetic constants of the catalytic cycle (with the possible exception of k_2), including the kinetic rate constants for association and dissociation of the inhibitor ($k_{a,I}$ and $k_{d,I}$ respectively); that is: $k_3 << [S]$* $k_1, k_3 << [I]k_{a,I}, k_3 << k_{d,I}$, *or, at least, $k_3 << [I]k_{a,I}+k_{d,I}$.* The last requirement implies that the inhibitor binds reversibly. It follows that irreversible and slowly binding inhibitors cannot be treated under the pseudo-equilibrium approximation and require a different approach.

The reason an enzyme cannot be studied under true equilibrium conditions (with the exception of special cases) and requires the pseudo-equilibrium approximation is that when the substrate is added to the mixture, it immediately starts being consumed. Thus, the time available for the system to reach its theoretical equilibrium condition is intrinsically limited by the duration of the assay and the consumption of the substrate. Only if every step of the catalytic cycle is significantly faster than the catalytic event may one assume that the reaction velocity is recorded in a system in which the concentration of each species approximates that expected under equilibrium. We encountered no such problem in the preceding Chapters 1 through 7, because in the absence of a catalytic event, one can wait for whatever time is required to reach the true thermodynamic equilibrium of the system.

Unfortunately, the requirements of the pseudo-equilibrium approximation are seldom checked and the pseudo-equilibrium approximation is applied as if it were granted. This may often be true with low-affinity inhibitors, but may fail in the case of high-affinity inhibitors whose dissociation from the enzyme can be slower than the catalytic event. The pseudo-equilibrium approximation is a rather generous one, and tolerates small deviations from its requirements, for example a dissociation rate constant of the enzyme-inhibitor complex only slightly larger than that of the catalytic event. However, it is contradicted and yields wrong results if the system is significantly non-conformant to its requirements.

Two very simple tests should be systematically carried out to verify that the system obeys the conditions for a pseudo-equilibrium analysis. (i) The rate of the catalyzed reaction should not depend on the order of addition of the components, and incubation of the enzyme with the inhibitor should have no effect on the reaction velocity. For practical reasons it is usually convenient that the enzyme is the last component to be added, but there is no real problem in repeating some determination adding the substrate as the last component. (ii) Dilution of the enzyme-inhibitor mixture should reduce the extent of inhibition.

To carry out these tests, a basic preliminary characterization of the system under study is required, to select the appropriate experimental conditions. The tests require that initial velocity measurements be carried out at constant substrate concentration, chosen to yield, say $V = 80\%\ V_{max}$. Next, one should select two inhibitor concentrations that, under the chosen experimental conditions, yield widely different levels of inhibition: for example, 25% and 75% (corresponding to 60% and 20% of V_{max}, respectively). The values of the desired velocities are indicative, not absolute, and the principle is that experimental conditions yielding $V \approx V_{max}$ or almost complete inhibition are to be avoided. Let the lower inhibitor concentration be I_1, and the higher I_2, and let their ratio be $R = I_2/I_1$.

To perform the test of rapid binding, the researcher prepares six samples, and records the initial velocity of the catalyzed reaction for each of them:

i) Sample #1: this sample contains buffer, enzyme, and substrate, added in this order, and the enzyme activity is recorded immediately after addition of the last component (the substrate). In this sample the inhibitor is absent and $V\#1 \approx 0.8\ V_{max}$ is expected.

ii) Sample #2: this sample is identical to the preceding one, except that it contains the inhibitor at concentration I_1. The activity is recorded immediately after addition of the substrate. If the inhibitor binds rapidly, $V\#2 \approx 0.6\ V_{max}$ is expected.

iii) Sample #3: is identical to sample #2, but the inhibitor concentration is I_2. If the inhibitor binds rapidly, $V\#3 \approx 0.2\ V_{max}$ is expected.

iv) Sample #4 is identical to sample #1, but the mixture containing the buffer and the enzyme at the final concentration of the assay is allowed to incubate for the chosen time (e.g., 20 min.). After the incubation, substrate is added and the enzyme activity is recorded. This sample is run to exclude that the enzyme may undergo inactivation during the incubation time, and ideally should have the same activity of sample #1.

v) Sample #5 is identical to sample #2, but the mixture containing the buffer, the inhibitor at concentration I_1 and the enzyme at the final concentration of the assay is allowed to incubate for the same time as sample #4. After the chosen incubation time, the substrate is added and the enzyme activity is recorded. If the inhibitor binds rapidly, the same activity as in sample #2 will be recorded ($V\#5 = V\#2$). A lower activity than in sample #2 suggests slow binding of the inhibitor, provided that no time dependent inactivation of the enzyme occurs.

vi) Sample #6 is identical to sample #5 except that the inhibitor concentration is I_2. If the inhibitor binds rapidly $V\#6 = V\#3$.

If one has doubts that the inhibitor may bind to components of the mixture other than the enzyme, further samples may be prepared that will contain every component of the mixture (including the substrate) except the enzyme, which will be added after the same incubation time as in samples #5 and #6.

If inactivation of the enzyme during the incubation time occurs (i.e., $V\#3 < V\#1$), all samples that were incubated with the inhibitor will display lower activity than those that were not. In this case rapid binding of the inhibitor causes $V\#2/V\#1 = V\#5/V\#4$ and $V\#3/V\#1 = V\#6/V\#4$, whereas $V\#2/V\#1 > V\#5/V\#4$ and $V\#3/V\#1 > V\#6/V\#4$ suggest slow binding of the inhibitor (i.e., greater inhibition in the samples that received incubation with the inhibitor).

If the inhibitor concentrations used are significantly higher than that of the enzyme, and binding is fast, this test strongly suggests that binding of the inhibitor is also reversible at least over the time scale of incubation. If, on the contrary, the inhibitor binds slowly to the enzyme, reversibility should be tested by dilution (see below).

To test reversibility of the binding of the inhibitor by dilution, one takes advantage of the same experimental conditions selected for the test of rapid binding, and prepares the following samples:

i) Sample #7: a mixture containing buffer and the enzyme at a concentration R times higher than required for the assay is prepared and left to incubate for the chosen time. At the end of the incubation, the sample is diluted R times with buffer in order to obtain the same enzyme concentration as in all the preceding samples, and a new incubation time is allowed, then substrate is added and the activity is recorded.

ii) Sample #8: a mixture containing buffer, the enzyme at a concentration R times higher than required for the assay, and the inhibitor at concentration I_2 is prepared and left to incubate for the chosen time. At the end of the incubation, the sample is diluted R times with buffer in order to obtain the same enzyme and inhibitor concentrations as in samples #2 and #5, and a new incubation time is allowed, then substrate is added and the activity is recorded.

If binding of the inhibitor is reversible, and the incubation time is sufficient, sample #8 should have the same activity as all samples containing the inhibitor at concentration I_1. Indeed, sample #8 differs from and #2 with #5 because of the history of the enzyme-inhibitor mixture but not because of its final composition, that is, the inhibitor concentration is I_1 in all of them, and the enzyme-inhibitor complex which formed in excess during the first incubation at I_2 must dissociate during the second incubation (after dilution). If on the contrary sample #8 remains as inhibited as sample #6 (or even more inhibited, because its total incubation with the inhibitor was longer) this suggests that dilution may not be able to induce dissociation. To decide whether the inhibitor binds irreversibly or its equilibration with the enzyme is much slower than anticipated a complete characterization of the time course of inactivation is required.

Finally, if time-dependent inactivation of the enzyme occurs, instead of comparing V#8 with V#5 and V#6 one should compare the ratios V#8/V#7 with V#5/V#4 and V#6/V#4.

The pseudo-equilibrium approximation holds if the inhibitor does not require incubation with the enzyme and if dilution promotes dissociation of the bound inhibitor, fails otherwise, as depicted in the flow-chart below (Figure 8.3). If either test fails, the reader should refer to the more sophisticated tests and experimental approaches described in Chapter 11.

In a standard steady-state experiment, *the velocity of the catalyzed reaction probes the concentration of the complex ES (or, in more complex reaction schemes, of the last complex before the catalytic step), which is used to reconstruct the binding polynomial. If the pseudo-equilibrium approximation holds, the binding polynomial will be that of a true steady-state mixture. If the pseudo-equilibrium approximation fails, steady-state experiments either fail to reconstruct a binding polynomial or reconstruct the binding polynomial of the mixture before addition of the last component.*

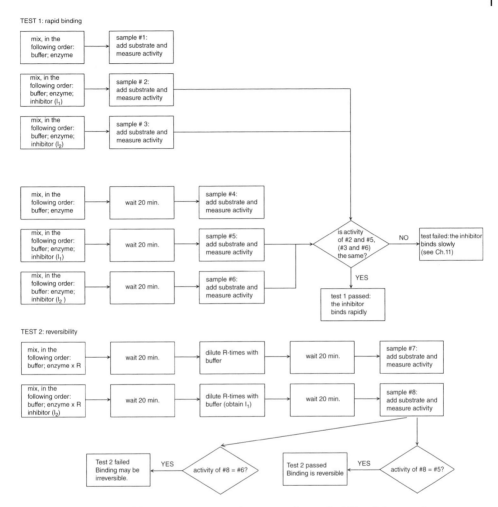

Figure 8.3 Flow chart of the experiments aimed at testing the applicability of the pseudo-equilibrium approximation.

Samples #2 and #5 contain exactly the same concentrations of substrate, enzyme, and inhibitor and differ only because of the incubation time; the same applies to samples #3 and #6. If the activity of sample #2 (#3) is greater than that of sample #5 (#6), this suggests slow binding. Sample #8 has been incubated with the inhibitor at the same concentration as samples #3 and #6, and then diluted in order to obtain the same inhibitor concentration as in samples #2 and #5. If the activity of sample #8 equals that of sample #5 this implies that dilution caused the inhibitor to dissociate and proves reversibility of binding. Samples #1, #4, and #7 serve as controls of the possible time-dependent inactivation of the enzyme (see text).

8.6 Inhibitors that Bind to the Same Site as the Substrate (Pure Competitive Inhibitors)

Some inhibitors compete with the substrate for the binding site and cannot form the three membered complex enzyme-substrate-inhibitor. These inhibitors are usually called competitive in biochemistry textbooks, but in view of possible confusions with other types of inhibitors we call them *pure competitive inhibitors*. Pure competitive inhibitors are common in biology, because minimal chemical changes may transform a substrate into a non-reactive analogue, which may behave as a pure competitive inhibitor.

One should be aware that we are using in this section a structural definition of the inhibitor, whereas competitive inhibitor is a functional definition. The two definitions do not necessarily overlap. *The thermodynamic relationship between an inhibitor and a substrate that compete for the same site on the enzyme is identical linkage (Section 1.8).* Competition between inhibitor and substrate may be observed also if the two molecules bind to different sites of the enzyme, because of negative heterotropic linkage (Sections 1.9 and 8.7). Thus, structural information should always be looked for.

The reaction scheme for a pure competitive inhibitor that obeys the requirements of the pseudo-equilibrium approximation is as follows:

$$EI + S \Leftrightarrow E + I + S \Leftrightarrow ES + I \rightarrow E + P + I$$

This system is described by the equilibrium constant K_I, the Michaelis-Menten constant K_M, and a kinetic constant, which for consistency with Section 8.1 we name k_3. We further assume that under the chosen experimental conditions $[E]_{tot}$ is much lower than both $[I]_{tot}$ and $[S]_{tot}$. The parameters of this model are:

$$K_M = [E][S]/[ES] = (k_2 + k_3)/k_1$$
$$K_I = [E][I]/[EI]$$

Under the pseudo-equilibrium condition it is unnecessary to determine the association and dissociation rate constants of the inhibitor ($k_{a,I}$ and $k_{d,I}$, respectively). Important and well-studied examples of pure competitive, reversible inhibitors of one-substrate enzymes are 5-phospho arabinoate, which inhibits glucose-6-phosphate isomerase (Seeholzer *et al.*, 1993), and benzamidine, which inhibits serine (and related) proteases (Markwardt *et al.*, 1968). Strictly speaking proteolytic enzymes (Beynon and Bond, 2001) use two substrates, peptide and water; however water, being present at high and constant concentration, may be neglected in the rate equations. Thus, these enzymes behave as single substrate ones. We do not include in this list such classical examples of pure competitive inhibitors as malonate, which inhibits succinate dehydrogenase, because the enzyme uses two substrates, and demands a different analysis (see Chapter 9).

The pseudo-equilibrium population of the various enzyme species, taking $[E]$ as the reference species, is as follows:

$$[EI] = [E][I]/K_I$$
$$[ES] = [E][S]/K_M$$

the resulting binding polynomial is:

$$[E]_{tot} = [E](1 + [I]/K_I + [S]/K_M) \qquad \text{(eqn. 8.9)}$$

and the fractional substrate saturation is:

$$[ES]/[E]_{tot} = ([S]/K_M)/(1 + [I]/K_I + [S]/K_M)$$
$$= [S]/\{K_M(1 + [I]/K_I) + [S]\}$$

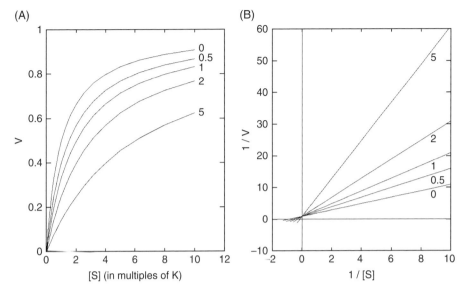

Figure 8.4 Pure competitive inhibition.

Panel A: Michaelis-Menten plots of velocity (expressed as a fraction of V_{max}) as a function of substrate concentration (in multiples of K_M measured in the absence of the inhibitor) at several concentrations of competitive inhibitor (indicated on the right of each curve, in multiples of K_I). Panel B: Lineweaver-Burk plots of the same data as in panel A. Notice that in this plot the highest values on the ordinate corresponds to the lowest velocity values.

The rate of conversion of substrate into product in the presence of inhibitor (V_I) is:

$$V_I = [ES]k_3 = [E]_{tot} k_3 [S] / \{ K_M (1 + [I] / K_I) + [S] \} \qquad \text{(eqn. 8.10)}$$

and we note that eqn. 8.10 reduces to 8.5 if [I]=0.

Eqn. 8.10 describes a rectangular hyperbola whose asymptote (V_{max} or k_{cat}) is the same as that recorded in absence of the inhibitor (eqn. 8.5; see Figure 8.4), while the apparent K_M, $K_{M app}$ is

$$K_{M app} = K_M (1 + [I] / K_I) \qquad \text{(eqn. 8.11)}$$

Thus, the pure competitive inhibitor changes the apparent K_M, but not V_{max} (or k_{cat}). Moreover, the $K_{M app}$ depends linearly on the inhibitor concentration, K_I, and K_M measured in the absence of the inhibitor. *In pure competitive inhibition, a plot of $K_{M app}$ versus [I] yields a straight line with slope K_M/K_I and intercept K_M.*

As evident from comparison of eqns. 1.12 and 8.10, or 8.11 and 1.15, *pure competitive inhibition is as a case of identical linkage* as already analyzed in Section 1.8.

An additional empirical parameter is often used to characterize the affinity of the inhibitor for the enzyme, **IC_{50}**, fully analogous to **$X_{1/2}$**, and defined as the free inhibitor concentration required to reduce the velocity of the catalyzed reaction to half its value in the absence of the inhibitor:

$$V_{[I] = IC50} = \frac{1}{2} V$$

If we solve eqn. 8.9 for IC_{50} we obtain:

$$[E]_{tot} k_3 [S]/\{K_M(1+IC_{50}/K_I)+[S]\} = 0.5[E]_{tot} k_3 [S]/(K_M+[S])$$
$$IC_{50} = K_I(1+[S]/K_M)$$

(eqn. 8.12)

Eqn. 8.12 has the same form as eqn. 8.11, as expected for identical linkage, and shows that IC_{50} of a pure competitive inhibitor increases linearly with [S] (see Table 1.1). Thus, a reported IC_{50} value for a pure competitive inhibitor is meaningless in the absence of information on [S]. Moreover, we observe that eqns. 8.11 and 8.12 are instances of the following general rule: IC_{50} is always a function of K_I, and $K_{M\,app}$ is always a function of K_M (see also Cheng and Prusoff, 1973).

8.7 Different Types of Heterotropic (Non-Competitive) Inhibitors

Heterotropic inhibitors bind to sites other than the substrate site, or at least the overlap between the binding sites of the inhibitor and substrate is incomplete. As a consequence, these inhibitors may form the three-membered complex EIS. The EIS complex is impossible with pure competitive inhibitors. Unfortunately, the absence of the EIS complex is an unreliable information about the inhibitor being pure competitive or heterotropic for the following reason. *The thermodynamic relationship between the substrate and a heterotropic inhibitor, if any exists, is heterotropic linkage, which may be positive or negative (Section 1.9). In the presence of strong negative heterotropic linkage, the population of the three-membered EIS complex may be very low or even negligible.* Thus, the EIS complex may be practically absent also in the case of heterotropic inhibitors. The typical reaction scheme of a heterotropic inhibitor is as follows:

$$E + S \xrightleftharpoons{K_M} ES \xrightarrow{k_3} E + P$$

with K_I, ^SK_I, and $EI + S \xrightleftharpoons{^IK_M} EIS$

The above scheme applies to every conceivable heterotropic inhibitor, but there is some confusion in the biochemistry textbook definitions of the different possibilities that may arise from the scheme. Cornish-Bowden (1995) adopts the definitions that we report in Table 8.1, with the modifications required by the different names used for the equilibrium constants in this book.

Mixed inhibition is possibly the most common of the cases listed in Table 8.1, all others being possible variants, and in view of its relevance, we shall name it heterotropic inhibition. However, for pedagogical purposes, we shall begin from the simpler case of pure non-competitive inhibition.

Pure non-competitive inhibition occurs if $K_M = {}^IK_M$ (which by necessity implies $K_I = {}^SK_I$). It implies absence of thermodynamic linkage between S and I. Pure non-competitive inhibition is conceptually very simple: a fraction of the enzyme is sequestered in the

Table 8.1 Different types of heterotropic inhibition.

Pure non-competitive	$^SK_I = K_I$; absence of linkage between S and I.
Mixed	^SK_I not too dissimilar from K_I; both EI and EIS attain significant population. Linkage between S and I may be positive or negative.
Competitive	$^SK_I \gg K_I$; population of EIS negligible. Extreme negative linkage between S and I.
Uncompetitive	$^SK_I \ll K_I$; population of EI negligible. Extreme positive linkage between S and I.

catalytically inactive EI complex, which can bind the substrate to form the still inactive complex EIS. The substrate binds to E and EI with the same affinity, thus it does not promote the release of the inhibitor. As a consequence, in a steady-state experiment, the concentration of active (i.e., uninhibited) enzyme is reduced and the system presents lower V_{max} than in the absence of the inhibitor. The fraction of uninhibited enzyme depends hyperbolically on inhibitor concentration, and so does V_{max}. The binding polynomial for a non-competitive enzyme under the pseudo-equilibrium approximation is:

$$
\begin{aligned}
[E]_{tot} &= [E]\left(1 + [S]/K_M + [I]/K_I + [S][I]/K_M K_I\right) \\
&= [E]\left(1 + [S]/K_M\right)\left(1 + [I]/K_I\right)
\end{aligned}
\qquad \text{(eqn. 8.13)}
$$

The velocity of the catalyzed reaction results:

$$
V = [ES]k_3 = \frac{[E]_{tot} k_3 [S]/K_M}{\left(1 + [S]/K_M\right)\left(1 + [I]/K_I\right)}
\qquad \text{(eqn. 8.14)}
$$

The maximum velocity in the presence of the inhibitor ($V_{max\,I}$) is obtained by assuming $[S] \rightarrow \infty$, which simplifies the above equation to:

$$
V_{max\,I} = [E]_{tot} k_3 / \left(1 + [I]/K_I\right)
$$

The apparent K_M is the substrate concentration required to obtain half the $V_{max\,I}$, that is:

$$
\begin{aligned}
& 2[E]_{tot} k_3 K_{M\,app} / K_M / \left(1 + K_{M\,app}/K_M\right)\left(1 + [I]/K_I\right) \\
& = [E]_{tot} k_3 / \left(1 + [I]/K_I\right)
\end{aligned}
$$

which yields:

$$
K_{M\,app} = K_M
$$

The above equation demonstrates that the pure non-competitive inhibitor does not affect K_M (Figure 8.5). Moreover, a plot of $V_{max,I}$ (or V at any fixed concentration of substrate) versus [I] is a hyperbola, whose midpoint (IC_{50}) equals K_I, as one can verify by solving for IC_{50} the equation:

$$
2[E]_{tot} k_3 [S]/K_M / \left(1 + [S]/K_M\right)\left(1 + IC_{50}/K_I\right) = [E]_{tot} k_3 [S]/K_M / \left(1 + [S]/K_M\right)
$$

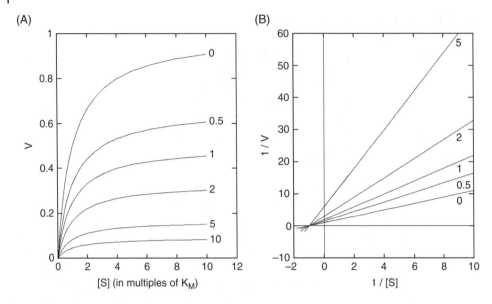

Figure 8.5 Pure non-competitive inhibition.
Panel A: Michaelis-Menten plots of velocity (in multiples of V_{max}) as a function of substrate concentration (in multiples of K_M) at several concentrations of non-competitive inhibitor (indicated at the right of each curve, in multiples of K_I). Panel B: Lineweaver-Burk plots of the same data as in panel A.

An alternative graphical method to estimate K_I is to plot the reciprocal of $V_{max\,I}$ as a function of [I]: one obtains a straight line with slope K_I/V_{max} and intercept $1/V_{max}$. Although a statistically much sounder procedure to determine K_I is to globally fit eqn. 8.14 to one's data, graphical methods have been widely resorted to in the past, and their study is necessary to read many classical enzymology papers.

Mixed inhibition represents a case of heterotropic linkage between the inhibitor and the substrate (see Section 1.9); thus, we call it heterotropic inhibition. Heterotropic linkage may be of the positive or negative type, that is, the inhibitor may either increase or decrease the affinity of the enzyme for the substrate. In both cases the EIS complex is inactive, thus, positive linkage does not imply relief of the inhibition; actually, the opposite is true, and positive linkage yields a stronger inhibition than negative linkage because an increase of substrate concentration causes an increase in the fraction of inhibited enzyme (Cornish-Bowden, 1986). The binding polynomial, under the pseudo-equilibrium approximation, results:

$$[E]_{tot} = [E]\left(1 + [S]/K_M + [I]/K_I + [S][I]/K_M{}^S K_I\right) \qquad \text{(eqn. 8.15)}$$

The velocity of the catalyzed reaction is:

$$V = [E]_{tot}\, k_3\,[S]/K_M / \left(1 + [S]/K_M + [I]/K_I + [S][I]/K_M{}^S K_I\right) \qquad \text{(eqn. 8.16)}$$

and

$$V_{max\,I} = [E]_{tot}\, k_3 / \left(1 + [I]/{}^S K_I\right)$$

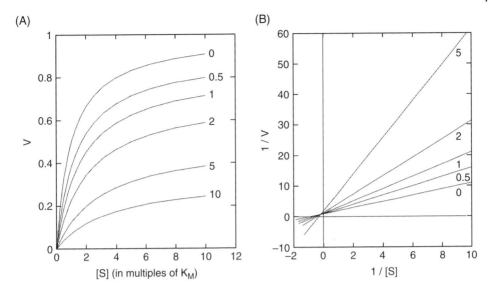

Figure 8.6 Heterotropic inhibition, mixed type.

Panel A: Michaelis-Menten plots of the velocity (in multiples of V_{max}) as a function of substrate concentration (in multiples of K_M measured in absence of the inhibitor) at several concentrations of the non-competitive inhibitor (indicated at the right of each curve, in multiples of K_I). The heterotropic interaction in this simulation is of the negative type, with $^SK_I = 5\,K_I$. Panel B: Lineweaver-Burk plots of the data in panel A.

The $K_{M\,app}$ results:

$$K_{M\,app} = K_M\left(1+[I]/K_I\right)/\left(1+[I]/\,^SK_I\right) = K_M\left(K_I+[I]\right)/\left(K_I+[I]K_I/\,^SK_I\right)$$

This equation may be compared with eqn. 1.20.

The above equations show that the uncompetitive inhibitor changes both V_{max} and $K_{M\,app}$ (Figure 8.6). In this case one must determine two binding constants for the inhibitor, K_I and ^SK_I. The statistically soundest method to determine these constants is to explore V over a large grid of concentrations of S and I and to globally fit the experimental data using eqn. 8.16. However, in many classical studies realized before personal computers became widespread, K_I and ^SK_I were determined from linear regression of the Michaelis-Menten parameters obtained at different concentrations of the inhibitor. Indeed *a plot of $1/V_{max\,I}$ versus [I] yields a straight line with slope $1/^SK_IV_{max}$ and intercept $1/V_{max}$ (where V_{max} is determined in the absence of the inhibitor). A plot of the inverse of the kinetic efficiencies ($K_{M\,app}/V_{max\,I}$) versus [I] yields a straight line with slope $K_M/V_{max}K_I$ and intercept the inverse of the kinetic efficiency measured in the absence of the inhibitor (K_M/V_{max}).*

The IC_{50} results:

$$2[E]_{tot}\,k_3[S]/K_M/\left(1+[S]/K_M+IC_{50}/K_I+[S]IC_{50}/K_M\,^SK_I\right)$$
$$=[E]_{tot}\,k_3[S]/K_M/\left(1+[S]/K_M\right)$$

which yields a hyperbolic dependence of IC_{50} on [S]:

$$IC_{50} = K_I\left(K_M+[S]\right)/\left(K_M+[S]K_I/\,^SK_I\right)$$

Two extreme cases of heterotropic inhibition deserve mention, namely those of extreme negative or positive heterotropic linkage between the substrate and the inhibitor.

Extreme negative heterotropic linkage causes the population of the ternary complex EIS to be negligible. This condition requires $^IK_M \gg K_M$ and $^SK_I \gg K_I$ and may cause the inhibitor to simulate the functional behavior of a pure competitive inhibitor, whence its name of competitive inhibition (Table 8.1). We do not particularly like this name, and prefer to name this case a competitive heterotropic inhibition for the following reasons. Energy conservation dictates that ^SK_I cannot be infinitely lower than K_I, thus EIS cannot be truly absent from the mixture. Consistently, the dependence of $K_{M\,app}$ on [I] cannot be described by a straight line, as it is in the case of pure competitive inhibition; rather, $K_{M\,app}$ depends hyperbolically on [I], as in mixed inhibition, but under practically accessible experimental conditions one only observes a small part of the hyperbola that may be indistinguishable from a straight line. In practice, ^SK_I is so large that the inhibitor behaves as a low-affinity ligand of ES (see Section 3.6). The binding polynomial and velocity equations for this type of inhibition may be derived from eqns. 8.15 and 8.16 by deleting the terms that contain ^SK_I.

Uncompetitive inhibition is a case of extreme positive heterotropic linkage, and causes the population of the complex EI to be negligible. Uncompetitive inhibition is uncommon, and Cornish Bowden in his classical review of the phenomenon lists essentially only one case, that of L-phenylalanine and alkaline phosphatase (Cornish-Bowden, 1986). Rather than competing with the substrate, the uncompetitive inhibitor strongly favors substrate binding; their relationship may be described as heterotropic cooperativity. Thermodynamic relationships are usually symmetric and this one is no exception; thus, the substrate favors the binding of the inhibitor. This has the paradoxical consequence that the higher the substrate concentration, the more inhibited the enzyme. The quantitative description of uncompetitive inhibition is opposite to the preceding one and requires $^IK_M \ll K_M$ and $^SK_I \ll K_I$; it is achieved by deleting the terms that contain K_I in eqns. 8.15 and 8.16. A consequence of this correction of eqns. 8.15 and 8.16, as one may easily verify, is that the kinetic efficiency of the enzyme is independent of the inhibitor, that is, $k_{cat,I}/K_{M\,app} = k_{cat}/K_M$. Thus a Lineweaver-Burk plot of a series of experiments carried out at different concentrations of the inhibitor yields a series of parallel straight lines, and a plot of $1/k_{cat,I}$ as a function of [I] yields a straight line with intercept $1/k_{cat}$ and slope $1/k_{cat}{}^SK_I$. As we observed in the case of competitive heterotropic inhibition, it is impossible that one species of a thermodynamic square (in this case, EI) is completely absent from the mixture, thus uncompetitive inhibition is a somewhat idealized case.

Cornish-Bowden (1986) remarked that naturally occurring uncompetitive inhibitors (and in general heterotropic inhibitors presenting strong positive linkage with the substrate) are extremely rare and suggested that enzymes that may be subject to this type of inhibition are counter selected. The reasoning goes as follows: every inhibitor increases the concentration of the enzyme's substrate because it reduces the rate at which the enzyme is able to use the substrate. The effect of competitive inhibitors (heterotropic or not) is self-limiting because the increase of substrate concentration releases the inhibition. This is a negative feedback loop that ultimately restores cell function. The affinity of uncompetitive inhibitors, on the contrary, increases as the substrate accumulates, and this causes a positive feedback cycle that ultimately blocks the cell metabolism and kills the cell.

(A) (B)

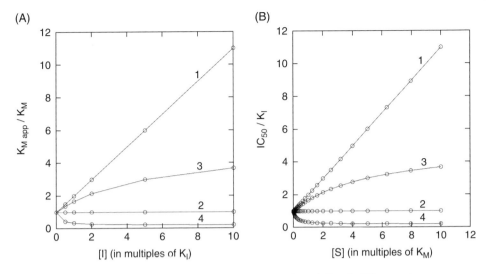

Figure 8.7 Dependence of $K_{M\,app}$ on inhibitor concentration (panel A) and of IC_{50} on substrate concentration (panel B). Curve 1 (both panels): competitive inhibition; curve 2: non-competitive inhibition; curve 3: uncompetitive inhibition, with negative linkage between I and S ($^SK_I = 5\,K_I$); curve 4: uncompetitive inhibition, with positive linkage between I and S ($^SK_I = K_I/5$).

    If one does not know which type of inhibition is being observed, the dependence of $K_{M\,app}$ on inhibitor concentration, and of IC_{50} on substrate concentration, distinguishes the cases of competitive, non-competitive, and uncompetitive inhibition (Figure 8.7; see also Table 1.1 and Figure 1.7).

    All the above relationships were derived under the pseudo-equilibrium approximation. If the pseudo-equilibrium approximation does not hold, both non-competitive and uncompetitive inhibitors will require an incubation with the enzyme before addition of the substrate. The duration of the incubation should be determined experimentally. The case of the slowly binding non-competitive inhibitor, after incubation, can be treated using the equations developed above for the pseudo-equilibrium case. Indeed, given that the substrate does not displace the non-competitive inhibitor, the equilibrium condition reached during the incubation is not altered by addition of the substrate or by the catalytic activity. The same is not true in the case of the uncompetitive inhibitor, which equilibrates in the incubation time according to K_I, but should re-equilibrate during the assay according (also) to ^SK_I. Thus in the latter case the experiment should be analyzed using the procedures for slowly equilibrating inhibitors (Chapter 11).

8.8 Heterotropic Regulation of Enzyme Activity

Heterotropic regulation of enzyme activity by an effector is conceptually analogous to uncompetitive inhibition, except that the enzyme-effector complex is not devoid of catalytic activity. This type of regulation is commonly observed in the case of allosteric, multi-subunit enzymes having more than a single catalytic site (e.g., Blangy *et al.*, 1968). However, the basic relationships of heterotropic enzyme regulation may be conveniently investigated in the simpler case of a monomeric enzyme having just one catalytic site for the substrate and one regulatory site for the heterotropic effector. An example is provided by the effect of proflavine on the Cys proteases papain and ficin (Hall and Anderson, 1974a, 1974b).

The reaction scheme is as follows:

$$\begin{array}{ccccc}
E + S & \overset{K_M}{\rightleftharpoons} & ES & \overset{k_3}{\longrightarrow} & E + P \\
+ & & + & & \\
X & & X & & \\
K_X \updownarrow & & {}^S K_X \updownarrow & & \\
EX + S & \underset{{}^X K_M}{\rightleftharpoons} & EXS & \overset{k_3'}{\longrightarrow} & EX + P
\end{array}$$

This case is analogous to that of uncompetitive or non-competitive inhibitors, depending on the effector X presenting heterotropic linkage with the substrate or not. The crucial difference between uncompetitive inhibition and heterotropic regulation is that the three-membered complex of the inhibitor (EIS) is inactive, whereas that of the effector (EXS) is active to a greater or lesser extent than the ES complex. Consistent with this premise, the binding polynomial, under the pseudo-equilibrium assumption, is identical to either eqn. 8.13 or 8.5. The rate of product formation, however, differs from the cases of non-competitive or uncompetitive inhibition, because the complex EXS has catalytic activity (and may even have higher activity than the two-membered complex ES, if $k_3' > k_3$).

In the absence of heterotropic linkage between X and S the velocity of the catalyzed reaction is as follows.

$$\begin{aligned}
V &= [ES]k_3 + [EXS]k_3' \\
&= [E]_{tot}[S] \left(K_x k_3 + [X]k_3' \right) / \left(K_M + [S] \right) \left(K_X + [X] \right)
\end{aligned} \qquad \text{(eqn. 8.17)}$$

In the presence of heterotropic linkage, the reaction velocity is as follows.

$$\begin{aligned}
V &= [ES]k_3 + [EXS]k_3' \\
&= [E]_{tot}[S] \left(K_x{}^X K_M k_3 + [X]K_M k_3' \right) / \left\{ K_M{}^X K_M \left([X] + K_X \right) + [S] \left([X]K_M + K_X{}^X K_M \right) \right\}
\end{aligned}$$

$$\text{(eqn. 8.18)}$$

In both cases the velocity (as well as V_{max}) depends hyperbolically on the concentration of the effector X and is limited between k_3 and k_3'.

References

Beynon, R. and Bond, J.S. (2001) Proteolytic Enzymes, 2nd ed. Oxford, UK: Oxford University Press.

Blangy, D., Buc, H. and Monod, J. (1968) Kinetics of the allosteric interactions of phosphofructokinase from Escherichia coli. J Mol Biol, 31: 13–35.

Cheng, Y.C. and Prusoff, W.H. (1973) Relationship between the inhibition constant (K_I) and the concentration of inhibitor which causes 50 per cent inhibition (I_{50}) of an enzymatic reaction. Biochem Pharmacol, 22: 3099–3108.

Cornish-Bowden, A. (1986) Why is uncompetitive inhibition so rare? FEBS Lett, 203: 3–6.

Cornish-Bowden, A. (1995) Fundamentals of Enzyme Kinetics. London, UK: Portland Press.

Hall, P.L. and Anderson, C.D. (1974a) Proflavine interactions with papain and ficin. I. Dye binding and its effects upon enzyme inactivation by N-alkylmaleimides. Biochemistry, 13: 2082–2087.

Hall, P.L. and Anderson, C.D. (1974b) Proflavine interactions with papain and ficin. II. Effects of dye binding upon reversible inhibition. Biochemistry, 13: 2087–2092.

Markwardt, F., Landmann, H. and Walsmann, P. (1968) Comparative studies on the inhibition of trypsin, plasmin, and thrombin by derivatives of benzylamine and benzamidine. Eur. J. Biochem, 6: 502–506.

Palmer, T. (1985) Understanding Enzymes, 2nd ed. Chichester, UK: Horwood Ltd.

Plaut, B. and Knowles, J.R. (1972) pH-dependence of the triose phosphate isomerase reaction. Biochem J, 129: 311–320.

Seeholzer, S.H. (1993) Phosphoglucose isomerase: a ketol isomerase with aldol C2-epimerase activity. PNAS USA, 90: 1237–1241.

9

Two-Substrate Enzymes and their Inhibitors

Two-substrate enzymes are very common, and represent the majority of enzymes (Cornish Bowden, 1995). Their catalytic properties are usually studied at constant concentration of one substrate, by systematically varying the concentration of the other. Under these experimental conditions one obtains two families of steady-state curves, one at variable concentration of the first substrate, the other at variable concentration of the second substrate. Each steady-state curve faithfully simulates that of single-substrate enzymes. However, the researcher must be aware that the similarity of the Michaelis-Menten curves of single- and two-substrate enzymes is superficial and the meaning of the parameters one obtains is not the same. Unfortunately, the catalytic cycle of two-substrate enzymes cannot be written once and for all, because several catalytic mechanisms are possible, depending on the order of substrate addition and product release, and the steady-state equations corresponding to different catalytic mechanisms differ.

We shall consider in this chapter two common catalytic mechanisms that have been considered in detail by Cleland (1967) and by Cheng and Prusoff (1973), with some of their variants. In one of these mechanisms the first substrate can be released before binding of the second and a chemically modified intermediate species of the enzyme is populated during the cycle, but formation of a ternary complex of the enzyme with its two substrates is not required. In the other mechanism, formation of the ternary complex is required, whereas chemical modification of the enzyme is not. Even if neither of these cases applies to an enzyme of interest to the reader, the information provided in this chapter should help him/her to derive a catalytic mechanism that is appropriate. Both types of catalytic mechanisms are common in biochemistry, and the order in which they are presented is dictated by pedagogical considerations, that is, we start with the mechanism whose quantitative description is simpler.

Together with each catalytic mechanism, we shall discuss the effect of pure competitive inhibitors. Heterotropic inhibitors, besides being less common, usually have simpler reaction schemes and do not seem equally stimulating. We shall rely on the pseudo-equilibrium approximation, since the kinetic treatment would yield very complicated equations not often required by the experimental data.

The study of two-substrate enzymes may appear complex, but one may take advantage of an experimental approach that is not accessible for single-substrate enzymes. Indeed, it is often possible to study the binding of one substrate in the absence of the other under true equilibrium conditions, provided that binding is associated with a signal. Thus, by incubating the enzyme with either of its substrates, the researcher can in some

Reversible Ligand Binding: Theory and Experiment, First Edition. Andrea Bellelli and Jannette Carey.
© 2018 John Wiley & Sons Ltd. Published 2018 by John Wiley & Sons Ltd.

cases determine a true equilibrium K_S for that substrate, with the *caveat* that in the presence of the other substrate the affinity may differ because of linkage. Moreover, incubation of the enzyme with only one of its substrates plus any inhibitor or other ligand allows the researcher to study their linkage under true equilibrium conditions, again provided that a signal due to binding is detected.

9.1   Two Basic Catalytic Mechanisms for Two-Substrate Enzymes

Two-substrate enzymes can operate according to several catalytic mechanisms, whose classification is largely due to the classic contributions of W.W. Cleland (e.g., Cleland, 1967). Cleland's classification takes into account the order of substrate addition and product release, and applies to every conceivable catalytic scheme. However, in this chapter we shall consider only two basic cases of two-substrate enzymes, both studied in the absence of their product(s), as is most commonly done in initial velocity experiments. Indeed, unless one is specifically interested in studying product inhibition, it is very unusual to record an enzymatic reaction in a system that contains controlled amounts of both substrate(s) *and* product(s).

A fundamental dichotomy in the classification of two-substrate enzymes is between those that do not require the formation of a ternary complex with both substrates (e.g., aminotransferases, oxidoreductases), and those that require such a complex (e.g., synthases). We shall describe the two mechanisms separately, under Sections 9.2 and 9.3 respectively.

An essential point to remark is that, although one can study the steady-state properties of two-substrate enzymes at fixed concentration of either substrate and obtain Michaelis-Menten-like hyperbolas, the parameters one obtains are always apparent parameters, whose magnitude depends on the concentration of the constant substrate. Thus, the reproducibility of these parameters depends crucially on the experimental conditions, which must be reported together with the parameter values.

In an effort to obtain concentration-independent estimates of the steady-state parameters, one may be tempted to determine K_M and k_{cat} for each substrate at saturating concentrations of the other, that is, to use a concentration of constant substrate much higher than its K_M. This approach should be adopted with some caution for two reasons. (i) Because the apparent affinity of a two-substrate enzyme for one substrate usually depends on the concentration of the other, a concentration that is saturating at low concentration of the other substrate may be non-saturating at high concentration, or vice versa. (ii) The catalytic mechanism determines how the concentration of the fixed substrate affects the Michaelis-Menten parameters of the variable substrate. Thus, important information on the catalytic mechanism is lost when using only saturating concentrations of the constant substrate.

In the following sections we shall pay little attention to the number of products and the order of their release. This is a consequence of the fact that we chose two very simple mechanisms as prototypes, which do not include the possibility of product inhibition. However, in the characterization of a real enzyme, the number of products and the order of their release may have great relevance.

The catalytic cycle of two-substrate enzymes can be remarkably complex and involve several intermediates. However, in the analysis of experimental data it is convenient to adopt the simplest possible reaction scheme and introduce more species only if required by a discrepancy between the data and the simulated time courses. The main reason to start with the simplest possible reaction mechanism is that the experimental data may not contain enough information to resolve all the steps of the catalytic cycle (usually because some intermediates are poorly populated), and a model requiring more intermediates than can be experimentally resolved yields unreliable estimates of the parameters. A minimal model, on the contrary, yields reliable estimates of apparent parameters, which may be very useful to describe the system at least from a phenomenological point of view. Given the above premises, the requirements of a minimum reaction scheme are as follows. (i) The minimum reaction scheme should be qualitatively consistent with the overall reaction of the enzyme. If the enzyme uses two substrates and releases one, two, or more products, the model should account for these facts; if an inhibitor is used, the model should be able to describe that, and so on. (ii) The minimum reaction scheme should quantitatively reproduce the experimental measurements with little or no systematic deviation. Large systematic deviations (i.e., non-random distribution of the fit residuals) indicate that a more complex reaction scheme is required.

9.2 Steady-State Parameters of Two-Substrate Enzymes that Do Not Form a Ternary Complex

Two-substrate enzymes that do not form a ternary complex are common. The two substrates need not be, and usually are not, present simultaneously in the active site. The two substrates may or may not share the same binding site. The catalytic cycle of these enzymes includes a chemically modified intermediate species of the enzyme (or its coenzyme). This type of mechanism, which was called a "ping-pong" mechanism by Cleland (1967), applies to several oxido-reductases (e.g., Saccoccia *et al.*, 2014; Bellelli *et al.*, 2000), aminotransferases, carboxylases, etc. All these enzymes may present exquisitely complex catalytic cycles, but in the following analysis we consider an oversimplified scheme, taking advantage of the fact that the catalytic cycle can be divided into two half-cycles:

$$E + XT \overset{K_{XT}}{\rightleftharpoons} EXT \xrightarrow{k_{XT}} ET + X$$
$$ET + Y \overset{K_Y}{\rightleftharpoons} ETY \xrightarrow{k_Y} E + YT$$

The catalytic scheme presented here is oversimplified because each of the enzyme-substrate complexes (EXT and ETY) may undergo internal chemical rearrangements before releasing the product; moreover, these reactions are usually reversible even though initial velocity studies allow one to neglect reversibility. In the scheme above, the enzyme catalyzes the irreversible transfer of a chemical group (or electrons) T from a donor substrate (XT) to an acceptor (Y), and the overall reaction catalyzed is:

$$XT + Y \Leftrightarrow X + YT$$

It is important to remark that the constants K_{XT} and K_Y in the above scheme are the equilibrium dissociation constants of the EXT and ETY complexes respectively, as would be measured under conditions where the transfer of group T does not occur. In other words, we are applying a pseudo-equilibrium approximation to the enzyme-substrate complexes. This is necessary, because, contrary to what happens with single-substrate enzymes, the reaction catalyzed in the half-cycle does not yield the initial state of the enzyme. For example, dissociation of EXT in the direction observed in initial velocity experiments yields ET, whereas the non-productive dissociation of EXT yields E.

A typical example is provided by the majority of oxido- reductases, in which the oxidized and reduced states of the enzyme alternate during the catalytic cycle:

$$E_{ox} + X_{red} \Leftrightarrow E_{ox}X_{red} \rightarrow E_{red} + X_{ox}$$
$$E_{red} + Y_{ox} \Leftrightarrow E_{red}Y_{ox} \rightarrow E_{ox} + Y_{red}$$

Oxidoreductases transfer electrons from the reducing substrate (X_{red}, which is converted to X_{ox}) to the oxidizing substrate (Y_{ox}, converted to Y_{red}). The first reaction of the scheme is often indicated as the reducing half-cycle, because the enzyme ends up in the reduced state, the second as the oxidizing half-cycle (e.g., Bellelli *et al.*, 2000). This scheme is robust, and can be applied to many oxidoreductases, even when more intermediate species are present in the catalytic cycle. When other intermediates are populated, the steady-state parameters one derives from the scheme are apparent ones, which combine or average some steps of the catalytic cycle. Yet, the above scheme is often able to describe the effect of the constant substrate, and to simplify the catalytic cycle of these important enzymes, whose more accurate description would be extremely demanding, if at all possible.

As an example of this type of analysis we quote the case of the flavoenzyme thioredoxin reductase (TR), in which the reducing substrate is NADPH and the oxidizing substrate is thioredoxin, where the simplified scheme proved sufficient to explain several puzzling results present in the literature (Saccoccia *et al.*, 2014). TR has a more complex catalytic scheme than the one depicted above, and both the reducing and oxidizing half-cycles have more intermediate species, for some of which the three dimensional structure has been solved by x-ray crystallography (Angelucci *et al.*, 2010). However, the reaction scheme proposed in the present section fulfills the requirements of a sufficient model.

Under the pseudo-equilibrium approximation, if applicable (see Section 8.5), the above scheme is fully described by two pseudo-equilibrium constants and two kinetic constants. To write the binding polynomial for the steady-state condition, we take into account that the rates of the two half-cycles must be equal to prevent changes in concentration of the reduced and oxidized species of the enzyme (i.e., $[EXT]k_{XT} = [ETY]k_Y$):

$$[E]_{tot} = [E]\left(1 + [XT]/K_{XT} + [XT]k_{XT}/K_{XT}k_Y + [XT]k_{XT}K_Y/[Y]K_{XT}k_Y\right) \quad \text{(eqn. 9.1)}$$

$$V = [E]_{tot}[XT][Y]k_{XT}k_Y/\left([Y]K_{XT}k_Y + [XT][Y]k_Y + [XT][Y]k_{XT} + [XT]k_{XT}K_Y\right)$$

$$\text{(eqn. 9.2)}$$

Properly speaking, this system does not lend itself to an obvious definition of the steady-state parameters, because the velocity depends on two variables. However, we can adapt to this case the definitions given in Section 8.1, and state that V_{max} is V at saturating concentration of the variable substrate, and K_M is the variable substrate concentration required to achieve half the V_{max}. The K_Ms and k_{cat}s we derive from the velocity function are as follows.

$$k_{cat\,XT,[Y]=const.} = [Y]k_{XT}k_Y / ([Y]k_Y + [Y]k_{XT} + k_{XT}K_Y) \qquad \text{(eqn. 9.3)}$$

$$k_{cat\,Y,[XT]=const.} = [XT]k_{XT}k_Y / ([XT]k_Y + [XT]k_{XT} + k_Y K_{XT}) \qquad \text{(eqn. 9.4)}$$

$$\begin{aligned} K_{M\,XT,[Y]=const.} &= K_{XT}[Y]k_Y / ([Y]k_Y + [Y]k_{XT} + k_{XT}K_Y) \\ &= K_{XT}k_{cat\,XT,[Y]=const.} / k_{XT} \end{aligned} \qquad \text{(eqn. 9.5)}$$

$$\begin{aligned} K_{M\,Y,[XT]=const.} &= K_Y[XT]k_{XT} / ([XT]k_{XT} + [XT]k_Y + k_Y K_{XT}) \\ &= K_Y k_{cat\,Y,[XT]=const.} / k_Y \end{aligned} \qquad \text{(eqn. 9.6)}$$

The above equations show that $k_{cat\,XT,\,[Y]=const.}$ presents a hyperbolic dependence on the concentration of constant substrate Y, and $k_{cat\,Y,\,[XT]=const.}$ presents a hyperbolic dependence on the concentration of constant substrate XT. The same dependence is presented by the K_Ms. Moreover, as the concentration of the constant substrate is increased to saturating levels, the k_{cat}s tend toward a common asymptote:

$$k_{cat,max} = k_{XT}k_Y / (k_Y + k_{XT}) \qquad \text{(eqn. 9.7)}$$

In contrast, at saturating concentrations of the constant substrate ($[Y]>>K_Y$ or $[XT]>>K_{XT}$, respectively), the K_Ms are directly proportional to the respective k_{cat}s, but with different proportionality constants:

$$K_{M\,XT,[Y]=saturating} = K_{XT}k_{cat,max} / k_{XT} \qquad \text{(eqn. 9.8)}$$

$$K_{M\,Y,[XT]=saturating} = K_Y k_{cat,max} / k_Y \qquad \text{(eqn. 9.9)}$$

Figure 9.1 reports the Michaelis-Menten and Lineweaver-Burk plots of the initial velocities one would observe at constant concentration of substrate Y, simulated using eqn. 9.2.

This catalytic scheme has some interesting and quite distinctive properties.

i) The kinetic efficiency of the variable substrate is independent of the concentration of the fixed substrate, that is,
$k_{cat\,Y,\,[XT]=const.}/K_{M\,Y,\,[XT]=const.} = k_Y/K_Y$ and $k_{cat\,XT,\,[Y]=const.}/K_{M\,XT,\,[Y]=const.} = k_{XT}/K_{XT}$.
As a consequence, the Lineweaver-Burk plots of $1/V$ versus $1/$[variable substrate] at different concentrations of the fixed substrate run parallel to each other, and do not have any intercept (see Figure 9.1B).

ii) The linkage between the two substrates is of the negative type, that is, an increase in the concentration of either substrate diminishes the apparent affinity of the other (Figure 9.2). In other words, the higher the concentration of the constant substrate, the higher the apparent K_M for the variable substrate (Figure 9.2B). This is consistent with the observation that both k_{cat} and K_M of the variable substrate tend toward zero at very low concentrations of the constant substrate. This linkage is based on the effect that each substrate concentration has on the relative populations of enzyme intermediates, and there is no ternary complex in which the protein can

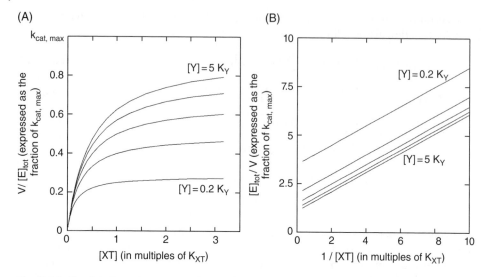

Figure 9.1 Simulated steady-state experiments for a two-substrate enzyme that does not form a ternary complex. Panel A: normalized rate of product formation (expressed as the fraction of $k_{cat,\,max}$) as a function of the concentration of variable substrate XT. Velocity data were simulated at five concentrations of constant substrate Y, over the range $0.2K_Y$, $0.5K_Y$, $1K_Y$, $2K_Y$, $5K_Y$. Parameters of the simulation: $k_{XT} = 1\,s^{-1}$, $k_Y = 1\,s^{-1}$, $k_{cat,\,max} = 0.5\,s^{-1}$, $K_{XT} = 1\,mM$, $K_Y = 1\,mM$. Panel B: Lineweaver-Burk plot of the same data as in panel A.

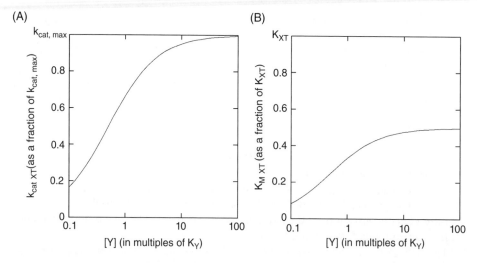

Figure 9.2 Dependencies of the Michaelis-Menten parameters for substrate XT obtained at constant concentration of substrate Y.

Panel A: hyperbolic dependency of k_{cat} of the variable substrate (XT) on the concentration of the constant substrate (Y). As expected, $k_{cat\,XT}$ approaches $k_{cat,\,max}$ as one raises the concentration of the constant substrate Y to saturating values. The parameters used for this simulation are the same as in Figure 9.1.

Panel B: hyperbolic dependency of K_M of the variable substrate on the constant substrate. As expected, $K_{M\,XT}$ cannot approach K_{XT} (see text).

directly regulate the affinity of each ligand. Thus, this linkage is unlike homotropic or heterotropic linkages that require a ternary complex, and is also unlike identical linkage since it does not require that the two substrates bind to the same site on the macromolecule. However, the linkage between the two substrates shares with identical linkage the absence of the ternary complex. We suggest that this type of linkage may be called indirect (because the two substrates do not bind at the same time nor do they share the same binding site) or statistical (because it is caused by each substrate changing the concentration of the enzyme intermediate capable of binding the other substrate).

iii) At least one K_M can never approach the corresponding equilibrium constant K_S (Cornish-Bowden, 1995). This is a consequence of the fact that $k_{cat\ max}$ may approximate the greater of the two catalytic constants of the reaction scheme, but not both. If the two catalytic constants are comparable, then neither K_M approaches the corresponding K_S (see eqns. 9.8 and 9.9; see also Figure 9.2B).

9.3 Competitive Inhibitors of Two-Substrate Enzymes That Do Not Form a Ternary Complex

Two-substrate enzymes that do not form a ternary complex may have several types of inhibitors, of which the competitive ones probably have the most interesting and peculiar behavior. Two cases of competitive inhibition are possible. If the binding site for the two substrates is the same, or the two sites are contiguous, competitive inhibitors may compete with both substrates. Conversely, if the two substrates bind to different sites, inhibitors compete with only one substrate. For example, transferases typically have contiguous or overlapping sites, whereas in oxidoreductases the binding sites for reducing and oxidizing substrates can be different.

The effect of an inhibitor that competes at the same time with both substrates of an enzyme presenting the ping-pong catalytic mechanism may be schematized as follows:

$$EI + XT \Leftrightarrow E + I + XT \Leftrightarrow EXT + I \rightarrow ET + X + I$$
$$ETI + Y \Leftrightarrow ET + I + Y \Leftrightarrow ETY + I \rightarrow E + YT + I$$

With respect to the mechanism considered in Section 9.2, we add two intermediates (EI and ETI) and two equilibrium constants to describe the dissociation of the enzyme inhibitor complex, one for EI, called K_I and one for ETI, called K_{IT}, leading to the following ligand-binding polynomial.

$$[E]_{tot} = [E](1 + [XT]/K_X + [I]/K_I + [XT]k_X/K_Xk_Y + [XT]k_XK_Y/[Y]K_Xk_Y$$
$$+ [XT][I]k_XK_Y/[Y]K_{IT}K_Xk_Y) \qquad \text{(eqn. 9.10)}$$

The rate equation is:

$$V = [E]_{tot}[XT][Y]k_Xk_YK_IK_{IT} / ([Y]K_Xk_YK_IK_{IT} + [XT][Y]k_YK_IK_{IT} + [I][Y]K_Xk_YK_{IT}$$
$$+ [XT][Y]k_XK_IK_{IT} + [XT]k_XK_YK_IK_{IT} + [XT][I]k_XK_YK_I)$$

$$\text{(eqn. 9.11)}$$

The rate equation contains the concentrations of three species (plus the total enzyme): XT, Y, and I. Experiments are typically carried out varying only one of these concentrations and keeping the other two constant. In particular, by alternately varying each substrate, one obtains Michaelis-Menten–type plots from which K_M and k_{cat} values can be derived for each substrate. The equations for k_{cat} and K_M for substrate XT, obtained at constant concentrations of substrate Y and inhibitor I, are as follows.

$$k_{cat\,XT,I,[Y]=const.} = k_{XT}[Y]k_Y K_{IT}/\left([Y]k_Y K_{IT} + k_{XT}K_Y K_{IT} + [Y]k_{XT}K_{IT} + [I]k_X K_Y\right)$$

(eqn. 9.12)

$$K_{M\,XT,I,[Y]=const.} = K_X k_{cat\,XT,I,[Y]=const.}\left([I]+K_I\right)/k_{XT}K_I$$

(eqn. 9.13)

We remark that $K_{M\,XT,\,[Y]=const.}$ is a function of the dissociation constant of the EXT complex (K_X), the ratio $k_{cat\,XT,I,\,[Y]=const.}/k_{XT}$, and the correction factor typical of competitive inhibition ($[I] + K_I)/K_I$ (see eqn. 8.11). We further remark that $K_{M\,XT,I,\,[Y]=const.}$ presents a different dependence on the concentration of the inhibitor than $k_{cat\,XT,I,\,[Y]=const.}$, hence, the kinetic efficiency $k_{cat\,XT,I,\,[Y]=const.}/K_{M\,XT,I,\,[Y]=const.}$ depends on the concentration of the inhibitor. Finally, if both the concentrations of XT and Y are raised to saturation, one obtains $k_{cat,max}$ because the inhibitor is fully displaced by competition with both substrates.

The corresponding equations for the other substrate are substantially identical, except that we need to exchange [Y] with [XT], K_X with K_Y, k_X with k_Y, and K_I with K_{IT}:

$$k_{cat\,Y,I,[XT]=const.} = k_Y[XT]k_X K_I/\left([XT]k_Y K_I + k_Y K_X K_I + [XT]k_{XT}K_I + [I]k_Y K_X\right)$$

(eqn. 9.14)

$$K_{M\,Y,I,[XT]=const.} = K_Y k_{cat\,Y,I,[XT]=const.}\left([I]+K_{IT}\right)/k_Y K_{IT}$$

(eqn. 9.15)

The same considerations as above apply to these equations.

Finally, from the rate equation (eqn. 9.11), we can derive IC_{50}, the inhibitor concentration required to reduce by half the rate of product formation at the same concentrations of the two substrates in the absence of the inhibitor:

$$IC_{50} = K_I K_{IT}\left\{[Y]k_Y\left([XT]+K_X\right)+[XT]k_X\left([Y]+K_Y\right)\right\}/\left([Y]k_Y K_X K_{IT}+[XT]k_X K_Y K_I\right)$$

(eqn. 9.16)

An interesting property of this equation is that IC_{50} is proportional to the product of the concentrations of both substrates. This is a consequence of the fact that the inhibitor competes with (is displaced by) either substrate, and is consistent with both $k_{cat\,XT,I,\,[Y]=const.}$ and $k_{cat\,Y,I,\,[X]=const.}$ being affected by the inhibitor, but $k_{cat\,max}$ being not. The reason for this behavior lies in the fact that in this (oversimplified) catalytic scheme the inhibitor binds to both states of the enzyme (E and ET), whereas each substrate binds to only one of them. Thus XT competes with I for E but not for ET, and even at saturating concentration XT does not displace I from the complex ETI. Figure 9.3 reports a series of simulated Michaelis-Menten and Lineweaver-Burk plots for this system.

In a number of oxido-reductases and other enzymes the reaction scheme for competitive inhibitors is different from the one considered above because the two substrates bind to different sites; hence, no inhibitor can compete with both substrates. For example,

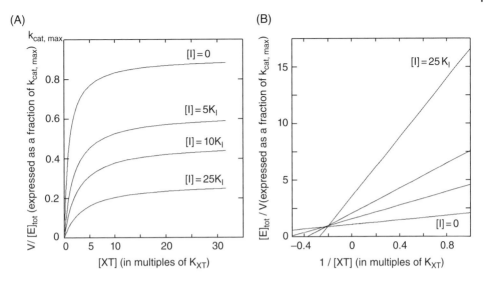

Figure 9.3 Simulated Michaelis-Menten (panel A) and Lineweaver-Burk plots (panel B) for an inhibitor that competes with the two substrates of an enzyme that does not form a ternary complex (see text). Simulation conditions: constant concentration of substrate Y = 5 K_Y; K_{IT} = 5 K_I. Notice that the inhibitor changes both $k_{cat\ XT,\ [Y]=const.}$ and $K_{M\ XT,\ [Y]=const.}$.

NADH- or NADPH-dependent reductases usually have a binding site for NADH/NADPH and another, non-overlapping site for the substrate to be reduced. A typical example is that of glutathione reductase or thioredoxin reductase that have a binding site for NADPH and a different site for the oxidized substrate (glutathione or thioredoxin, respectively). Electrons are transferred internally between the two sites. In these cases the inhibitor competes with one substrate and behaves as a non-competitive inhibitor in experiments in which the concentration of the competing substrate is kept constant. However, as explained below, the non-competitive inhibition observed in this case has peculiar characteristics. The reaction scheme is as follows.

$$EI + XT \Leftrightarrow E + I + XT \Leftrightarrow EXT + I \rightarrow ET + X + I$$
$$ET + Y \Leftrightarrow ETY \rightarrow E + YT$$

In this scheme we do not write, and can safely ignore, the possible ETI/ETYI complexes, because, if they form at all, we may assume at least to a first approximation that these species are fully active toward substrate Y, and essentially indistinguishable from the complexes ET/ETY. In order to describe this scheme, we need only one equilibrium constant for the enzyme-inhibitor complex, K_I. The binding polynomial we obtain contains five species.

$$[E]_{tot} = [E]\left(1 + [XT]/K_X + [I]/K_I + [XT]k_X/K_X k_Y + [XT]k_X K_Y/[Y]k_Y K_X\right) \quad \text{(eqn. 9.17)}$$

The rate equation for initial velocity measurements is

$$V = [E]_{tot}[XT][Y]k_X k_Y K_I / ([Y]K_X k_Y K_I + [XT][Y]k_Y K_I + [I][Y]K_X k_Y$$
$$+ [XT][Y]k_X K_I + [XT]k_X K_Y K_I) \quad \text{(eqn. 9.18)}$$

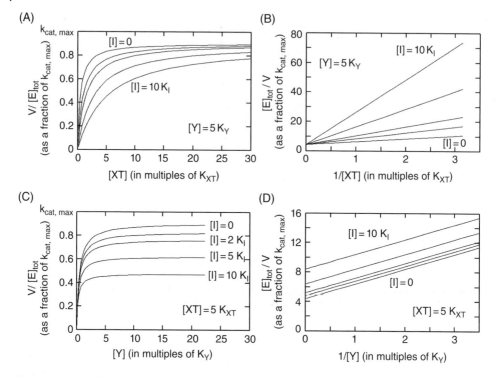

Figure 9.4 Simulations of the rate of product formation according to eqn. 9.18.

Panel A: V as a function of the concentration of the variable substrate XT, at [I]=0,1,2,5,10 K_I. The concentration of substrate Y is constant, at 5-fold K_Y. Panel B: Lineweaver-Burk plots of the same data as in panel A.

Panel C: V as a function of the concentration of the variable substrate Y, at [I]=0,1,2,5,10 K_I. The concentration of substrate XT is constant, at 5-fold K_{XT}. Panel D: Lineweaver-Burk plots of the same data as in panel C.

Two simulated data sets for this type of inhibition are reported in Figure 9.4; the former is a series of determinations of V measured at constant concentration of substrate Y (Figures 9.4A and 9.4B), the latter a series of determinations of V measured at constant concentration of substrate XT (Figures 9.4C and 9.4D). In both series four concentrations of the inhibitor were tested and compared with the plots obtained in the absence of the inhibitor.

When the reaction rate is determined at constant concentration of the non-competing substrate Y, Michaelis-Menten and Lineweaver-Burk plots identical to those that characterize pure competitive inhibition in one-substrate enzymes (Section 8.6) are obtained. At constant concentration of the competing substrate XT, Michaelis-Menten and Lineweaver-Burk plots reminiscent of those that characterize uncompetitive inhibition in one-substrate enzymes (Section 8.7) are obtained.

An example of an enzyme-inhibitor system that obeys eqn. 9.18 is provided by thioredoxin reductase and dimethylarsonous diiodide (Lin *et al.*, 1999). After long incubation times the inhibitor becomes covalently bound to the enzyme, thus we are interested here only in the initial, rapid and reversible binding process. Dimethylarsonous diiodide competes with the oxidizing substrate of the enzyme (thioredoxin or DTNB), but not with the

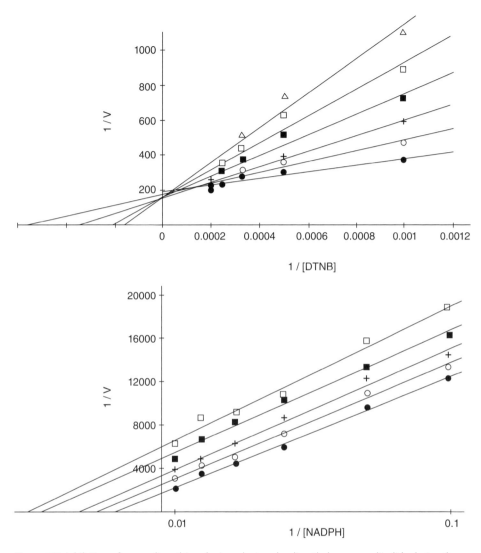

Figure 9.5 Inhibition of mouse liver thioredoxin reductase by dimethylarsonous diiodide during the fast, reversible binding process (original data by Lin *et al.*, 1999; redrawn and reanalyzed by Saccoccia *et al.*, 2014). The enzyme has two substrates, DTNB (or thioredoxin) and NADPH; the inhibitor competes with the former, but not with the latter. *Source:* Adapted from Lin *et al.* (1999) and Saccoccia *et al.* (2014).

reducing substrate NADPH, and yields the Lineweaver-Burk plots reported in Figure 9.5, which the reader may compare with the simulations reported in Figure 9.4B and 9.4D.

From eqn. 9.18, we derive the Michaelis-Menten parameters for experiments carried out at constant concentration of the non-competing substrate Y.

$$k_{cat\,XT,I,[Y]=const.} = k_{XT}[Y]k_Y / ([Y]k_Y + k_{XT}K_Y + [Y]k_{XT}) \qquad (\text{eqn. } 9.19)$$

$$K_{M\,XT,I,[Y]=const.} = K_{XT}k_{cat\,XT,I,[Y]=const.}([I]+K_I)/k_{XT}\,K_I \qquad (\text{eqn. } 9.20)$$

These equations are essentially identical to the ones we found above except that the terms containing K_{IT} have disappeared, and the inhibitor behaves as a simple competitive inhibitor of a one-substrate enzyme in that it does not affect k_{cat} but changes K_M (compare eqn. 9.3 with eqn. 9.19).

The opposite experiment, in which the concentration of the competing substrate XT is kept constant, and that of the non-competing one is varied, yields a very different picture.

$$k_{cat\,Y,I,[XT]=const.} = k_Y[XT]k_{XT}K_I/([XT]k_YK_I + k_YK_{XT}K_I + [XT]k_{XT}K_I + [I]k_YK_{XT})$$

(eqn. 9.21)

$$K_{M\,Y,I,[XT]=const.} = K_Yk_{cat\,Y,I,[XT]=const.}/k_Y$$

(eqn. 9.22)

Here again we have the same equations obtained above, modified because the terms containing K_{IT} have disappeared. But in this case the inhibitor appears in the equation for k_{cat} (eqn. 9.21), as it would in a mixed-heterotropic inhibition. Moreover, K_M is directly proportional to k_{cat}, which is reminiscent of uncompetitive inhibition as it implies that *the inhibitor increases the apparent affinity of the enzyme for the non-competing substrate.* This is a case of statistical linkage, which depends on the fact that increasing the concentration of non-competing substrate favors conversion of ET to E, the species that is inhibited. That is, the higher the concentration of non-competing substrate, the faster the conversion of the enzyme to the state to which the inhibitor binds, and consequently the higher the saturation of the enzyme with the inhibitor. We may better explain this case by stating that the reduction of $K_{M\,Y,I,\,[XT]=const.}$ caused by the inhibitor is not due to an effective increase in affinity of the enzyme for the non-competing substrate (as in uncompetitive inhibition), but to the fact that the asymptote of the Michaelis-Menten hyperbola measures the fraction of uninhibited enzyme. This fraction decreases as the non-competing substrate concentration is increased.

While the functional relationship between the inhibitor and the competing substrate is an identical linkage, that between the inhibitor and the non-competing substrate is statistical, that is, it depends on a shift of the populations of ligation intermediates, rather than on a change in the affinity of the enzyme for the substrate induced by the binding of the inhibitor.

Finally, the inhibitor concentration required to reduce the rate of product formation to half its value in the absence of inhibitor is:

$$IC_{50} = K_I\{[Y]k_Y([XT]+K_{XT}) + [XT]k_{XT}([Y]+K_Y)\}/[Y]k_YK_{XT}$$

(eqn. 9.23)

We stated more than once that the statistically soundest method to analyze one's experimental data and extract the best estimate of the binding parameters is to globally fit the appropriate equation to the entire data set that has been collected. However, since a graphical representation of the data is usually necessary for their presentation, graphical methods are still employed to derive K_I values from steady-state parameters determined at different inhibitor concentrations. Moreover, given the complexity of the equations that describe the functional behavior of two-substrate enzymes, the textbook analysis, which is developed for one-substrate enzymes, is often used. Thus, it is of interest to explore what happens if one applies the equations developed in Sections 8.6 and 8.7 to the inhibitors considered in this section.

To graphically estimate K_I from a series of experiments carried out at constant concentration of the non-competing substrate (Figures 9.4A and 9.4B; upper panel of Figure 9.5), one may adopt the procedure we described in Section 8.6 and indeed a plot of $K_{M XT,I, [Y]=const.}$ versus $[I]$ yields a straight line with intercept $K_{M XT,I, [Y]=const.}$ and slope $K_{M XT,I,[Y]=const.}/K_I$. Thus, in this case the simplistic approach yields the correct result.

The other approach is to treat a series of experiments carried out at constant concentration of the competing substrate as a case of uncompetitive or mixed inhibition for a single-substrate enzyme and apply the procedure we described in Section 8.7. However, in this case, a plot of $1/k_{cat Y,I, [XT]=const.}$ versus $[I]$ yields a straight line with intercept $1/k_{cat Y, [XT]=const.}$ and slope $K_{XT}/[XT]k_{XT}K_I$, much different from the expected $1/k_{cat Y, [XT]=const.}K_I$. If one tries to estimate K_I from the ratio intercept/slope, one obtains a wrong estimate of K_I (Saccoccia *et al.*, 2014).

9.4 Steady-State Parameters of Two-Substrate Enzymes Forming a Ternary Complex

The active site of this type of enzyme has room for both substrates simultaneously, and the catalyzed reaction requires the presence of both. Due to the complex architecture of the active site, these enzymes may be prone to product inhibition. In some cases non-productive complexes may be formed containing one substrate and one product. However, these complications can be neglected in initial velocity studies, where the product(s) are absent or negligible, and the reaction mechanism may be simplified as follows.

$$
\begin{array}{c}
\quad K_X \nearrow EX + Y \underset{}{\overset{X K_Y}{\rightleftharpoons}} \\
E + X + Y \qquad\qquad\qquad EXY \xrightarrow{\ k_p\ } E + P \\
\quad K_Y \searrow EY + X \underset{}{\overset{Y K_X}{\rightleftharpoons}}
\end{array}
$$

This mechanism is a case of heterotropic linkage between the two substrates (Section 1.9). Under the pseudo-equilibrium assumption the mechanism is described by three equilibrium constants and one kinetic constant (k_p). If the binding of either substrate is associated with a detectable signal, one may determine the equilibrium constant directly, under conditions where catalysis cannot occur. In this case it is a rational choice to determine K_X in the absence of substrate Y, and K_Y in the absence of substrate X. The remaining equilibrium constant ($^X K_Y$ or $^Y K_X$) can be determined in a steady-state experiment, and the fourth from linkage relationships. However, if the binding of the substrate(s) is not associated with a detectable signal, and all the parameters must be determined under steady-state conditions from the signal associated with product formation or substrate consumption, it is a more rational choice to adopt as independent parameters the constants $^X K_Y$, $^Y K_X$, k_p and either K_X or K_Y.

The pseudo-equilibrium binding polynomial of the enzyme is:

$$
[E]_{tot} = [E]\left(1 + [X]/K_X + [Y]\, ^Y K_X/K_X\, ^X K_Y + [X][Y]/K_X\, ^X K_Y\right) \qquad \text{(eqn. 9.24)}
$$

and the initial velocity of product formation is:

$$
V = [E]_{tot}\, k_p [X][Y] / \left(K_X\, ^X K_Y + [X]\, ^X K_Y + [Y]\, ^{YI} K_X + [X][Y]\right) \qquad \text{(eqn. 9.25)}
$$

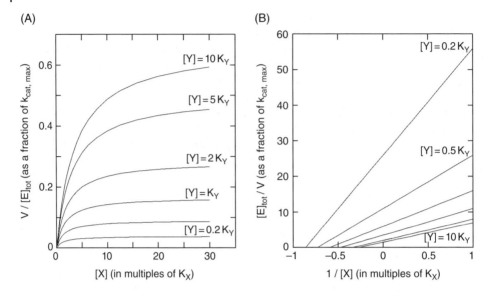

Figure 9.6 Michaelis-Menten (panel A) and Lineweaver-Burk (panel B) plots of the rate of product formation simulated according to eqn. 9.25. $^{X}K_Y = 5K_Y$; $^{Y}K_X = 5K_X$; variable substrate is X, and six concentrations of the constant substrate Y are explored: $0.2K_Y$, $0.5K_Y$, K_Y, $2K_Y$, $5K_Y$, and $10K_Y$. In this mechanism the k_{cat} and K_M vary as a function of the constant substrate concentration, but with different laws (see eqns. 9.26 and 9.28), hence the kinetic efficiency exhibits a dependence on the constant substrate concentration.

A set of simulated steady-state experiments for an enzyme obeying eqn. 9.25 is reported in Fig. 9.6.

The k_{cat} for each substrate at constant concentration of the other result.

$$k_{cat\,X,[Y]=const.} = k_p[Y]/\left(^{X}K_Y + [Y]\right) \qquad \text{(eqn. 9.26)}$$

$$k_{cat\,Y,[X]=const.} = k_p[X]/\left(^{Y}K_X + [X]\right) \qquad \text{(eqn. 9.27)}$$

We note that the two k_{cat}s are identical except that they exchange [X] and [Y], $^{X}K_Y$ and $^{Y}K_X$. Moreover, we observe that both k_{cat}s show a hyperbolic dependence on the concentration of constant substrate, with asymptote $k_{cat,max} = k_p$ at saturating concentration of constant substrate (See Figure 9.7).

The K_M, that is, the concentration of variable substrate required to achieve half the V_{max} ($=[E]_{tot}k_{cat}$), result.

$$K_{M\,X,[Y]=const.} = \left(K_X\,^{X}K_Y + [Y]\,^{Y}K_X\right)/\left([Y] + ^{X}K_Y\right) \qquad \text{(eqn. 9.28)}$$

$$K_{M\,Y,[X]=const.} = ^{X}K_Y\left([X] + K_X\right)/\left([X] + ^{Y}K_X\right) \qquad \text{(eqn. 9.29)}$$

Both K_Ms present a hyperbolical dependence on concentration of the fixed substrate and their asymptotes are:

$$\lim K_{M\,X,[Y]\to\infty} = ^{Y}K_X; \quad \lim K_{M\,X,[Y]\to 0} = K_X$$

$$\lim K_{M\,Y,[X]\to\infty} = ^{X}K_Y; \quad \lim K_{M\,Y,[X]\to 0} = K_X\,^{X}K_Y/^{Y}K_X = K_Y;$$

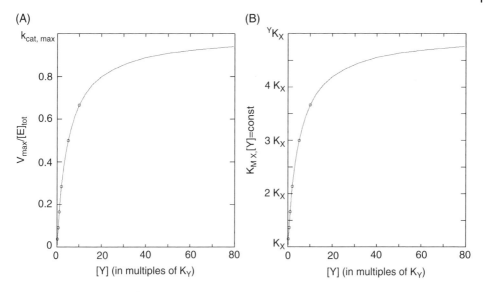

Figure 9.7 Dependency of $k_{cat\,X,\,[Y]=const.}$ (panel A) and $K_{M\,X,\,[Y]=const.}$ (panel B) on the concentration of substrate Y. Same data and parameters as in Figure 9.6. Points represent the values of $k_{cat\,X,\,[Y]=const.}$ (panel A) and $K_{M\,X,\,[Y]=const.}$ for the curves in Figure 9.6; lines are drawn according to eqns. 9.26 or 9.28.

The equations that describe the K_Ms may seem uninformative. However, if we substitute ^YK_X with $K_X{}^XK_Y/K_Y$ (or ^XK_Y with $K_Y{}^YK_X/K_X$), they convert to the equation for $X_{1/2}$ described in Section 1.9 for heterotropic linkage, as the reader might have expected. The reason the two-substrate enzyme is treated as in eqns. 9.24 to 9.29 rather than as heterotropic linkage in Section 1.9 is two-fold (i) The enzyme requires both its substrates to catalyze the reaction, which in turn provides the observed signal. Thus, it is not always possible to determine directly the binding of either substrate in the absence of the other. (ii) ^YK_X is $K_{M\,X}$ at saturating concentration of Y and ^XK_Y is $K_{M\,Y}$ at saturating concentration of X. Thus, these two parameters are easy to determine in steady-state experiments. By contrast K_X and K_Y, the equilibrium constants for each substrate in absence of the other, cannot be determined in a steady-state experiment, because in absence of either substrate catalysis cannot occur. It is unfortunate that the jargon used in enzymology is different from that used to describe ligand binding experiments, because this difference obscures the obvious fact that the same equations are used and the same results are obtained.

There are two important differences between this catalytic mechanism and the one considered in Section 9.2

i) This mechanism is compatible with a heterotropic linkage between the two substrates, which may be positive or negative. In the case of enzymes that do not form the ternary complex, linkage between substrates is statistical and negative (eqns. 9.5 and 9.6).

ii) The dependence of K_M for the variable substrate on the fixed substrate concentration is hyperbolic but it does not pass through the origin of the axes. $K_{M\,X}$ is limited between K_X and ^YK_X, and $K_{M\,Y}$ between K_Y and ^XK_Y (see above).

Therefore, collection of a set of Michaelis-Menten curves for one substrate at several concentrations of the other may allow the researcher to detect or exclude the formation of a ternary complex. This is a compelling reason for exploring the effect of the concentration of fixed substrate, rather than limiting oneself to the saturating condition.

The preceding mechanism simplifies under limiting cases.

i) In the *absence of linkage* (i.e., $K_Y = {}^X K_Y$ and $K_X = {}^Y K_X$) K_M of the varied substrate is independent of the concentration of the constant substrate.

ii) When the order of substrate addition is *obligate*, the second substrate binds only if the first substrate is already bound and the intermediate with the second substrate is never populated. The reaction scheme simplifies as follows.

$$E + X + Y \Leftrightarrow EX + Y \Leftrightarrow EXY \rightarrow E + P(s)$$

For the sake of consistency we describe this system using the two pseudo-equilibrium constants K_X and ${}^X K_Y$ and the kinetic constant k_p. This mechanism presents an extreme degree of positive heterotropic linkage in the sense that the equilibrium dissociation constant K_Y is so high that substrate Y does not bind appreciably to the enzyme in the absence of X. The k_{cat} and K_M are as in the parent case except that all terms in which K_Y appears as a divisor are neglected:

$$k_{cat\,X,[Y]=const.} = k_p[Y]/\left({}^X K_Y + [Y]\right)$$

$$k_{cat\,Y,[X]=const.} = k_p$$

$$K_{M\,X,[Y]=const.} = K_X /\left(1 + [Y]/{}^X K_Y\right)$$

$$K_{M\,Y,[X]=const.} = {}^X K_Y \left(K_X + [X]\right)/[X]$$

It may appear paradoxical that $k_{cat\,Y,\,[X]=const.}$ does not depend on the concentration of the constant substrate X. The reason for this effect is that, within the limits of the compulsory substrate addition model, the positive heterotropic interaction between the two substrates is so strong that if the concentration of substrate Y is increased sufficiently, its effect in promoting the formation of the complex EXY causes substrate X to bind, and indeed $K_{M\,X,\,[Y]=const.}$ tends to zero as [Y] increases. Thus the paradox stems from the fact that setting an infinitely large K_Y implies an infinite increase in affinity for Y in the presence of X.

9.5 Competitive Inhibitors of Two-Substrate Enzymes Forming a Ternary Complex

Two cases are possible. The inhibitor may compete with both substrates at the same time, or it may compete with only one of them. The former case is described by the following reaction scheme:

$$EI \underset{K_I}{\overset{}{\rightleftharpoons}} I + E + X + Y \begin{array}{c} \overset{K_X}{\nearrow} EX + Y \overset{{}^X K_Y}{\nwarrow} \\ \underset{K_Y}{\searrow} EY + X \underset{{}^Y K_X}{\nearrow} \end{array} EXY \xrightarrow{k_p} E + P$$

The binding polynomial of the above catalytic scheme under the pseudo-equilibrium approximation is:

$$[E]_{tot} = [E]\left(1 + [I]/K_I + [X]/K_X + [Y]^Y K_X/K_X{}^X K_Y + [X][Y]/K_X{}^X K_Y\right)$$

The initial velocity is:

$$V = [E]_{tot} k_p [X][Y]K_I / \left(K_X K_I{}^X K_Y + [I]K_X{}^X K_Y + [X]K_I{}^X K_Y + [Y]^Y K_X K_I + [X][Y]K_I\right)$$

The Michaelis-Menten steady-state parameters are as follows.

$$k_{cat\, X,I,[Y]=const.} = k_p [Y]/\left({}^X K_Y + [Y]\right)$$

$$K_{M\, X,I,[Y]=const.} = K_{M\, X,[Y]=const.} + K_X{}^X K_Y [I]/K_I \left([Y] + {}^X K_Y\right)$$

$$k_{cat\, Y,I,[X]=const.} = k_p [X]/\left({}^Y K_X + [X]\right)$$

$$K_{M\, Y,I,[X]=const.} = K_{M\, Y,[X]=const.} + K_X{}^X K_Y [I]/K_I \left([I] + {}^Y K_X\right)$$

The IC$_{50}$ results:

$$IC_{50} = K_I \left[\left(1 + [X]/K_X\right) + [Y]/K_Y \left(1 + [X]/{}^Y K_X\right)\right]$$

The roles of the substrates in this catalytic scheme are perfectly symmetric, that is, the equations that describe the steady-state parameters at variable concentration of X and constant concentration of Y are identical to those that describe the steady-state parameters at variable concentration of Y and constant concentration of X, except that they exchange [X] and [Y], K_X and K_Y, and ${}^Y K_X$ and ${}^X K_Y$. Thus, we report in Figure 9.8, the steady-state rates of product formation simulated under only one condition (i.e., at constant concentration of Y). The plot of this type of enzyme and inhibitor system is unremarkable because it is essentially identical to that of a pure competitive inhibitor of a single-substrate enzyme, and, as expected, the competitive inhibitor does not change k_{cat}, but increases the apparent K_M.

The alternative case to be considered is that the inhibitor replaces only one of the substrates but not the other. Competition is thus observed with only one of the substrates, according to the catalytic mechanism:

$$\begin{array}{c} {}^X K_I \\ EIX \rightleftharpoons EX + Y \; {}^X K_Y \\ {}^I K_X \updownarrow \qquad \qquad \searrow K_X \qquad \searrow \\ EI \rightleftharpoons I + E + X + Y \qquad\qquad EXY \xrightarrow{k_p} E + P \\ K_I \qquad\qquad\qquad \searrow \qquad \nearrow \\ K_Y \; EY + X \; {}^Y K_X \end{array}$$

The binding polynomial for the above scheme, under the pseudo-equilibrium condition is:

$$[E]_{tot} = [E]\left(1 + [I]/K_I + [I][X]/K_I{}^I K_X + [X]/K_X + [Y]^Y K_X/K_X{}^X K_Y + [X][Y]/K_X{}^X K_Y\right)$$

The initial velocity is:

$$V = [E]_{tot} k_p [X][Y]K_I{}^I K_X / (K_X{}^X K_Y K_I{}^I K_X + [I]K_X{}^X K_Y{}^I K_X + [I][X]K_X{}^X K_Y \\ + [X]^X K_Y{}^I K_X K_I + [Y]^Y K_X K_I{}^I K_X + [X][Y]K_I{}^I K_X)$$

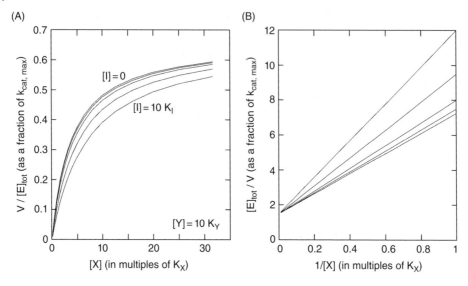

Figure 9.8 Simulation of the Michaelis-Menten (panel A) and Lineweaver-Burk (panel B) plots of a two-substrate enzyme in the presence of an inhibitor that competes with both substrates. Simulation parameters as in Figure 9.6, plus five inhibitor concentrations equalling 0, K_I, $2K_I$, $5K_I$ and $10K_I$.

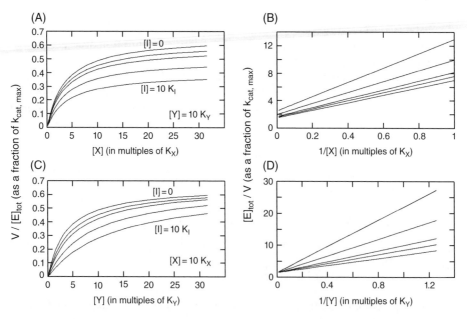

Figure 9.9 Michaelis-Menten and Lineweaver-Burk plots for a two-substrate enzyme that forms a ternary complex in the presence of an inhibitor that competes with only one of the substrates (see text). Simulation parameters as in Figure 9.6, $^IK_X = 5K_X$; inhibitor concentrations: 0, K_I, $2K_I$, $5K_I$, $10K_I$. Panels A and B: rates of product formation at constant concentration of the competing substrate Y. Panels C and D: rates of product formation at constant concentration of the non-competing substrate X. Notice that in experiments at variable concentration of the competing substrate (panels C and D) one obtains plots identical to those of competitive inhibitors of single-substrate enzymes.

A set of simulated steady-state experiments at different concentrations of the inhibitor is reported in Figure 9.9.

The Michaelis-Menten steady-state parameters for this system are as follows.

$$k_{\text{cat X,I,}[Y]=\text{const.}} = k_p [Y] K_I{}^I K_X / ([I] K_X{}^X K_Y +{}^X K_Y{}^I K_X K_I + [Y] K_I{}^I K_X)$$

$$k_{\text{cat Y,I,}[X]=\text{const.}} = k_p [X] / ({}^Y K_X + [X]) = k_{\text{cat Y,}[X]=\text{const.}}$$

$$k_{\text{M X,I,}[Y]=\text{const.}} = {}^I K_X \left\{ K_X{}^X K_Y (K_I + [I]) + [Y]{}^Y K_X K_I \right\} / \left\{ {}^X K_Y ([I] K_X + K_I{}^I K_X) + [Y] K_I{}^I K_X \right\}$$

$$K_{\text{M Y,I,}[X]=\text{const.}} = {}^X K_Y ([X] + K_X)(K_I{}^I K_X + [I] K_X) / K_I{}^I K_X ({}^Y K_X + [X])$$

References

Angelucci, F., Dimastrogiovanni, D., Boumis, G., Brunori, M., Miele, A.E., Saccoccia, F. and Bellelli A. (2010) Mapping the catalytic cycle of Schistosoma mansoni thioredoxin glutathione reductase by x-ray crystallography. J Biol Chem, 285: 32557–32567.

Bellelli, A., Morpurgo, L., Mondovi, B. and Agostinelli, E. (2000) The oxidation and reduction reactions of bovine serum amine oxidase. A kinetic study. Eur J Biochem, 267: 3264–3269.

Cleland, W.W. (1967) Enzyme kinetics. Annu Rev Biochem, 36: 77–112. PMID: 18257716

Cheng, Y.C. and Prusoff, W.H. (1973) Relationship between the inhibition constant (KI) and the concentration of inhibitor which causes 50 per cent inhibition (I50) of an enzymatic reaction. Biochem Pharmacol, 22: 3099–3108.

Cornish-Bowden, A. (1995) Fundamentals of Enzyme Kinetics. London, UK: Portland Press.

Lin, S., Cullen, W.R. and Thomas, D.J. (1999) Methylarsenicals and arsino-thiols are potent inhibitors of mouse liver thioredoxin reductase. Chem Res Toxicol, 12: 924–930.

Saccoccia, F., Angelucci, F., Boumis, G., Carotti, D., Desiato, G., Miele, A.E. and Bellelli, A. (2014) Thioredoxin reductase and its inhibitors. Curr Protein Pept Sci, 15: 621–646.

10

Beyond the Steady State: Rapid Kinetic Methods for Studying Enzyme Reactions

There are two main conditions in which the pseudo-equilibrium approximation of enzyme kinetics is insufficient to describe the experimental data. (i) Some enzymes, and some enzyme-inhibitor pairs, do not reach steady-state conditions during initial velocity experiments. Thus, these experiments test a pre-steady-state condition to which the steady-state approximations do not apply. (ii) The researcher aims explicitly at resolving the rate constants of the single steps of the catalytic cycle and has access to the instrumentation necessary to study rapid kinetic reactions (usually a stopped flow instrument).

Whatever the case, solving pre-steady-state kinetics is a demanding enterprise that usually requires more information than the signal provided by conversion of substrate into product, and is best carried out on enzymes having chromogenic coenzymes whose signal monitors the interconversion of different intermediates of the catalytic cycle. Detection of a signal related to the enzyme, rather than to substrate or product, requires much higher enzyme concentrations than those employed in steady-state experiments; as a consequence, the reaction velocities are higher than in typical steady-state experiments. Thus, the use of rapid kinetic methods is mandatory for this type of study.

It is impossible to generalize a method for pre-steady-state studies involving chromogenic substrates and enzyme chromophores because the number of variables is too large. As a consequence, we shall analyze only one case, taken from the experience of one of the authors, with the intent of examining the theoretical complexities and practical requirements of this type of study.

10.1 Structural and Catalytic Properties of Copper-Containing Amine Oxidases

Copper-containing amine oxidases are ubiquitous enzymes that catalyze the oxidative deamination of amine substrates usually derived from the decarboxylation of amino acids. They play an important dissimilatory role given that their substrates may be biologically active or toxic. The reaction catalyzed by amine oxidases is as follows.

$$R - CH_2NH_2 + O_2 + H_2O \rightarrow R - CHO + NH_3 + H_2O_2$$

Reversible Ligand Binding: Theory and Experiment, First Edition. Andrea Bellelli and Jannette Carey.
© 2018 John Wiley & Sons Ltd. Published 2018 by John Wiley & Sons Ltd.

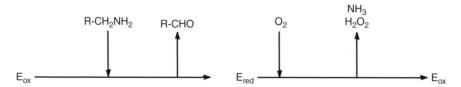

Figure 10.1 Catalytic scheme of a copper containing amine oxidase: order of substrate addition and product release.

The reaction is essentially irreversible, and is made completely irreversible by addition of traces of catalase, which also prevents the nonspecific inactivation of the enzyme by hydrogen peroxide. The enzyme uses two substrates (neglecting water) and releases three products, in a precise order (see scheme in Figure 10.1). Catalysis requires formation of covalent adducts with specific residues of the enzyme active site and does not require formation of a ternary complex; the catalytic mechanism is thus a Ping-Pong Bi-Ter (Figure 10.1).

The enzyme is a homodimeric metalloprotein, but the two subunits react independently in the catalytic mechanism and there is no need to consider the dimeric structure in this context. The active site contains a copper ion coordinated by three His residues, and a post-translationally modified Tyr residue (tri-hydroxy phenylalanine quinone, TPQ), that is essential for catalysis and has a distinctive absorbance spectrum. The catalytic scheme may be divided into oxidative and reductive half-cycles (Figure 10.2). The chemical intermediates of the reductive half cycle are the TPQuinone (a), TPKetimine (b), TPQuinolamine (c), and TPAminoresorcinol (d).

The oxidative half cycle may follow two distinct pathways. Either intermediate (d) reacts directly with oxygen to yield the resting oxidized enzyme (a) or it isomerizes (reversibly) to the TPsemiQuinolamine radical (e) in which an electron is transferred to the Cu ion (Dooley *et al.*, 1991; Turowski *et al.*, 1993).

Intermediate (e) reacts with oxygen at a much faster rate than (d). The TPsemi-Quinolamine radical never exceeds 50% of the available active sites, and may be much less.

10.2   Experimentally Accessible Information on Copper-Amine Oxidases

Copper-amine oxidases constitute quite a fortunate case for enzymologists, thanks to the rich spectroscopic features of their cofactors. These can be supplemented with the use of artificial chromogenic substrates (aromatic amines). Moreover, important additional information comes from the comparison of enzymes from different sources. For example, copper-amine oxidases from plants have a much greater yield of the semi-quinolamine radical intermediate than copper-amine oxidases of animal origin, and this facilitates attribution of rate constants for reoxidation to the two oxygen-reactive enzyme derivatives.

Several experiments are possible that lead to a consistent picture. The experiments fall into two groups: those that probe mostly or exclusively one step of the catalytic cycle

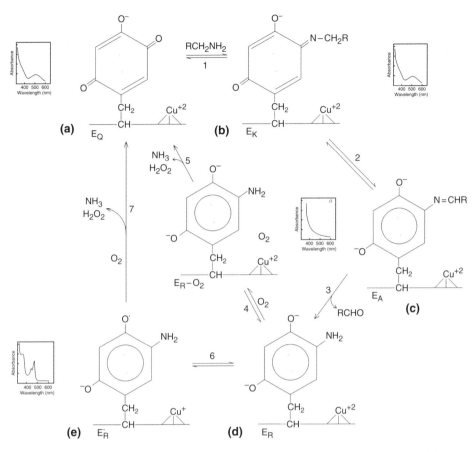

Figure 10.2 The catalytic cycle of Lentil seedling amine-oxidase. Schematic representation of the active site, showing the TPQ group and the adjacent copper ion. Absorbance spectra of TPQ in its different functional states: spectrum (i) is characteristic of the fully oxidized intermediates (a) and (b) of the upper panel; spectrum (ii) is characteristic of the reduced intermediates (c) and (d); spectrum (iii) is characteristic of intermediates containing the semiquinolamine radical (e). Source: Adapted from Medda *et al.* (1998).

(e.g., single turnover experiments, as in cases i–iii below), and those that probe all steps, under steady-state or pre-steady-state conditions (cases iv–v).

i) The rate constant for oxidation by oxygen can be measured by rapidly mixing in a stopped flow apparatus the reduced enzyme (obtained by adding a stoichiometric amount of the amine substrate to a deoxygenated solution of the enzyme) with an oxygen-containing solution, both under conditions in which the semiquinolamine radical is present (e.g., at neutral pH) and under conditions where it is absent or scarce (e.g., at acidic or basic pH). This experiment ideally is less than a single-turnover, because the enzyme runs only through the oxidative half-cycle, and not through the reductive half-cycle. If a slight excess of the amine substrate is added to prevent reoxidation by trace amounts of oxygen, a fraction of the total enzyme may cycle more than once, but the reaction of interest (reoxidation) can easily be resolved by computational methods. The enzyme concentration required for this type of experiment is high, given that the absorbance signal is due to the TPQ cofactor, whose extinction coefficients are in the order of $4,000$–$8,000\,mM^{-1}cm^{-1}$, depending on the derivative.

ii) One may measure the reduction time course by rapidly mixing the oxidized enzyme with amine substrate in the stopped flow. This experiment is carried out under single turnover conditions, by using a stoichiometric amount of the amine. Care must be taken to protect the samples from contaminating oxygen. This requirement is less stringent under experimental conditions or with enzymes that do not form the semiquinolamine radical, because reoxidation of the aminoresorcinol derivative is relatively slow and can be easily resolved from reduction and cycling by computational methods.

iii) The rate of interconversion of reduced aminoresorcinol and semiquinolamine radical can be studied by T jump. This is possible because the two species are in equilibrium and their fully reversible interconversion has a significant ΔH. This experiment is carried out on the reduced enzyme, in the presence of excess amine substrate and in the absence of oxygen (Turowski *et al.*, 1993].

iv) One can collect standard Michaelis-Menten type hyperbolas under steady-state conditions at variable concentrations of either the amine or oxygen substrate, while keeping constant the concentration of the other. Oxygen consumption and H_2O_2 production provide convenient signals, or one may use artificial chromogenic amine substrates (e.g., p-dimethylamino methyl benzylamine).

v) One may follow the pre-steady-state and steady-state in the stopped flow, starting from the oxidized or reduced enzyme and arranging the oxygen and amine concentrations in order to obtain the oxidized or the reduced enzyme at the end of the reaction (Figure 10.3). The analysis of this experiment is demanding because it requires minimization of all the kinetic parameters, using numerical simulation of the time courses; however, the experiment is easy to run and highly informative (Bellelli *et al.*, 2000).

10.3 From Kinetic Constants to Steady-State Parameters

A meaningful interpretation of the data collected in the experiments described in the preceding section demands that the kinetic single-step parameters and steady-state ones be integrated. This is a laborious task even for a relatively simple kinetic scheme like the one depicted in Figure 10.2. We shall consider the simplest possible analysis, which, however, is time-consuming and error-prone, and then we shall consider one shortcut, the King and Altman graphical method. The advantage of the standard kinetic analysis is that one is always perfectly aware of its chemical significance, whereas graphical shortcuts rely on procedures that, although yielding the same result with less effort, are somewhat less transparent as to the chemical meaning of the terms they use to build up the equations.

Usually, experimental design and analysis go hand in hand and constitute an iterative process in which analysis of the first experiment suggests the second, which in turn allows a refinement of the analysis. However, for the description of this example we shall assume that the catalytic mechanism depicted in Scheme 10.2 can be taken for granted,

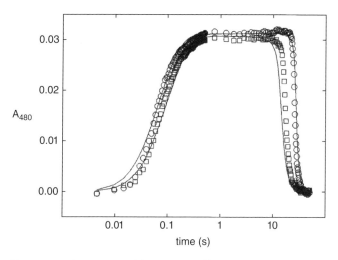

Figure 10.3 Time course of the reaction of bovine serum amine oxidase (BSAO) with benzylamine. The experiment was run in a stopped-flow apparatus, by rapidly mixing a 33.4 µM solution of BSAO in the presence of 1.25 mM benzylamine with an equal volume of buffer containing 280 (circles) or 150 µM O_2 (squares). Concentrations after mixing: reduced BSAO (species ER in the scheme) 16.7 µM; benzylamine 625 µM; O_2 140 or 75 µM. Potassium phosphate buffer pH 7.2; T = 25 °C. The time courses present three clearly separated processes: the pre-steady state rapid oxidation of the enzyme, marked by an increase of absorbance; the steady state process, without changes of absorbance; and the exhaustion of oxygen, with return of the enzyme to the reduced state. Lines were calculated by numerical integration of the set of differential equations reported in the text, and a non-linear least-squares minimization routine was used to find the best estimates of the eight kinetic rate constants, as reported in Table 10.1. Source: Reproduced with permission from Bellelli *et al.* (2000).

and that all the experiments described in Section 10.2 have been carried out. The first step is to critically analyze the catalytic mechanism and see whether some steps cannot be resolved experimentally and must be considered together. Combining two or more steps means that they will be assigned apparent rate constants, rather than intrinsic ones. It is also possible that the analysis of experimental data will require additional steps to be considered, for example because a reaction that we expect to be second order is apparently first order in the experiment, implying one more reaction intermediate.

Inspection of the crude data makes it evident that if we limit ourselves to stopped flow experiments, we shall not be able to resolve the reversible interconversion of aminoresorcinol and the semiquinolamine radical (which by T jump has been assigned a rate constant of $20,000\,s^{-1}$, see Turowski *et al.*, 1993); thus, combination of the two reduced derivatives into one species is justified. Moreover, the enzyme whose kinetics we present below, Cu-AO from bovine serum, populates the semiquinolamine radical to a very minor extent, if at all. Thus, the following scheme proved sufficient to describe the experimental data.

$$E_Q + RNH_2 \Leftrightarrow E_K + H_2O \qquad\qquad \text{(reaction 10.1)}$$

$$E_K \Leftrightarrow E_A \qquad\qquad \text{(reaction 10.2)}$$

$$E_A + H_2O \rightarrow E_R + RCHO \qquad\qquad \text{(reaction 10.3)}$$

$$E_R + O_2 \Leftrightarrow E_R - O_2 \tag{reaction 10.4}$$

$$E_R - O_2 \rightarrow E_Q + H_2O_2 + NH_3 \tag{reaction 10.5}$$

Where E_Q (standing for enzyme-TPquinone) corresponds to species (a) in Figure 10.2; E_K (enzyme-TPketimine) to species (b); E_A (enzyme-TPaldimine) to species (c); E_R (enzyme-reduced) to the sum of species (d) and (e); and E_R-O_2 to a labile complex of the (sum of) reduced species with oxygen. The reversible oxygen complex of the reduced enzyme, E_R-O_2, was not predicted from the chemical scheme but was added at a later stage of the analysis, in order to fit the time course of reoxygenation, which could not be satisfactorily simulated using a simple second-order reaction.

The above set of kinetic chemical equations fully accounts for the observable steps of the catalytic cycle, and for substrate consumption and product formation. Two reactions of this scheme have been assumed to be irreversible for chemical reasons: reaction 3 because the equilibrium constant strongly favors the aldehyde product over the amine substrate; and reaction 5 because of the large redox potential of O_2.

Once the simplification of the catalytic reaction scheme has been carried out, it is necessary to assign tentatively the spectroscopic features of each species. In the present case this is helped by the fact that several species can be obtained under static conditions (see Figure 10.2), yet this requires some guesswork for those species that are populated only transiently, in mixtures with other species.

The next step is to write the set of differential kinetic equations that describe the system. We assign the following kinetic rate constants: k_1 and k_{-1} (reaction 1, forward and reverse, respectively); k_2 and k_{-2} (reaction 2); k_3 (reaction 3); k_4 and k_{-4} (reaction 4); k_5 (reaction 5). In spite of our attempts to simplify the system as far as possible, the scheme requires 8 rate constants, some of which are very difficult or impossible to measure directly, but are necessary for consistency with the chemical reactions. The differential equations are:

$$\delta[E_Q]/\delta t = [E_R - O_2]k_5 + [E_K]k_{-1} - [E_Q][RNH_2]k_1$$
$$\delta[E_K]/\delta t = [E_Q][RNH_2]k_1 + [E_A]k_{-2} - [E_K](k_{-1} + k_2)$$
$$\delta[E_A]/\delta t = [E_K]k_2 - [E_A](k_3 + k_{-2})$$
$$\delta[E_R]/\delta t = [E_A]k_3 + [E_R - O_2]k_{-4} - [E_R][O_2]k_4$$
$$\delta[E_R - O_2]/\delta t = [E_R][O_2]k_4 - [E_R - O_2](k_{-4} + k_5)$$
$$\delta[RNH_2]/\delta t = [E_K]k_{-1} - [E_Q][RNH_2]k_1$$
$$\delta[O_2]/\delta t = [E_R - O_2]k_{-4} - [E_R][O_2]k_4$$
$$\delta[RCHO]/\delta t = [E_A]k_3$$
$$\delta[H_2O_2]/\delta t = \delta[NH_3]/\delta t = [E_R - O_2]k_5$$

It is necessary that the set of equations describes all the chemical species involved, or at least all the chemical species that are required to account for the signal. For example, in the scheme above it is necessary to account for all the enzyme intermediates and all the reagents that are consumed in the course of the reaction. Moreover, if one uses a chromogenic amine substrate whose aldehyde generates an absorbance signal, it is

Table 10.1 Kinetic parameters of bovine serum amine oxidase.

k_1	k_{-1}	k_2	k_{-2}	k_3	k_4	k_{-4}	k_5
9×10^{-4}	250	1.2	0.01	3.3	3.6×10^5	33	20

First-order rate constants (k_{-1}, k_2, k_{-2}, k_3, k_{-4}, k_5) in s^{-1}; second order rate constants (k_1, k_4) in $M^{-1}s^{-1}$ (from Bellelli *et al.*, 2000).

necessary to account for the appearance of this product. By contrast, because the reaction is essentially irreversible, it is unnecessary to account for the two spectroscopically silent products NH_3 and H_2O_2 (although in this example they were calculated, they were not used further).

Computer simulation of the reaction further requires (i) a set of initial conditions (the concentration of substrates and enzyme immediately after mixing and the initial redox state of the enzyme, either E_Q or E_R) and (ii) the extinction coefficient of each species at the observation wavelength. In order to find the best estimates of the kinetic rate constants, the variance between the numerically simulated time courses and the experimental points is calculated, and a non-linear minimization routine is used to improve the initial guess on the kinetic rate constants. The results of the minimization procedure for two different substrate concentration are reported in Figure 10.3, and the corresponding parameters are listed in Table 10.1.

The kinetic parameters obtained from the above analysis can be used to calculate the steady-state parameters. This is an important test of the entire procedure because the steady-state parameters (Table 10.2) are usually known with great accuracy. We shall describe two ways to derive K_MS and $k_{cat}S$ from the kinetic differential equations. It is important to stress that the kinetic model should be kept to the minimum complexity compatible with fitting the experimental data; there is no point in adding reaction steps that cannot be experimentally resolved.

The direct method to derive the steady-state parameters is similar to that we apply to define the binding polynomial for the pseudo-equilibrium approximation. We equate to zero all the differential kinetic equations pertaining to the enzyme intermediates, per the definition of the steady state, and express the concentration of all enzyme intermediates as a function of the kinetic constants, the substrates' concentrations, and the intermediate chosen as the reference species. Any enzyme species may be used as reference and a little experience will usually suggest the one that yields the most straightforward solution of the problem. One would obtain the same result whatever species is used as reference, but not with the same ease.

For the scheme above we define the rate of substrate transformation as $\delta[RCHO]/\delta t$, because the signal of the chromogenic substrate is generated in this step. Moreover, this step is irreversible, and this simplifies the rate expression. Finally, using E_A, the species whose concentration limits the rate-determining step, as the reference species leads to a relatively simple derivation of the steady-state population of all other forms of the enzyme. To simplify the equations we define:

$$k_1' = k_1[RNH_2]$$
$$k_4' = k_4[O_2]$$

With these notations we imply nothing about the substrate or product concentrations since in our differential equations these are always associated with the respective rate constant. It is important to stress that k_1' and k_4' are not pseudo-first order constants, because during the simulations of the time courses the concentrations of RNH_2 and O_2 are recalculated at each integration step, whereas in the calculations required to obtain the steady-state parameters these concentrations were assumed as constant and equal to the one used in the steady-state experiments. We also remark that internal reversible isomerization steps can be included in the analysis only if they are associated with a signal and can be experimentally determined by the instrument used: for example, this is the case for the reversible interconversion between EK and EA. Otherwise, they should be grouped together as we did with the semiquinolamine radical and aminoresorcinol derivatives.

We proceed to define the concentrations of all other enzyme species as a function of $[E_A]$:

$$[E_K] = [E_A](k_3 + k_{-2})/k_2$$

$$[E_Q] = \{[E_K](k_{-1} + k_2) - [E_A]/k_{-2}\}/k_1'$$

$$= [E_A](k_3 k_{-1} + k_3 k_2 + k_{-2} k_{-1})/k_2 k_1'$$

$$[E_R - O_2] = ([E_Q]k1' - [E_K]k_{-1})/k_5 = [E_A]k_3/k_5$$

$$[E_R] = [E_R - O_2](k_{-4} + k_5)/k_4' = [E_A](k_3 k_{-4} + k_3 k_5)/k_4' k_5$$

We can now write an expression that, for lack of a better name, we may call the steady-state kinetic binding polynomial:

$$[E]_{tot} = [E_Q] + [E_K] + [E_A] + [E_R] + [E_R - O_2]$$

$$= [E_A]\{(k_3 k_{-1} + k_3 k_2 + k_{-2} k_{-1})k_4' k_5 + (k_3 + k_{-2})k_1' k_4' k_5 + k1' k_2 k_4' k_5$$

$$+ (k_3 k_{-4} + k_3 k_5)k_1' k_2 + k_1' k_2 k_3 k_4'\}/k_1' k_2 k_4' k_5$$

(eqn. 10.1)

In this expression the parentheses are unnecessary, but we keep them for clarity, to identify the groups of constants that correspond to the concentration of every enzyme species.

The rate of product formation is:

$$V = [E_A]k_3 = [E]_{tot} k_1' k_2 k_3 k_4' k_5 / \{(k_3 k_{-1} + k_3 k_2 + k_{-2} k_{-1})k_4' k_5$$

$$+ (k_3 + k_{-2})k_1' k_4' k_5 + k_1' k_2 k_4' k_5 + (k_3 k_{-4} + k_3 k_5)k_1' k_2 + k_1' k_2 k_3 k_4'\}$$

(eqn. 10.2)

From the rate equation, we can calculate k_{cat}s and K_Ms:

$$k_{cat\ RNH2,[O2]=const.} = k_2 k_3 k_4[O_2]k_5 / (k_3 k_4[O_2]k_5 + k_{-2}k_4[O_2]k_5 + k_2 k_4[O_2]k_5$$

$$+ k_2 k_3 k_4[O_2] + k_2 k_3 k_{-4} + k_2 k_3 k_5)$$

$$K_{M\ RNH2,[O2]=const.} = k_{cat\ RNH2,[O2]=const.}(k_{-1}k_{-2} + k_{-1}k_3 + k_2 k_3)/k_1 k_2 k_3$$

$$k_{cat\ O2,\ [RNH2]=const.} = k_1[RNH_2]k_2 k_3 k_5 / (k_1[RNH_2]k_2 k_3 + k_1[RNH_2]k_2 k_5$$

$$+ k_1[RNH_2]k_{-2}k_5 + k_1[RNH_2]k_3 k_5 + k_{-1}k_{-2}k_5 + k_{-1}k_3 k_5 + k_2 k_3 k_5)$$

$$K_{M\ O2,[RNH2]=const.} = k_{cat\ O2,[RNH2]=const.}(k_5 + k_{-4})/k_4 k_5$$

Table 10.2 Steady-state parameters of bovine serum amine oxidase.

	$k_{cat\,O2}$ $[RNH_2]=5\,mM$	$K_{M\,O2}$ $[RNH_2]=5\,mM$	$k_{cat\,BzNH2}$ $[O_2]=0.27\,mM$	$K_{M\,BzNH2}$ $[O_2]=0.27\,mM$
Calc.	0.60	4.5	0.82	1900
Exp.	0.65	5.7	1.23	1700

Steady-state parameters of bovine serum amine oxidase for benzylamine as calculated from rapid kinetic experiments (Calc.) or directly determined (Exp.). Experimental conditions as in Figure 10.3. K_{cat}s in s^{-1}; K_Ms in μM. Steady-state parameters for benzylamine (BzNH$_2$) were determined in air-equilibrated buffer ([O$_2$]=0.27 mM); for O$_2$ at 5 mM benzylamine (from Bellelli et al., 2000).

10.4 The Method of King-Altman to Derive Steady-State Parameters

Finding the sets of kinetic constants that describe the steady-state population of each intermediate in the catalytic cycle can be accomplished using graphical methods. The first and most classical of these methods was described by King and Altman (King and Altman, 1956).

To apply this method one should first represent the whole catalytic cycle with its eventual alternative paths in graphical form (scheme I in Figure 10.4).

The next step is to isolate one intermediate of the cycle at a time and determine the polynomial of pseudo-first order rate constants that describe its steady-state concentration. We start with the TPquinone intermediate (E_O, species a). We eliminate one side of the polygon that represents the catalytic cycle at a time and we check whether or not we can draw one or more paths that connect all other intermediates of the cycle to the selected species (a). The cycle of CuAO has five intermediates and five reactions, thus we consider five possibilities. Of these, possibilities II and III are discarded because of the irreversible reaction 3, which prevents some intermediate to be connected to (a). Possibilities IV, V, and VI are valid because paths may be drawn that connect all intermediates to (a), and we write down, in form of products, the rate constants characteristic of these paths: $k_{-1}k_{-2}k_4k_5$; $k_{-1}k_3k_4k_5$; and $k_2k_3k_4'k_5$. The polymonial that represents the concentration of species E_O under steady-state conditions is thus: ($k_{-1}k_{-2}k_4k_5 + k_{-1}k_3k_4k_5 + k_2k_3k_4'k_5$).

We repeat the procedure for the other species and obtain the following polynomials (or monomials): $E_K = (k_4k_5k_1k_{-2} + k_3k_4k_5k_1)$; $E_A = k_4k_5k_1k_2$; $E_R = k_{-4}k_1k_2k_3$; $E_{R-O2} = k_1k_2k_3k_4$.

We can now reconstruct the steady-state kinetic polynomial using species E_A as the reference species:

$$[E]_{tot} = [E_A]\left(k_{-1}k_{-2}k_4k_5 + k_{-1}k_3k_4k_5 + k_2k_3k_4'k_5\right) + \left(k_4k_5k_1k_{-2} + k_3k_4k_5k_1\right)$$
$$+ k_4k_5k_1k_2 + k_{-4}k_1k_2k_3 + k_1k_2k_3k_4 / k_4k_5k_1k_2.$$

The reader may verify that the polynomial thus obtained is identical to eqn. 10.1.

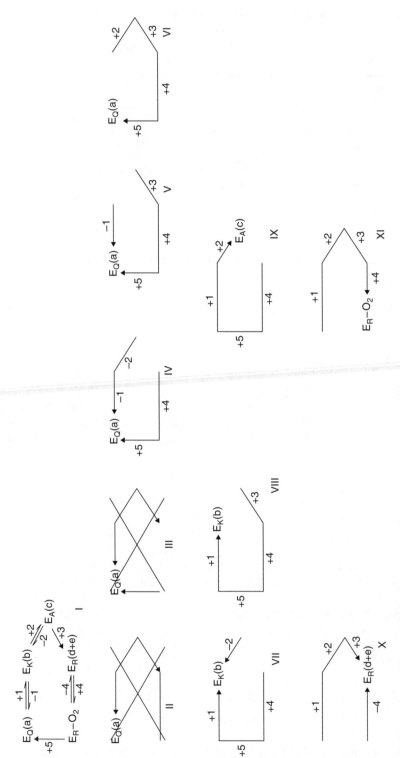

Figure 10.4 Application of the King-Altman method to the case of bovine serum CuAO (see text).

References

Bellelli, A., Morpurgo, L., Mondovi, B. and Agostinelli, E. (2000) The oxidation and reduction reactions of bovine serum amine oxidase. A kinetic study. Eur J Biochem, 267: 3264–3269.

Dooley, D.M., McGuirl, M.A., Brown, D.E., Turowski, P.N., McIntire, W.S. and Knowles, P.F. (1991) A Cu(I)-semiquinone state in substrate-reduced amine oxidases. Nature, 349: 262–264.

King, E.L. and Altman, C. (1956) A schematic method of deriving the rate laws for enzyme-catalyzed reactions. J. Phys Chem, 60: 1375–1378.

Medda, R., Padiglia, A., Bellelli, A., Sarti, P., Santanche, S., Finazzi Agro', A. and Floris, G. (1998) Intermediates in the catalytic cycle of lentil (Lens esculenta) seedling copper-containing amine oxidase. Biochem J, 332: 431–7.

Turowski, P.N., McGuirl, M.A. and Dooley, D.M. (1993) Intramolecular electron transfer rate between active-site copper and topa quinone in pea seedling amine oxidase. J Biol Chem, 268: 17680–17682.

11

Slowly Binding and Irreversible Enzyme Inhibitors

The catalytic mechanism of an enzyme can usually be described as an unbranched cycle, and this causes the reaction to enter a steady-state condition that will be more or less closely described by the pseudo-equilibrium approximation. Indeed, in an unbranched cycle the enzyme does not get trapped in non-productive dead ends, and will be freed from a substrate that dissociates slowly or not at all by converting it to the product and dissociating the product instead. In some cases the enzyme may present resting or inactivated states and, depending on the experimental conditions, may require some type of activation before measuring its catalytic activity, but even in these cases one may find conditions in which a steady state is attained.

Addition of an inhibitor creates a branch in the catalytic cycle in which the enzyme can be trapped, as may occur when testing irreversible or slowly equilibrating inhibitors. These cases are important because irreversible or slowly equilibrating inhibitors may be of great interest for biotechnological and pharmacological applications.

Irreversible binding is usually associated with the formation of a covalent bond between the inhibitor and a residue in the enzyme active site. The study of high-affinity, irreversible, and slowly binding enzyme inhibitors presents complications that are often overlooked; as a consequence, puzzling information abounds in the literature. Several possible cases must be considered, keeping in mind that the categories of high affinity, irreversible, covalent and slowly binding, though largely overlapping, remain separate, and no one implies another. This chapter begins with a classification of these types of inhibitors; each type is then treated separately, but the classification is not exclusive and the same inhibitor may belong to two or more classes. All these inhibitors share one property: they do not obey the equations derived in Chapters 8 and 9, and may prevent the enzyme from entering true steady-state conditions. Often these inhibitors require incubation of variable duration with the enzyme in the absence of the substrate for their inhibitory properties to be manifest.

A preliminary description of tests aimed at identifying irreversible and slowly binding inhibitors was presented in Section 8.5, to provide the information required to distinguish them from rapidly binding, reversible inhibitors. However, in Chapter 8 we did not consider slowly equilibrating and irreversible inhibitors, and were satisfied with their identification. In this chapter we present more refined tests and a basic treatment of slowly equilibrating and irreversible inhibitors.

Reversible Ligand Binding: Theory and Experiment, First Edition. Andrea Bellelli and Jannette Carey.
© 2018 John Wiley & Sons Ltd. Published 2018 by John Wiley & Sons Ltd.

11.1   Definitions and Classifications

Enzyme inhibitors are ligands capable of preventing catalytic activity. They obey all laws governing ligand-binding, as explored in Chapter 1, but may present a specific complication: namely, the approach to equilibrium in the final enzyme-inhibitor-substrate mixture must occur within the time window of the activity assay. *An inhibitor whose binding requires a time window comparable to or longer than that of the activity assay is defined as slowly binding inhibitor, and its binding is tested in the assay under non-equilibrium conditions* (Section 11.3). Slowly binding inhibitors require an incubation with the enzyme in the absence of substrate. If they present a thermodynamic linkage with the substrate, even the incubation does not yield a pseudo-equilibrium mixture in the activity assay, as detailed below.

High-affinity inhibitors *have K_I lower than the enzyme concentrations used in the assay*, and their treatment parallels that reported for high-affinity ligands in Section 3.7. Even though the high-affinity inhibitor may compete structurally with the substrate (i.e., it may bind to the active site of the enzyme), competition may not be functionally apparent, and the Michaelis-Menten and Lineweaver-Burk plots may resemble those of non-competitive inhibitors. The reason for this peculiarity is two-fold. (i) If the free inhibitor concentration is low with respect to that of the EI complex, the substrate-induced release of a tiny fraction of the bound inhibitor will significantly increase its free concentration in solution and prevent further competition. (ii) Because the association rate constant is usually diffusion limited, high affinity may also imply slow dissociation of the enzyme inhibitor complex. Thus, the inhibitor, if competitive, may not have time to dissociate during the assay.

The limit of high affinity is irreversibility and the distinction between an *irreversible* inhibitor and a high-affinity one rests on whether dissociation of the enzyme-inhibitor complex may be achieved or not in practice, for example, by dialysis. *Irreversible inhibitors once bound do not dissociate, their K_I cannot be defined, and their IC_{50} equals zero.* Irreversible inhibitors can be distinguished from high-affinity inhibitors only in especially designed tests and experiments because the procedures required to dissociate a high-affinity enzyme-inhibitor complex are incompatible with the duration of the assay of enzyme activity.

Irreversible inhibitors usually form a *covalent bond* with some residue of the active site. Because the substrate cannot compete with an irreversible inhibitor, Michaelis-Menten and Lineweaver-Burk plots identical to those of reversible pure non-competitive inhibitors may be obtained, even when the irreversible inhibitor binds to the active site and prevents the formation of the ternary complex EIS.

Irreversible enzyme inhibitors are easily recognized if they bind rapidly to the enzyme (Section 11.4). If however, as often occurs, the irreversible inhibitor binds slowly to the enzyme (Section 11.5), or if the irreversible complex forms with probability lower than unity (Section 11.6), the inhibitor may be confused with a non-competitive one. Thus, the systematic test of reversibility of binding is recommended. There are several types of irreversible inhibitors, classified according to the mechanism of their reaction with the enzyme, as detailed in Table 11.1.

If the inhibitor binds slowly to the enzyme, has high affinity, or is irreversible *it is always advisable to fully analyze the time course of binding,* by sampling the extent of inhibition as a function of the incubation time. The time course of binding of a

Table 11.1 Some important definitions.

Slow-binding inhibitor	Reaches its equilibrium condition with the enzyme over a time longer than that required for the activity assay.
High-affinity inhibitor	Has a dissociation constant (K_I) comparable to or lower than the enzyme concentration used in the assay.
Irreversible inhibitor	Once bound to the enzyme does not dissociate significantly upon dialysis or dilution (caution: the dialysis buffer might require special components, see text). K_I cannot be defined, and IC_{50} is zero.
Covalent inhibitor	Forms a covalent bond with some residue of the active site or with the coenzyme (if present).
Mechanism-based/suicide substrate	A variant of the covalent inhibitor, it is converted by the enzyme to a reactive form that has a finite probability of forming a covalent bond with the enzyme. Requires the enzyme activity to react: for example, in the case of a two-substrate enzyme it may require the presence of the partner substrate.

reversible inhibitor asymptotically tends to an inhibitor concentration-dependent fraction of residual activity, whereas in the case of an irreversible inhibitor the residual activity tends to zero as the incubation time is increased, unless the inhibitor is sub-stoichiometric.

Some covalent inhibitors may bind rapidly to the target enzyme and form an initial reversible complex, often without requiring incubation; this complex slowly evolves to an irreversible one over long incubation times. An example is provided by methylarsonous diiodide and thioredoxin reductase (Lin *et al.*, 1999), that we discussed in Section 9.3 (see Figure 9.5). This case can be tricky because incubation increases the fraction of inhibited enzyme only slowly and the binding time course is strongly biphasic; thus, the irreversible binding process may escape notice.

11.2   Test of Reversibility of Binding

Irreversible inhibitors cannot be dealt with using the equations presented in Chapters 8 and 9, and require a different approach. These inhibitors simulate the behavior of pure non-competitive ones, thus, when non-competitive inhibition is observed it is advisable that reversibility is tested. Probably, the majority of irreversible inhibitors bind covalently to the enzyme, at the active site or elsewhere, and many of them may have important applications in biotechnology and pharmacology. However, the two categories of covalent binding and irreversible inhibition do not completely overlap. The reason to stress the importance of the nature of the chemical bond(s) between the enzyme and the inhibitor is that, as will be shown below, it may influence the experimental conditions required for testing reversibility.

Two important considerations (which provide indications for possible control experiments) are as follows.

i) *Covalent bond formation may be, and usually is, slow* because of the high activation energies of bond breakage and formation. Thus, long incubation times may be required, and many irreversible inhibitors bind slowly to the target enzyme.

ii) *Covalent binding does not necessarily mean irreversible, even though the two conditions are often associated in practice.* Indeed, if the free energy change associated with formation of a covalent bond may be very large (thus suggesting irreversible binding), the reaction may be associated with breakage of other covalent bonds. The overall ΔG, which results from the difference of the energies of the bonds that are formed and broken, may be small. An interesting example is that of alkylating agents reacting with active site Cys residues, whose reaction scheme is:

$$E - SH + IX \Leftrightarrow E - S - I + H^+ + X^- \qquad \text{(eqn. 11.1)}$$

In this reaction the inhibitor I forms a covalent bond with the sulfur of a Cys residue (S-I), but at the same time the I-X bond is broken, and X^- (or HX) is released. Thus, irreversibility, although likely, is not certain and must be tested, as discussed in Section 1.11.

If the reaction of the inhibitor with the enzyme produces a leaving group (see eqn. 11.1, above) and is reversible, the equilibrium constant for the dissociation of these inhibitors may be dimensionless or may be expressed as M^{-1} rather than M. For example, the equilibrium constant for the dissociation of the E-S-I complex from eqn. 11.1 is:

$$K_I = \left([E - SH][IX] \right) / \left([E - S - I][H^+][X^-] \right) \qquad \text{(eqn. 11.2)}$$

which has the units of the reciprocal of a molarity. If the reaction is carried out in a buffered solution, the term $[H^+]$ is constant and K_I is dimensionless. As a consequence, *even if the reaction of these inhibitors is reversible, they may not dissociate upon dilution, and dilution may actually increase the fraction of the enzyme-inhibitor complex.* A further consequence is that *if reversibility is tested by dialysis, one should add to the dialysis medium a high, fixed concentration of the leaving group* (HX or X^- in this example), otherwise dialysis will increase complex formation rather than dissociate it (Section 1.11). Finally, we remark that in a reaction scheme similar to eqn. 11.1 the concept of IC_{50} is hard to define, because the inhibitor concentration required to inhibit 50% of the enzyme depends on the enzyme concentration. If this concept is to be maintained in a meaningful way, it is strongly advisable to measure IC_{50} in the presence of excess of X^- (or HX), in order to make constant the concentration of this component. An example of covalent reversible inhibitors is provided by nitrosothiols, which are able to transfer a NO group onto a Cys residue (necessary for catalysis), as described in Section 1.11.

Under suitable experimental conditions reversible covalent inhibitors do not differ from typical reversible inhibitors. For example, CO is a reversible covalent inhibitor of cytochrome oxidase whose binding is fast and does not yield any leaving group. If, as is often the case, the covalent inhibitor binds slowly with respect to the duration of the assay, it should be incubated with the enzyme and treated as described in Section 11.3.

Reversibility does not require testing if the inhibitor competes with the substrate, and obeys the equations developed in Section 8.5, or if it presents heterotropic linkage with the substrate: that is, *competition proves reversibility.* This consideration, however, applies only to the time window explored during the assay.

An exception to the rule that competition proves reversibility occurs if the binding mechanism entails a two-step reaction scheme in which an initially reversible complex is formed rapidly and then converts to the irreversible complex:

$$E + I \Leftrightarrow EI \rightarrow EI^* \qquad \text{(eqn. 11.3)}$$

In this case, competition with substrate proves reversibility of the first reaction, but does not exclude irreversibility of the second reaction, and one can often separate the two processes, taking advantage of the fact that the assay of the catalytic activity is faster than the time required by the conversion of EI to EI*. The initial process observed after short or no incubation should be treated as a case of reversible inhibition, whereas the end product, obtained after prolonged incubation of the enzyme with the inhibitor should be treated as an irreversible complex. Competition with the substrate may only be observed before the EI* complex is formed.

Irreversibility is immediately detected in the case of rapidly binding irreversible inhibitors (Section 11.4), because the total inhibitor concentration required to fully inhibit the enzyme equals the enzyme concentration itself (or is a constant multiple of it). If binding is slow, however, one may easily confuse incomplete binding due to kinetic reasons (insufficient incubation time) with incomplete saturation due to the enzyme-inhibitor mixture having reached an equilibrium condition. Thus, reversibility tests are specially indicated in the case of slowly binding inhibitors. The problem is compounded by the fact that to speed up binding, the researcher may be tempted to increase the concentration of the inhibitor, which then greatly exceeds the concentration of the enzyme, obscuring the fact that binding is a stoichiometric titration.

There are three fundamental tests of the reversibility of binding (see Sections 1.11 and 8.5): kinetics, dilution, and dialysis.

The time course of binding of an irreversible inhibitor always tends to the complete loss of activity of the enzyme, at least if the inhibitor concentration equals or exceeds the amount required by the stoichiometry of the reaction (Figure 11.1). Moreover, if the mechanism of the reaction is simple and allows one to estimate both the association and dissociation rate constants from a rapid mixing experiment (see Figure 2.3), the reversible inhibitor has a non-zero dissociation constant, whereas the dissociation constant of an irreversible inhibitor is zero. An important note of caution is the following: the dissociation rate constant of a high-affinity inhibitor in a rapid mixing experiment may be indistinguishable from zero.

The kinetic test of reversibility is straightforward, but has demanding requirements: (i) The binding time course should be slow enough to allow one to test the enzyme activity at the desired time intervals. (ii) The reaction mechanism should not include long-lived inhibited intermediates that would complicate the picture. (iii) High-affinity, reversible inhibitors may have kinetic dissociation constants close to zero, thus leading to possible confusion with irreversible inhibitors.

To test reversibility by dilution one applies the procedure described in Section 8.5, except that different incubation times should be tested, and a full time course of enzyme inactivation should be recorded.

To test reversibility by dialysis the researcher should demonstrate that the activity of a sample of the inhibited enzyme can be fully recovered after the inhibitor is removed by extensive dialysis. As explained above, if the reaction of the inhibitor with the enzyme produces a leaving group, both the dialysis and the dilution tests should be carried out in the presence of a high and fixed concentration of the free leaving group, otherwise a false negative result will be obtained (Section 1.11).

We close this section on a note of caution: it is not uncommon to find in the literature irreversible enzyme inhibitors that are misclassified as reversible non-competitive, or vice versa. In some cases it is the chemical nature of the inhibitor that should raise the

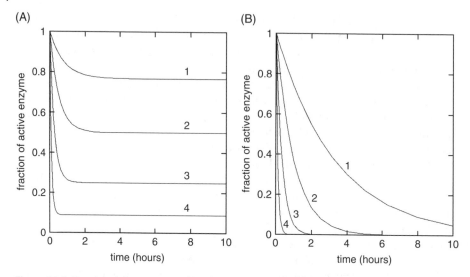

Figure 11.1 Simulated time courses of binding of a reversible (panel A) and an irreversible inhibitor (panel B).

Panel A: Time courses of the reaction of a reversible, slowly binding inhibitor with the target enzyme, at four different concentrations of the inhibitor. The time courses tend asymptotically to the equilibrium condition; the apparent rate constant is a function of the inhibitor concentration and both the association and dissociation rate constants.

Panel B: Time courses of the reaction of an irreversible, slowly binding inhibitor with the target enzyme, at four different concentrations of the inhibitor. The time courses tend asymptotically to zero residual activity; the apparent rate constant is a function of the inhibitor concentration and the association rate constant, while the dissociation rate constant is zero.

suspicion, in other cases discrepancies in IC_{50} among different laboratories. The most common reason this error occurs is that binding of the irreversible inhibitor requires a long incubation, which the researcher has not fully explored. If the incubation time allowed is insufficient for the reaction to reach completion, some residual activity will be recorded, and something resembling an IC_{50} will be estimated by the standard assay (see Section 3.4 and Figure 3.2). This parameter will exhibit a dependence on the incubation time: as already noticed, one will obtain a time-dependent IC_{50}. The time-dependent IC_{50} of an irreversible inhibitor is a scarcely meaningful parameter, which corresponds to the free inhibitor concentration required so that the arbitrarily chosen incubation time is the $t_{1/2}$ of the reaction time course. Clearly, this does not conform to the definition of IC_{50} given in Section 8.5 and we suggest that it is discarded altogether and replaced by a more properly defined $t_{1/2}$.

A simulation of the typical (but wrongly interpreted) experiment is presented in Figure 11.2. The simulation assumes an inhibitor that binds irreversibly to the enzyme in a (slow) bimolecular reaction carried out at variable inhibitor concentration but always at $[I]_{tot} \gg [E]_{tot}$. The fraction of active enzyme is tested by the steady-state catalytic assay after a fixed incubation time at different concentrations of inhibitor. Under the assumptions of the simulation (irreversible binding by the simple bimolecular process E+I --> EI, Figure 11.2A), the fraction of the inhibited enzyme is an exponential function of the inhibitor concentration, which can be misinterpreted as a hyperbolic

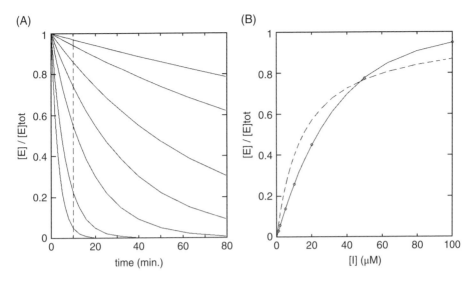

Figure 11.2 Simulation of the binding of an irreversible enzyme inhibitor.

Panel A: time courses of the irreversible combination of inhibitor with enzyme, simulated under the assumption of pseudo-first order using the equation $[E]_t = [E]_{tot}\, e^{-k[I]t}$. All time courses in due time tend to zero; k is $3\times10^3\ M^{-1}min^{-1}$. The dashed line represents the incubation time chosen (10 min.). Panel B: the points represent the fraction of the inhibited enzyme after the chosen incubation time, as a function of the inhibitor concentration (taken from panel A). The continuous line represents the exponential function that correctly describes the data ($[EI]=[E]_{tot}\,(1 - e^{-k10[I]})$). The dashed line represents a "binding" hyperbola with $K_d = 15\ \mu M$; even though its shape is different from the exponential, experimental errors and other possible deviations from ideal behavior may induce the researcher to accept the hyperbola as a description of the experimental data.

one (Figure 11.2B). We suspect that at least some (wrong) time-dependent IC_{50} have been obtained because of the misinterpretation of exponentials for hyperbolas (Saccoccia *et al.*, 2014; Angelucci *et al.*, 2016).

11.3 Slowly Equilibrating Competitive Inhibitors

If the binding or (more frequently) the dissociation of the inhibitor is slow with respect to the catalytic event, the pseudo-equilibrium approximation cannot be applied (Morrison, 1982; Morrison and Walsh, 1988). This is often the case with high-affinity inhibitors, whose dissociation from the enzyme may be slow, and with covalent irreversible inhibitors, whose binding may have a high activation barrier.

However, in what follows it will be assumed that the slowly-binding inhibitor considered is reversible and is not a high-affinity one, as happens, for example, in the case of some inhibitors of cholinesterase (Masson and Lushchekina, 2016).

If the slowly-equilibrating inhibitor is of the pure non-competitive type (i.e., if it binds to a site different from that of the substrate, and has no thermodynamic linkage with it), the researcher only needs to carefully determine the incubation time required by the enzyme and the inhibitor to reach the equilibrium condition before adding the substrate and recording the activity, which probes the fraction of uninhibited enzyme.

The case of the competitive (either purely competitive or mixed), slowly equilibrating inhibitor is more complex, and cannot be solved by incubation of the enzyme with the inhibitor because the enzyme-substrate-inhibitor mixture should find an equilibrium condition which is not the same as that reached by the enzyme-inhibitor mixture at the end of the incubation. Indeed, the substrate should induce a decrease or increase of the fraction of the enzyme-inhibitor complex that was formed during the incubation, but the assay occurs over too short a time to allow this. Thus, the slowly equilibrating competitive inhibitor may functionally behave as a non-competitive inhibitor. Irreversible inhibitors can bind slowly to the enzyme, but their case is conceptually different, because they cannot be displaced by the substrate over any time window. Thus, irreversible inhibitors require still a different treatment, as detailed in Sections 11.4 and 11.5. However, slowly equilibrating and irreversible inhibitors may share some superficial similarities, for example, they both behave as pure non-competitive inhibitors, even though they may bind at the same site as the substrate.

Two cases of slow-binding, reversible, competitive inhibitors will be considered, namely (i) inhibitors whose binding occurs in the same time window as the assay of the catalytic activity (usually a couple of minutes, at most), and (ii) inhibitors whose binding is significantly slower than the time window of the assay.

The basic characteristics of this type of inhibitors are: (i) incubation of the enzyme with the inhibitor in the absence of substrate is required for the inhibitor to exert its full effect; and (ii) the activity of the inhibited enzyme can be restored by dialysis carried out under the appropriate experimental conditions (see above). Inhibitors requiring incubation with the enzyme are slowly binding, and inhibitors that cannot be removed by dialysis are irreversible. The two properties are not mutually exclusive.

An inhibitor that associates and dissociates slowly with respect to the time window of the activity assay fails the test of the pseudo-equilibrium approximation (Section 8.5), since the samples in which incubation of the enzyme with the inhibitor has been allowed before addition of the substrate present a greater extent of inhibition than those where incubation was not carried out. The classical textbook equations for enzyme inhibitors are derived under the pseudo-equilibrium approximation, and are invalid if this approximation fails. If they are applied nonetheless, they yield wrong estimates of the affinity of the inhibitor. In particular, if the slow-binding inhibitor is not allowed an incubation with the enzyme, or if the incubation is too short, initial velocity measurements underestimate its affinity for the enzyme, because the transformation of the substrate starts in a mixture in which little enzyme-inhibitor complex has formed (Figure 11.3). By contrast, if the enzyme and inhibitor are incubated before adding the substrate, the initial velocity measurements overestimate the affinity of the inhibitor for the enzyme. This is because the amount of the enzyme-inhibitor complex formed in the absence of the substrate exceeds the amount expected at steady state, where competition between the inhibitor and the substrate should occur but does not occur due to the short duration of the activity assay. Thus, we formulate the following take home message: *if the inhibitor requires incubation with the enzyme, or its effect changes as a function of the incubation time, then initial velocity measurements are not carried out under pseudo-equilibrium conditions, and the textbook equations cannot be applied.*

The quantitative description of an inhibitor whose binding time course overlaps with the duration of the assay of enzyme activity requires a kinetic approach, and depends on

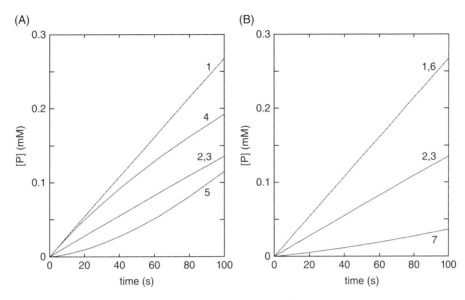

Figure 11.3 Effect of the slowly binding pure competitive inhibitor.

The time courses of product formation were simulated using a numerical integration routine adapted to eqns. 11.4 through 11.9 under the following conditions: $[E]_{tot} = 10^{-6}$ M; $[S] = 10^{-3}$ M; $[I]$ (where present) $= 10^{-4}$ M; $k_1 = 10^6$ M^{-1}s^{-1}; $k_2 = 100$ s^{-1}; $k_3 = 3$ s^{-1}; $k_{ass.I}$ and $k_{diss.I}$ differ in each time course (see below), but their ratio is constant and leads to $K_I = 10^{-5}$ M. With these parameters, the theoretical (pseudo-)equilibrium fraction $[EI]/[E]_{tot}$ results 0.91 in the absence of the substrate (i.e., at the end of the incubation), and 0.5 in the presence of the substrate. The starting condition, in the absence of incubation is $[E] = 10^{-6}$ M, $[ES] = 0$, $[EI] = 0$. If the enzyme is pre-incubated with the inhibitor before adding the substrate for a time long enough to reach the equilibrium the starting condition is $[E] = 0.1 \times 10^{-6}$ M, $[EI] = 0.9 \times 10^{-6}$ M.

Panel A: time course 1: no inhibitor; 2: $[I] = 10^{-4}$ M, no pre-incubation of the enzyme with the inhibitor; $k_{ass.I} = 10^6$ M^{-1}s^{-1}, $k_{diss.I} = 10$ s^{-1}; 3: $[I] = 10^{-4}$ M, with incubation, same kinetic constant as time course 2; 4: $[I] = 10^{-4}$ M, no pre-incubation of the enzyme with the inhibitor; $k_{ass.I} = 10^3$ M^{-1}s^{-1}, $k_{diss.I} = 0.01$ s^{-1}; 5: $[I] = 10^{-4}$ M, with incubation, same kinetic constant as time course 4.

Panel B: time courses 1, 2, and 3 as in panel A. Time course 6: $[I] = 10^{-4}$ M, no pre-incubation of the enzyme with the inhibitor; $k_{ass.I} = 10^2$ M^{-1}s^{-1}, $k_{diss.I} = 10^{-3}$ s^{-1}; 7: $[I] = 10^{-4}$ M, with incubation, same kinetic constant as time course 6.

Time courses 2 and 3 obey the pseudo-equilibrium approximation, hence incubation of the enzyme with the inhibitor has no effect. Time courses 4 and 5 do not obey the pseudo-equilibrium approximation, but their kinetic constants are such that the pseudo-equilibrium condition is reached during the assay, hence their upward or downward bending. In the case of time courses 6 and 7 $k_{ass.I}$ and $k_{diss.I}$ are so much lower than k_3 that no significant binding or release of the inhibitor occurs during the assay.

the order of addition of the components of the mixture. The kinetic differential equations for this type of inhibitor are as follows.

$$\delta[E]/\delta t = [EI]k_{diss.I} + [ES](k_2 + k_3) - [E]([I]k_{ass.I} + [S]k_1) \qquad \text{(eqn. 11.4)}$$

$$\delta[EI]/\delta t = [E][I]k_{ass.I} - [EI]k_{diss.I} \qquad \text{(eqn. 11.5)}$$

$$\delta[ES]/\delta t = [E][S]k_1 - [ES](k_2 + k_3) \qquad \text{(eqn. 11.6)}$$

$$V = \delta[P]/\delta t = [ES]k_3 \qquad\qquad\qquad \text{(eqn. 11.7)}$$

$$\delta[S]/\delta t = [ES]k_2 - [E][S]k_1 \qquad\qquad\qquad \text{(eqn. 11.8)}$$

$$\delta[I]/\delta t = [EI]k_{diss.I} - [E][I]k_{ass.I} \qquad\qquad\qquad \text{(eqn. 11.9)}$$

The first three equations sum to zero, as required by mass conservation.

The time evolution of this system goes through a pre-steady-state process, and may never reach the true steady state, because of substrate consumption before equilibration. Both the pre-steady-state and steady-state time courses may be simulated by numerical integration of the above equations. It is important to keep in mind that there is no safe method to directly determine the steady-state parameters of such a system, since the initial velocity is measured during the pre-steady-state process, under non-steady state conditions.

Since binding of the inhibitor occurs during the course of the assay, one observes a characteristic bending of the time course of product formation. The bending is downward (i.e., velocity diminishes) if the enzyme has not been incubated with the inhibitor and the inhibitor binds during the assay. The bending is upward (i.e., the velocity increases) if the enzyme has been incubated with the inhibitor and the inhibitor dissociates during the assay as a consequence of the competition with the substrate (Figure 11.3A). The fact that velocity of product formation depends on the incubation clearly demonstrates that the pseudo-equilibrium approximation does not apply.

The case in which a bent time course of product formation is recorded is highly informative, because it contains information about all the kinetic constants of the reaction. One should check that bending is not caused by other events (e.g., exhaustion of the substrate or product inhibition) and is absent in the absence of the inhibitor. The best way to study this type of inhibitor is to collect an extended data set in which the inhibitor concentration and the duration of incubation are systematically varied and to globally fit the experimental data with time courses calculated using eqns. 11.4 through 11.9. The procedure demands some computational power, but is well within the capabilities of modern personal computers, and the only requirement is availability of a software capable of minimizing the kinetic parameters while using a numerical integration routine to calculate the time courses.

An interesting observation is the following. Unless very low concentrations of inhibitor and substrate are used, the bent time courses of inhibition (lines 4 and 5 in Figure 11.3A) obey the kinetic law of ligand replacement reactions (Section 2.6). In the absence of incubation the inhibitor replaces the substrate (which binds faster), whereas in the presence of incubation the substrate replaces the inhibitor. This time course is different from that of binding of the inhibitor during the incubation, which is a case of reversible association (Section 2.2) instead of a replacement.

Inhibitors that associate with or dissociate from the enzyme over time windows significantly longer than the duration of the assay are easier to analyze. In these cases, the enzyme is uninhibited or almost so if not incubated with the inhibitor, and over-inhibited if incubated (Figure 11.3B).

The important property of this type of inhibitor is that their complex with the enzyme does not significantly dissociate during the assay as a consequence of competition with the substrate. Thus the assay probes the fraction of uninhibited enzyme at the

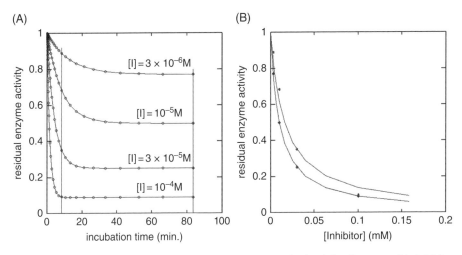

Figure 11.4 Simulation of the time course of the incubation of a slowly binding reversible inhibitor with an enzyme.

Panel A: Time courses of the formation the enzyme-inhibitor complex as simulated using eqn. 2.2, $k_i = 100\,M^{-1}s^{-1}$ and $k_{-i} = 10^{-3}\,s^{-1}$, at four inhibitor concentrations. Vertical lines indicate the reaction times 8.3 and 83 min.

Panel B: The points at 8.3 and 83 min from the time courses in panel A are plotted as a function of the inhibitor concentration and fitted to a hyperbola. Notice that while the points taken after 83 min incubation are at equilibrium and perfectly fit the hyperbola calculated for $K_I = k_{-i}/k_i = 10^{-4}$ M, those taken after 8.3 min have not yet reached the equilibrium condition and deviate not only from the expected K_I, but also from the ideal hyperbolic behavior.

end of the incubation. The first step of the characterization of this type of inhibitor is to record the time courses of the binding of the inhibitor to the enzyme at several inhibitor concentrations. This is done by testing the enzyme activity (at saturating substrate concentration) at different time intervals during the incubation (see Figure 11.4A). In some cases several tens of minutes may be required.

Once the time courses of binding of the inhibitor have been collected, their end points can be plotted as a function of the concentration of the inhibitor to obtain the binding hyperbola of the inhibitor. Indeed, in this case the V_{max} probes the fraction of uninhibited enzyme, which depends hyperbolically on the inhibitor concentration, and the midpoint of the hyperbola is K_I (Figure 11.4B). Moreover, quantitative analysis of the binding time courses reveals the kinetic constants for binding and dissociation, and may possibly reveal some clue to the binding mechanism. This type of inhibitor cannot be easily distinguished from a pure non-competitive slowly equilibrating one. It is distinguished from an irreversible one (Section 11.5) because the time course of binding of the irreversible inhibitor tends to zero residual activity (Figure 11.1). Moreover, while direct observation of competition of the substrate with the inhibitor (Figures 8.3, and 8.7) proves reversibility, absence of competition proves nothing because it occurs in slowly binding, reversible competitive inhibitors, in reversible, non-competitive inhibitors irrespective of their binding being rapid or slow (Figure 8.6), and in irreversible inhibitors. Thus, in the case of inhibitors that do not compete with the substrate, it is important to check the reversibility of the enzyme-inhibitor complex by dialysis, as detailed above.

In the absence of observable competition between the slowly binding inhibitor and the substrate, the researcher has no clue whether the respective binding sites overlap or not. To ascertain this point further experimental evidence is required. The binding site of the inhibitor may be directly observed by solving the structure of the enzyme-inhibitor complex by x-ray crystallography. If this approach fails, dedicated experiments may be designed. In the case of two-substrate enzymes, the researcher has access to a straightforward experiment: he/she checks the extent of inhibition after a long enough incubation of the enzyme with the inhibitor in the absence and in the presence of varying concentrations of either substrate, starting the reaction with the addition of the other substrate. This is an experiment of ligand replacement (Section 1.8), and if competition of either substrate with the inhibitor occurs, the researcher will record less inhibition in samples incubated in the presence of the competing substrate(s).

Single-substrate enzymes are more awkward to study, because the above experiment cannot be carried out. There are however possible substitutes, whose design is left to the creativity of the researcher and to the characteristics of the enzyme: for example, if a rapidly binding competitive inhibitor is available, the researcher may incubate the enzyme in the presence of both inhibitors before testing the enzyme activity (using high substrate concentrations). The fraction of the enzyme bound to the rapidly reacting inhibitor will be partially reactivated due to competition with the substrate.

A common misconception is that slowly binding inhibitors may be appropriately described by adding a time parameter to IC_{50} under the simplistic assumption that IC_{50} is an apparent parameter and may thus be redefined at will. For example, one may create the parameter "IC_{50} after 10 min. incubation." We consider this practice erroneous and strongly discourage its use. IC_{50} is analogous to $X_{1/2}$, and is defined as the free inhibitor concentration required to inhibit enzyme activity to 50% of that observed under the same experimental conditions in the absence of the inhibitor. As such, IC_{50} is a thermodynamic parameter, related to the equilibrium dissociation constant of the inhibitor, K_I (Cheng and Prusoff, 1973), and is to be determined after the equilibrium condition between I and E has been reached, ideally at infinite time (in practice after at least 5 half-times). If one introduces a time function into the concept of the IC_{50}, one changes its very nature from that of a thermodynamic parameter to that of an arbitrary kinetic parameter of quite complex meaning. Figure 11.4 illustrates the conceptual error implicit in a time dependent IC_{50}, and demonstrates that it not only amounts to an insufficiently equilibrated mixture, but that the degree of equilibration reached with the same incubation time varies with the concentration of the inhibitor.

Finally, we remark that, in order to speed up binding and shorten the incubation time, one may be tempted to increase the inhibitor concentration. Indeed, as shown in Figure 11.4A, if the association reaction of the inhibitor is second order, increasing the inhibitor concentration makes the reaction faster. This practice should be discouraged: there is no substitute to waiting till the enzyme plus inhibitor system has reached its equilibrium condition.

11.4 Rapidly Binding Irreversible Inhibitors

Rapidly binding irreversible inhibitors require minimal or no incubation with the enzyme to bind and to form an irreversible complex. In these cases one finds that the total concentration of the inhibitor required to cause a 50% loss of the enzyme activity equals, within the experimental error, half the total enzyme concentration. This is not the IC_{50} of the inhibitor, as IC_{50} is defined as the *free* inhibitor concentration required to half the reaction velocity. If one tries to calculate the free inhibitor concentration under these conditions, one finds it to be zero within the experimental error. The binding of the inhibitor conforms to a stoichiometric titration (Section 3.7; Figure 3.1A), and allows one to determine the stoichiometric ratio I:E required to completely inhibit the enzyme (Figure 11.5). It is important to remark that in this experiment an irreversible inhibitor cannot be distinguished from a high-affinity one. One is thus prompted to test the reversibility of binding, as explained above (Sections 8.5 and 11.2). If no activity can be recovered after extensive dialysis under appropriate conditions, binding of the inhibitor may be considered irreversible.

A typical example of rapidly binding irreversible inhibitors is provided by organic derivatives of mercury, which rapidly and irreversibly inhibit enzymes having a reactive Cys residue in the active site (see Section 1.11). This type of inhibitor has marginal interest in biotechnology and pharmacology because of its low specificity and high toxicity. However, it may have interest in toxicology. These inhibitors form an irreversible Hg-S adduct (often quite rapidly) in a second-order process, and the reaction time course may often be followed by absorbance spectroscopy, rather than by assay of the residual activity. For example, the classical thiol reagent p-hydroxy mercury benzoate reacts with Cys residues in a rapid second order reaction, easily followed in the stopped flow at 240 nm (Gibson, 1973).

IC_{50} and equilibrium constant K_I of an irreversible inhibitor are zero, and the inhibitor apparently behaves as a non-competitive inhibitor because it lowers the V_{max} and leaves the K_M unchanged, but does so at total concentrations lower than or equal to the total concentration of the enzyme, that is, the inhibitor stoichiometrically titrates the enzyme (Figure 11.5). One may expect that the stoichiometric ratio I:E is unity, but this

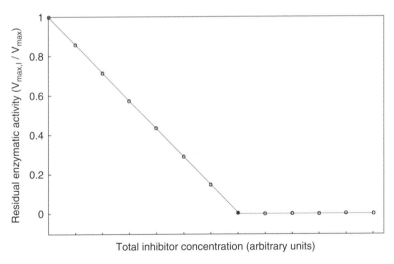

Figure 11.5 Titration of an enzyme with a rapidly binding irreversible inhibitor. The closed point marks the stoichiometric equivalence (see text). Because the irreversible inhibitor lowers the total concentration of the active enzyme, its effect is evident as a reduction of the apparent V_{max} ($V_{max,I}$), which reaches zero when all the enzyme has been inactivated.

is not always the case. For example, organic mercurials react with every exposed Cys residue, and if the enzyme has exposed Cys residues not essential for the catalytic activity, these exert an apparent protective effect on the activity and increase the stoichiometric ratio one determines in an experiment like that reported in Figure 11.5. Two fundamental points in any case remain, that is, (i) reversibility tests fail; and (ii) the total inhibitor concentration required to halve the enzyme activity is a multiple of the total enzyme concentration rather than a function of the K_I.

Usual enzyme concentrations employed in activity assays are in the nanomolar range, thus K_I values in the low nanomolar range or below should be regarded with suspicion and should prompt an accurate test of reversibility.

11.5 Slowly Binding Irreversible Inhibitors

Irreversible inhibitors that bind slowly to the enzyme are common. These require incubation with the enzyme before addition of the substrate, but except for this requirement they essentially behave as the rapidly binding irreversible inhibitors described in the preceding section and their IC_{50} is zero. In the simplest possible case, the inhibitor binds to the enzyme in a second order irreversible reaction and what the researcher can measure is the kinetic rate constant. Thus, the characterization of the time course of incubation required for complex formation (at several inhibitor concentrations) is essential (see Figure 11.1B).

As in the case of slowly binding reversible inhibitors (Section 11.3) the researcher may be tempted to increase the inhibitor concentration in order to reduce the incubation time. This practice is even more discouraged than in the case of reversible inhibitors, as it obscures the fact that the total inhibitor concentration required to inhibit the enzyme is a multiple of the enzyme concentration. Moreover, if the researcher has not fully characterized the association time course, and uses an insufficient incubation time, a residual activity is recorded from which the researcher is induced to calculate a non-zero IC_{50} (see Figure 11.2). Ideally, one should use incubation times long enough to obtain a stoichiometric titration analogous to that reported in Figure 11.5. This, however, may be impractical because some covalent irreversible inhibitors may require incubation of several hours duration under stoichiometric conditions.

The information that fully describes the binding of an irreversible inhibitor is: (i) the binding time course and its kinetic rate constant; (ii) the mechanism of the reaction, and its intermediates (see below); and (iii) the demonstration that reversibility tests consistently fail. If possible the binding site should be identified by x-ray crystallography of the enzyme-inhibitor complex or by other means. Irreversible inhibition resembles the non-competitive one, because the substrate cannot displace the inhibitor, but, as already noticed, this observation is poorly informative.

Although slowly binding irreversible inhibitors do not conceptually differ from rapidly binding ones (Section 11.4), their characterization is prone to errors, and it may often happen that these inhibitors are erroneously assigned a non-zero IC_{50}. This may occur because of (i) insufficient incubation time; (ii) higher than unity stoichiometric ratio; (iii) errors in the calculation of the free inhibitor concentration or use of the total inhibitor concentration to determine the IC_{50}. If irreversible binding is demonstrated by dilution or dialysis (Sections 1.11 and 11.2), the IC_{50} should be assumed to be zero, and contrary evidence should be carefully scrutinized.

In some cases the inhibitor rapidly forms an intermediate non-covalent complex that is already inhibited, but reversible. The reversible complex slowly forms a covalent enzyme-inhibitor bond and becomes irreversible. An example of this type of behavior is provided by methylarsonous diiodide, a covalent inhibitor of thioredoxin reductase (Lin *et al.*, 1999). This case is interesting and allows the researcher to determine an equilibrium constant K_I for the reversible complex and a first order kinetic constant for the conversion to the irreversible complex. The reaction scheme is as follows.

$$EI^* + S \leftarrow EI + S \Leftrightarrow E + I + S \Leftrightarrow ES + I \rightarrow E + I + P$$

If sufficient time is allowed, all the enzyme will be trapped into the irreversible complex EI^*. However, if the reaction mixture is prepared without incubation of the enzyme with the inhibitor, it is often observed that the inhibitor behaves as if it were reversible. This is because during the time of the assay little or no irreversible EI^* complex is formed. We shall consider the most interesting and informative case, that is, that the inhibitor is a pure competitive one before it becomes irreversible. The hallmark of such a system is that without incubation one records Michaelis-Menten and Lineweaver-Burke plots typical of pure competitive inhibition, with invariant k_{cat} and increased K_M. With a long incubation the activity disappears, and inhibition becomes non-competitive and irreversible. The transition between the two types of inhibition is gradual and depends on the duration of the incubation. This quite complex behavior is fully explained by an equilibrium constant K_I, which describes the formation of the reversible complex, that is:

$$K_I = [E][I]/[EI]$$

and a first-order kinetic rate constant $k_{inact.}$, which describes the monomolecular conversion of the reversible complex EI to the irreversible complex EI^*. These parameters can be estimated from a careful characterization of the time course of inactivation during the incubation of the enzyme with the inhibitor (Figure 11.6).

If the concentration of the inhibitor significantly exceeds that of the enzyme, the time course of formation of the irreversible complex obeys a pseudo first-order law.

$$\delta[EI^*]/\delta t = [EI]k_{inact.} = [E]_{tot} k_{app}$$

with

$$k_{app} = [I]k_{inact.}/([I]+K_I)$$

Thus, the concentrations of the different species present after an incubation of duration t result

$$[EI^*]_t = [E]_{tot}\left(1 - e^{-kapp.t}\right)$$
$$[E]_t = [E]_{tot} e^{-kapp.t} K_I/([I]+K_I)$$
$$[EI]_t = [E]_{tot} e^{-kapp.t} [I]/([I]+K_I)$$

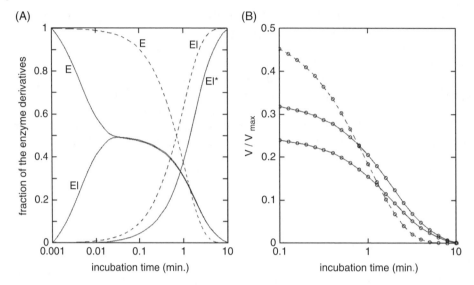

Figure 11.6 Simulated time courses of enzyme inactivation by irreversible inhibitors.

Panel A: Simulation of binding of irreversible inhibitors. Continuous lines: reaction scheme $E+I<==>EI-->EI^*$, $K_I = 10^{-4}$ M; $k_{inact.} = 1$ min^{-1}; enzyme concentration 1 uM. Dashed lines: simple irreversible bimolecular reaction, $k = 10^4$ M^{-1}min^{-1}, other conditions as in the preceding case. Notice that the simulation is carried out in the absence of substrate.

Panel B: Simulation of the enzyme activity, as it would be recorded during the time courses of the reaction with the inhibitor reported in panel A. Conditions as in panel A, substrate concentration equal to the K_M recorded in the absence of the inhibitor. Dashed line: simple irreversible bimolecular binding of the inhibitor (compares with the dashed lines in panel B). Upper continuous line: the inhibitor, while the complex is reversible, is of the pure competitive type. Lower continuous line: the inhibitor, while the complex is reversible is of the pure non-competitive type. Notice that, if the incubation time is long enough, the activity is zero in all cases; non-zero activity is recorded only if the incubation time is too short to allow completion of the reaction.

When the enzyme activity is measured, one finds that both enzyme-inhibitor complexes are inactive. If in the reversible complex the inhibitor competes with the substrate, the fraction of the active enzyme will be higher than that calculated with the above equations, because of the effect of the substrate in promoting the dissociation of the enzyme-inhibitor complex.

In some cases the above scheme is insufficient and it is necessary to replace K_I with the two kinetic rate constants of binding and release of the inhibitor k_i and k_{-i}, plus the monomolecular conversion rate constant for inactivation $k_{inact.}$.

Very interesting cases are provided by the irreversible, covalent inhibitors of enzymes having a reactive Cys residue at their active site, of which we present two examples below. These enzymes are inhibited by thiol reagents: for example, iodoacetamide, mercury, gold, and other metals and their organo-metallic derivatives, alkylating agents, and so on. The reaction mechanisms are different for different inhibitors. The majority of these inhibitors are irreversible, but due to the peculiar reactivity of thiols, this point deserves some caution, and a phenomenon that for lack of a better name we call *pseudo-reversibility* may occur and will be discussed below.

Auranofin is a gold-containing inhibitor of the enzyme thioredoxin reductase. This enzyme catalyzes the NADPH-dependent reduction of thioredoxin and has two crucial

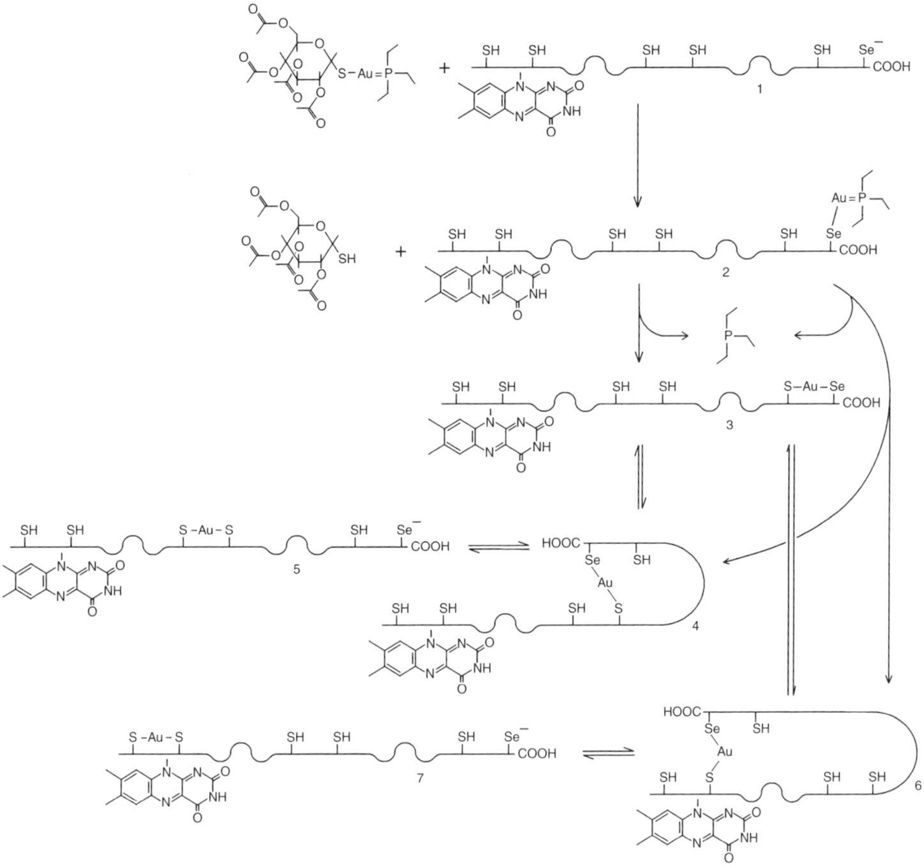

Figure 11.7 Proposed mechanism of gold transfer from auranofin to thioredoxin reductase. Auranofin is a bidentate complex of gold, whose ligands are acetoxy-thioglucose and triethyl phosphine. A seleno-Cys residue of the target enzyme displaces acetoxy-thioglucose, presumably by forming a transient tridentate complex; then an adjacent Cys residue displaces triethyl phosphine, forming a bidentate complex in the C-terminal Cys-Sec couple (protein adduct 3). Since the C-terminal Cys-Sec couple is redox active, and is responsible for the electron transfer from the active site (which contains a FAD and a Cys couple) to thioredoxin, this adduct is already inactive. The enzyme contains other redox-active Cys couples and the gold ion can be transferred to these from the flexible C-terminal arm (to form complexes 5 and 7, both observed by X-ray crystallography). From Angelucci *et al.*, 2009.

redox sites: a Cys-Cys couple in front of the FAD, and a SeC-Cys couple at the C-terminus. The electrons flow from NADPH to FAD to the Cys-Cys couple to the SeC-Cys couple to thioredoxin. Chemical modification of either or both the Cys-Cys or the SeC-Cys couple inactivates the enzyme.

The formula of auranofin is reported in Figure 11.7, together with the presumed reaction mechanism. Some important considerations on this mechanism are as follows. (i) Gold is stripped from its ligands by preferential reaction with Se, probably via a transient 3-coordinate unstable derivative; once bound to the SeC-Cys couple it can be transferred to the Cys-Cys couple. (ii) The enzyme inhibitor complex is irreversible under the usual experimental conditions, because of the very favorable geometry of the Cys-Au-Cys or SeC-Au-Cys complex. (iii) Nonetheless, it is possible that addition of high concentrations of acetoxy-thioglucose and triethyl phosphine may partially remove gold from the active site(s) of the protein; thus, it is hard to exclude that a K_I might be measured. If so, it has the dimensions of M^{-1} rather than M. Dialysis removes acetoxy-thioglucose and triethyl phosphine, thus one cannot expect it to induce the dissociation of the metal ion (and indeed it does not). (iv) Auranofin (or, to be more precise, gold) presents pseudo-reversibility because reduced thioredoxin (the product of the reaction catalyzed by the enzyme), or other thiols, may remove gold from the enzyme and restore its activity; this is not true thermodynamic reversibility as the original state of the inhibitor (auranofin) is not reformed.

Another very interesting irreversible inhibitor of thioredoxin reductase is the compound methylarsonous diiodide (Lin *et al.*, 1999; Saccoccia *et al.*, 2014). This compound rapidly forms an initial non-covalent and fully reversible complex. While the enzyme-inhibitor complex is reversible, competition between the inhibitor and the chromogenic substrate DTNB is observed (at constant NADPH concentration). Since DTNB is reduced by the Cys-Cys redox site, in close proximity to the FAD, competition probes the binding site of the inhibitor. Competition is not observed with NADPH, at constant concentration of DTNB; under these experimental conditions an atypical case of uncompetitive inhibition is recorded instead (Figure 9.5). This is explained as an effect of the thermodynamic linkage between the two substrates of the enzyme (Saccoccia *et al.*, 2014). If methylarsonous diiodide is incubated with thioredoxin reductase, the binding becomes irreversible, and neither competition nor linkage are observed any more.

11.6 Mechanism-Based Inhibitors

Mechanism-based enzyme inhibitors (also called suicide substrates) are molecules that are transformed by the enzyme, like substrates, but form a reactive intermediate that may or may not react with amino acid residues of the catalytic site and inactivate the enzyme (Walsh, 1984). Since the probability that the reactive intermediate forms a covalent bond with the enzyme is lower than unity, the average inhibitor:enzyme stoichiometric ratio required for inactivation is significantly higher than one. For example, inactivation may occur, on average, every 10 or 20 or more cycles. These inhibitors not only require an incubation with the enzyme, but also require that during the incubation the enzyme is fully active, and transforms the inhibitor. For example, in the case of a two-substrate enzyme, the other substrate must be added to allow the enzyme to activate the mechanism based inhibitor.

Examples of this type of inhibitor are serpins (Gettins and Olson, 2016) and 1,4-diamino 2-butyne, which inhibits copper-containing amine oxidases (Frébort *et al.*, 2000).

Mechanism based inhibitors usually bind covalently and irreversibly to the enzyme and cannot be removed by dialysis; thus, the enzyme stays inhibited even when the excess free inhibitor is removed.

The characterization of mechanism based inhibitors requires: (i) to determine the average number of turnovers required for inhibition; and (ii) to determine the reaction mechanism, and possibly the nature of the reactive intermediate formed during the catalytic cycle. If the number of cycles required to inactivate the enzyme is high enough, one may even determine K_M and k_{cat} values for these inhibitors, pertinent to the non-inactivating cycle (Figure 11.8).

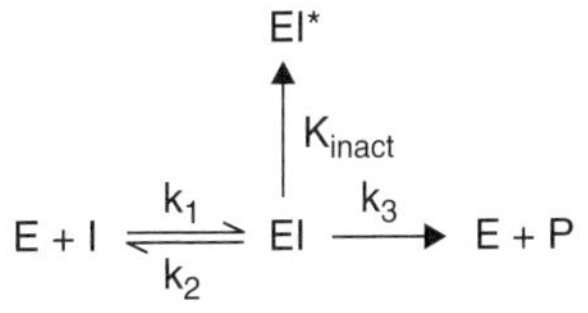

Figure 11.8 Typical cycle of a mechanism based inhibitor. The average number of cycles required for inactivation is $1+k_{cat}/k_{inact}$.

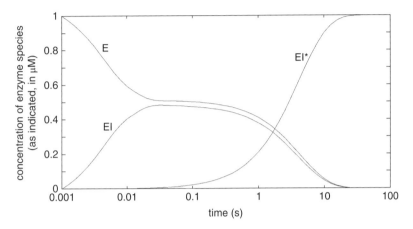

Figure 11.9 Simulation of the time course of inactivation of an enzyme reacting with a suicide substrate as in Figure 11.8.

Conditions: $k_1 = 10^6\,M^{-1}s^{-1}$; $k_2 = 100\,s^{-1}$; $k_3 = 5\,s^{-1}$; $k_{inact} = 0.5\,s^{-1}$; $[E]_{tot} = 10^{-6}\,M$; $[I] = 10^{-4}\,M$. Lines are obtained by numerical integration of the differential kinetic equations presented in the text.

The quantitative description of this system is as follows. Under pseudo-first order conditions (i.e., $[I]_{tot} >> [E]_{tot}k_3/k_{inact}$) the time course of formation of the irreversibly inactivated enzyme obeys an exponential law, with apparent rate constant

$$k_{app.} = [I]k_{inact}/([I]+K_M)$$

$$[EI^*]/[E]_{tot} = e^{k_{app.}t}$$

The overall time course of product formation exactly parallels that of enzyme inactivation and at any time the ratio $[P]/[EI^*]$ equals k_3/k_{inact}.

The initial velocity of product formation is:

$$V = [E]_{tot}k_3[I]/([I]+K_M)$$

The above equation assumes that the disappearance of the functional enzyme due to the formation of the inactivated EI^* complex is negligible with respect of the duration of the assay. This assumption strongly depends on the condition $k_3 >> k_{inact}$ and may fail; if this happens, the simplest strategy is to numerically integrate the following set of kinetic differential equations (Figure 11.9).

$$\delta[E]/\delta t = [EI](k_2+k_3)-[E][I]k_1$$

$$\delta[EI]/\delta t = [E][I]k_1-[EI](k_2+k_3+k_{inact})$$

$$\delta[EI^*]/\delta t = [EI]k_{inact}$$

$$\delta[I]/\delta t = [EI]k_2-[E][I]k_1$$

$$\delta[P]/\delta t = [EI]k_3$$

The determination of the parameters of this system is easy if the reaction provides a suitable signal (e.g., the absorbance or fluorescence of EI* differs from that of E+I or EI; or the absorbance or fluorescence of P differs from that of I); more complex if the signal is the residual activity of the enzyme at different incubation times in the presence of the inhibitor. An extreme case occurs if the reaction of the enzyme with the inhibitor is fast and does not yield any detectable signal. In this case the experiment one may resort to is the determination of the residual activity of the enzyme after completion of the reaction with a non-saturating amount of the inhibitor. Indeed, the amount of the inhibitor required to completely inactivate the enzyme in this mechanism is:

$$[I]_{saturating} = [E]_{tot}(k_3 + k_{inact})k_{inact}$$

Thus, the system behaves as a stoichiometric titration with molar ratio $I : E = (k_3+k_{inact})/k_{inact} : 1$ (see Section 3.7 and Figure 3.3A), and a plot of the residual enzyme activity as a function of the total inhibitor concentration yields a straight line from whose slope the ratio k_3/k_{inact} can be determined.

References

Angelucci, F., Sayed, A.A., Williams, D.L., Boumis, G., Brunori, M., Dimastrogiovanni, D., Miele, A.E., Pauly, F. and Bellelli, A. (2009) Inhibition of Schistosoma mansoni thioredoxin-glutathione reductase by auranofin: structural and kinetic aspects. J Biol Chem, 284: 28977–28985.

Angelucci, F., Miele, A.E., Ardini, M., Boumis, G., Saccoccia, F. and Bellelli A. (2016) Typical 2-Cys peroxiredoxins in human parasites: Several physiological roles for a potential chemotherapy target. Mol Biochem Parasitol, 206: 2–12.

Cheng, Y.C. and Prusoff, W.H. (1973) Relationship between the inhibition constant (KI) and the concentration of inhibitor which causes 50 per cent inhibition (I50) of an enzymatic reaction. Biochem Pharmacol, 22: 3099–3108.

Frébort, I., Sebela, M., Svendsen, I., Hirota, S., Endo, M., Yamauchi, O., Bellelli, A., Lemr, K. and Pec, P. (2000) Molecular mode of interaction of plant amine oxidase with the mechanism-based inhibitor 2-butyne-1,4-diamine. Eur J Biochem, 267: 1423–1433.

Gettins, P.G. and Olson, S.T. (2016) Inhibitory serpins. New insights into their folding, polymerization, regulation and clearance. Biochem J, 473: 2273–2293.

Gibson, Q.H. (1973) p-Mercuribenzoate as an indicator of conformation change in hemoglobin. J Biol Chem, 248: 1281–1284.

Lin, S., Cullen, W.R. and Thomas, D.J. (1999) Methylarsenicals and arsino-thiols are potent inhibitors of mouse liver thioredoxin reductase. Chem Res Toxicol, 12: 924–930.

Masson, P. and Lushchekina, S.V. (2016) Slow-binding inhibition of cholinesterases, pharmacological and toxicological relevance. Arch Biochem Biophys, 593: 60–68. doi: 0.1016/j.abb.2016.02.010.

Morrison, J.F. (1982) The slow-binding and slow, tight-binding inhibition of enzyme-catalysed reactions. Trend Biochem Sci, 7: 102–106.

Morrison, J.F. and Walsh, C.T. (1988) The behavior and significance of slow-binding enzyme inhibitors. Adv Enzymol Relat Areas Mol Biol, 61: 201–301.

Saccoccia, F., Angelucci, F., Boumis, G., Carotti, D., Desiato, G., Miele, A.E. and Bellelli, A. (2014) Thioredoxin Reductase and its inhibitors. Curr Protein Pept Sci, 15: 621–646.

Walsh, C.T. (1984) Suicide substrates, mechanism-based enzyme inactivators: recent developments. Annu Rev Biochem, 53: 493–535.

Index

Reversible Ligand Binding: Theory and Experiment, First Edition. Andrea Bellelli and Jannette Carey.
© 2018 John Wiley & Sons Ltd. Published 2018 by John Wiley & Sons Ltd.